Non-Newtonian Fluids for Industrial Applications

Scrivener Publishing
100 Cummings Center, Suite 541J
Beverly, MA 01915-6106

Publishers at Scrivener
Martin Scrivener (martin@scrivenerpublishing.com)
Phillip Carmical (pcarmical@scrivenerpublishing.com)

Non-Newtonian Fluids for Industrial Applications

Modeling and Simulations

Edited by

Dhananjay Yadav

Mukesh Kumar Awasthi

and

Harith Mohamed Al-Azri

Scrivener Publishing

WILEY

This edition first published 2026 by John Wiley & Sons, Inc., 111 River Street, Hoboken, NJ 07030, USA and Scrivener Publishing LLC, 100 Cummings Center, Suite 541J, Beverly, MA 01915, USA

For more information about Scrivener publications please visit www.scrivenerpublishing.com.

Wiley Global Headquarters
111 River Street, Hoboken, NJ 07030, USA

For details of our global editorial offices, customer services, and more information about Wiley products visit us at www.wiley.com.

The manufacturer's authorized representative according to the EU General Product Safety Regulation is Wiley-VCH GmbH, Boschstr. 12, 69469 Weinheim, Germany, e-mail: Product_Safety@wiley.com.

Library of Congress Cataloging-in-Publication Data

ISBN 9781394356225

Cover image: Generated with AI using Adobe Firefly
Cover design by Russell Richardson

Set in size of 11pt and Minion Pro by Manila Typesetting Company, Makati, Philippines

Printed in the USA

10 9 8 7 6 5 4 3 2 1

Contents

Contributing Author List

D. D. Ganji
Babol Noshirvani University of Technology, Babol, Iran

Mukesh Kumar Awasthi
Department of Mathematics, Babasaheb Bhimrao Ambedkar University, Lucknow, India

Atul Kumar Shukla
Department of Mathematics, Ganga Singh College, Chapra, Bihar, India

Dhananjay Yadav
Department of Mathematical & Physical Sciences, University of Nizwa, Oman

K. Jyothi
G. Pulla Reddy Engineering College, Kurnool, Andhra Pradesh, India

Yeddula Rameswara Reddy
JNTUA College of Engineering Pulivendula, Andhra Pradesh, India

Ramachandra Reddy Vaddemani
K.S.R.M College of Engineering, Kadapa, Andhra Pradesh, India

Raghunath Kodi
St. John's College of Engineering and Technology, Yemmiganur, Andhra Pradesh, India

Sahin Ahmed
Rajiv Gandhi University, Arunachal Pradesh, India

Bikash Das
Rajiv Gandhi University, Arunachal Pradesh, India

Anil Nangkar
Rajiv Gandhi University, Arunachal Pradesh, India

Nava Jyoti Hazarika
Tyagbir Hem Baruah College, Assam, India

Eny Tayang
Rajiv Gandhi University, Arunachal Pradesh, India

Damodara Reddy Annapureddy
JNTUACE Pulivendula, Andhra Pradesh, India

Sunil
NIT Hamirpur, Himachal Pradesh, India

Sweta Sharma
NIT Hamirpur, Himachal Pradesh, India

Deepak Kumar
NIT Hamirpur, Himachal Pradesh, India

Poonam Sharma
NSCBM Govt. College, Hamirpur, Himachal Pradesh, India

Akanksha Thakur
NIT Hamirpur, Himachal Pradesh, India

Reeta Devi
Govt. College Nagrota Bagwan, Kangra, Himachal Pradesh, India

Abdul Faiz Ansari
University of Lucknow, Lucknow, India

Sameera Iqram
University of Lucknow, Lucknow, India

Piyush Jaiswal
University of Lucknow, Lucknow, India

Vinod Y.
Davangere University, Davangere, India

Mohd. Asif
Integral University, Lucknow, India

Pushap Lata Sharma
Himachal Pradesh University, Shimla, India

Praveen Lata
Himachal Pradesh University, Shimla, India

Ajit Kumar
Himachal Pradesh University, Shimla, India

G. C. Rana
Govt. College, Dhaneta, Himachal Pradesh, India

Aswin Kumar Rauta
S.K.C.G. Autonomous College, Paralakhemundi, Odisha, India

S. M. Sachhin
Davangere University, Davangere, India

G. M. Sachin,
Davangere University, Davangere, India

K. R. Harshitha
Davangere University, Davangere, India

U. S. Mahabaleshwar
Davangere University, Davangere, India

Priyanka Chandra
SR University, Warangal, Telangana, India

Raja Das
VIT, Vellore, Tamil Nadu, India

Abdulhalim Musa Abubakar
Modibbo Adama University, Yola, Nigeria

Suleiman A. Wali
Modibbo Adama University, Yola, Nigeria

Abubakar Mohammed
Modibbo Adama University, Yola, Nigeria

Vivek Kumar Pandey
Chitkara University, Punjab, India

Gandhimathi G.
SRM TRP Engineering College, Tiruchirappalli, India

Chellaswamy C.
SRM TRP Engineering College, Tiruchirappalli, India

Geetha T. S.
SA Engineering College, Chennai, India

Awad M. M.
Mansoura University, Egypt

Aim & Scope

This book aims to comprehensively explore the behavior, mathematical modeling, and simulation of non-Newtonian fluids in various industrial and applied domains. Non-Newtonian fluids exhibit complex rheological properties—such as shear-thinning, shear-thickening, viscoelasticity, and yield stress—that make them essential in modern applications ranging from biomedical engineering and 3D printing to environmental engineering and heat transfer systems. The book combines foundational theory with advanced computational approaches, offering insights into current research and practical applications in diverse industrial contexts.

The book is structured across 16 chapters, each addressing a critical aspect of non-Newtonian fluid dynamics. **Chapter 1** introduces the fundamentals of non-Newtonian fluids, highlighting their types, behaviors, and significance across various sectors including medical, food, and environmental industries. **Chapter 2** provides a detailed discussion of viscoelastic fluid models, presenting several classical and modern constitutive equations, including Maxwell, Oldroyd, Power-law, and Walters' B models. **Chapter 3** focuses on computational fluid dynamics (CFD) applied to Carreau nanofluids with Marangoni boundary conditions, offering simulations using FEM to analyze velocity and temperature fields. **Chapter 4** explores heat and mass diffusion in Casson fluid over a stretching sheet embedded in a non-Darcy porous medium, emphasizing the effects of viscous dissipation and ohmic heating.

Chapter 5 analytically investigates CNT-based non-Newtonian nanofluids in porous vertical channels, highlighting thermal regulation and the role of porosity and magnetic fields. **Chapter 6** studies the effects of chemical reactions and diffusive heating in unsteady Maxwell fluid flows through exponentially stretching porous sheets using numerical methods. **Chapter 7** examines the influence of couple stresses on thermal convection in Navier–Stokes–Voigt fluids within porous media, using energy methods and stability analysis. **Chapter 8** investigates rotating ferrofluid convection in porous media under non-equilibrium conditions, applying Galerkin methods and analyzing subcritical convection onset.

Chapter 9 analyzes Casson fluid flow in a porous rotating channel with seepage, providing analytical solutions using similarity transformations and highlighting engineering applications. **Chapter 10** presents a study of thermosolutal convection in Walters' B' nanofluid under varying boundary conditions, considering rotational effects and stability thresholds. **Chapter 11** deals with two-phase flow characteristics over a stretching sheet, applying Runge–Kutta and shooting methods to solve transformed boundary-layer equations. **Chapter 12** models Maxwell fluid flow influenced by thermophoresis and Brownian motion over a permeable stretching sheet with variable magnetic field and porous media.

Chapter 13 discusses the impact of Arrhenius activation energy and viscosity ratio on Casson fluid dynamics in porous surfaces with magnetic fields, focusing on heat and mass transfer. **Chapter 14** explores non-Newtonian fluid behavior over exponentially stretching surfaces with thermal sources/sinks using finite-element-based CFD analysis. **Chapter 15** addresses the environmental engineering applications of non-Newtonian fluids, covering sediment transport, wastewater treatment, and biofilm behavior using CFD techniques. **Chapter 16** focuses on the role of non-Newtonian fluids in additive manufacturing and 3D printing, analyzing shear-thinning and viscoelastic behavior for process optimization.

Together, the chapters form a cohesive reference that bridges theoretical fluid mechanics and practical industrial design, offering readers insights into modeling techniques, simulation methods, and real-world case studies involving non-Newtonian fluids. This book is intended for researchers, engineers, and graduate students in applied mathematics, mechanical engineering, chemical engineering, and fluid mechanics.

Editors
Dhananjay Yadav
Mukesh Kumar Awasthi
Harith Mohamed Al-Azri

Preface

The book titled *Non-Newtonian Fluids for Industrial Applications: Modeling and Simulations* presents a comprehensive overview of the theoretical and practical aspects of non-Newtonian fluid dynamics, emphasizing its wide-ranging industrial applications. Non-Newtonian fluids, unlike their Newtonian counterparts, exhibit complex rheological behaviors such as shear-thinning, shear-thickening, and viscoelasticity. These properties are pivotal in understanding flow and heat transfer characteristics in multiple fields including environmental engineering, manufacturing, biofluid mechanics, and material processing.

Chapter 1 introduces the fundamental concepts of non-Newtonian fluids and their critical importance in diverse fields such as food processing, medicine, and environmental science. **Chapter 2** explores viscoelastic fluid models in detail, presenting classical and modern constitutive equations and their relevance in contemporary applications. **Chapter 3** uses computational fluid dynamics (CFD) to study Carreau nanofluids with Marangoni effects and highlights the impact of thermal radiation and magnetic parameters using FEM in Mathematica. **Chapter 4** presents a numerical analysis of Casson fluid flow over a stretching sheet embedded in a variable porous medium using the bvp4c MATLAB routine, offering insights into heat and mass transfer behavior. **Chapter 5** analytically examines CNT-based nanofluid flow in vertical porous channels, addressing effects such as porosity, magnetohydrodynamics, and nanoparticle dispersion through convergent series solutions. **Chapter 6** studies unsteady Maxwell fluid flow with diffusive heating and chemical reactions, employing a Runge–Kutta-based shooting method for numerical simulation of mixed convection phenomena.

Chapter 7 analyzes the effect of couple stresses on thermal convection in a Navier–Stokes–Voigt fluid through linear and nonlinear stability analyses relevant to geotechnical and polymer processing industries. **Chapter 8** investigates convective heat transfer in rotating ferrofluids under non-equilibrium conditions using Galerkin methods, with applications in biomedical, filtration, and cooling systems. **Chapter 9** provides an analytical solution for Casson fluid flow in a porous rotating channel, offering useful insights for biomedical and industrial fluid systems by applying similarity transformations to Darcy–Brinkman equations. **Chapter 10** examines rotating Walters's B' nanofluid convection in porous media under different boundary conditions, elucidating stabilizing and destabilizing parameters through numerical simulations in MATLAB. **Chapter 11** focuses on the two-phase boundary layer flow over a stretching sheet and incorporates particle electrification and mass transfer, solved using Runge–Kutta and shooting techniques.

Chapter 12 discusses thermophoresis and Brownian motion effects on Maxwell fluid in porous media under magnetic influence, emphasizing energy and concentration distributions using BVP4c numerical methods. **Chapter 13** explores the role of Arrhenius activation energy and viscosity ratio in Casson fluid dynamics across porous surfaces with magnetic fields, presenting implications in biomedical and energy systems. **Chapter 14** employs CFD using a variational finite element method to model Casson fluid flow over an exponentially stretching surface with source/sink and magnetic effects, analyzing key thermal performance metrics. **Chapter 15** delves into environmental applications of non-Newtonian fluids in sediment transport, wastewater treatment, and biofilm modeling, supported by CFD and case studies. **Chapter 16** discusses the vital role of non-Newtonian fluids in additive manufacturing and 3D printing, detailing flow properties like viscoelasticity and their impact on process control and product quality.

This book serves as an essential resource for students, researchers, and professionals seeking to understand and apply advanced modeling and simulation techniques to complex fluids in real-world industrial contexts. By combining theoretical foundations with computational methods and application-driven case studies, this work aims to foster deeper insights and innovative solutions in the ever-expanding domain of non-Newtonian fluid dynamics.

Editors
Dhananjay Yadav
Mukesh Kumar Awasthi
Harith Mohamed Al-Azri

Acknowledgement

We extend our sincere gratitude to all the contributing authors whose dedication and expertise have significantly shaped the success of this book, *Non-Newtonian Fluids for Industrial Applications: Modeling and Simulations*, to be published by Wiley–Scrivener Publishing. Your scholarly contributions spanning diverse topics—from viscoelastic fluid models, porous media dynamics, and Maxwell and Casson fluid flows, to environmental and industrial applications—have enriched this volume and created a valuable resource for the scientific and engineering communities.

We are deeply thankful to the editorial and production team at Wiley–Scrivener for their continued support, professionalism, and guidance throughout the development of this project. Special thanks go to the reviewers and advisors who provided critical insights, which helped refine the quality and focus of the book.

We acknowledge the support of our respective institutions and departments, whose infrastructure and encouragement enabled us to carry out this editorial endeavor successfully.

Finally, we extend our appreciation to our families and friends for their patience and encouragement throughout this journey. Their unwavering support and understanding have been invaluable during the rigorous process of compiling and editing this book.

This book is a result of collaborative efforts by researchers across the globe, and we hope it will serve as a cornerstone for future research, education, and industrial innovation in the field of non-Newtonian fluid dynamics.

Editors
Dhananjay Yadav
Mukesh Kumar Awasthi
Harith Mohamed Al-Azri

1

Introduction to Non-Newtonian Fluids

D. D. Ganji

Babol Noshirvani University of Technology, Babol, Iran

Abstract

Understanding non-Newtonian fluids is essential due to their widespread presence and applications across various industries and natural phenomena. In contrast to Newtonian fluids, which maintain a constant viscosity regardless of shear rate, non-Newtonian fluids display complex behaviors such as shear thinning, shear thickening, and viscoelasticity. These unique properties are crucial for designing and optimizing products and processes in the food, cosmetics, pharmaceuticals, and materials science industries. For instance, the shear-thinning property of ketchup allows for easy pouring from a bottle while retaining thickness on a plate, enhancing user experience. In the medical field, the non-Newtonian nature of blood is crucial for developing accurate blood flow models, which can improve the design of medical devices and treatments for cardiovascular diseases. Furthermore, understanding non-Newtonian fluids is vital in environmental engineering, where accurately modeling the behavior of natural substances like mud and lava is essential for predicting and mitigating natural disasters. The study of non-Newtonian fluids also extends to advanced technologies, including developing nanofluids for enhanced heat transfer in cooling systems. A comprehensive understanding of non-Newtonian fluids enables scientists and engineers to innovate and improve various applications, leading to better products, safer processes, and more effective solutions to complex problems. This knowledge drives technological advancements and strengthens our ability to tackle real-world health, safety, and environmental sustainability challenges.

Keywords: Non-Newtonian fluids, shear thinning, shear thickening, viscoelasticity, blood flow modeling, environmental engineering, nanofluids, rheology

Email: ddg_davood@yahoo.com; ORCID: 0000-0002-4293-5993

Dhananjay Yadav, Mukesh Kumar Awasthi and Harith Mohamed Al-Azri (eds.) Non-Newtonian Fluids for Industrial Applications: Modeling and Simulations, (1–28) © 2026 Scrivener Publishing LLC

1.1 Overview

1.1.1 Definition

A non-Newtonian fluid is a special fluid that adjusts its viscosity[1] (μ) in response to applied shear stress (τ) or shear rate. This is unlike Newtonian fluids, which maintain a constant viscosity regardless of the applied stress. The relationship between shear stress and fluid flow is defined below:

$$\tau \propto \frac{du}{dy}, \tag{1.1}$$

where du/dy is known as the shear strain rate or velocity gradient, and the dynamic viscosity (μ) is the fluid property that relates shear stress (τ) and fluid flow (du/dy). So, Equation 1.1 can be modified by considering dynamic viscosity as follows:

$$\tau = \mu \frac{du}{dy}, \tag{1.2}$$

where Equation 1.2 can be used for both Newtonian and non-Newtonian fluids. But with this difference in the Newtonian fluids, as it is demonstrated in Figure 1.1, μ (relationship between shear stress and shear rate) is linear, which means that viscosity remains unchanged when subjected to a force. Typical examples of Newtonian fluids include crude oil (has a viscosity of approximately $100\ \text{mPa}\cdot\text{s}$ at 15.56°C), olive oil (has a viscosity of approximately $84\ \text{mPa}\cdot\text{s}$ at 15.56°C), glycerol (has a viscosity of approximately $1.5\ \text{mPa}\cdot\text{s}$ at 15.56°C), water (has the viscosity of approximately $1.12\ \text{mPa}\cdot\text{s}$ at 15.56°C and approximately 0.2816 $0.2816\ \text{mPa}\cdot\text{s}$ at 37.78°C), and air (has a viscosity of approximately $1.12\ \text{mPa}\cdot\text{s}$ at 15.56°C).

However, according to Figure 1.2, in non-Newtonian fluids, it is not, which means their viscosity can either increase or decrease when subjected to a force. Typical examples of non-Newtonian fluids include ketchup, which becomes runnier when shaken (shear-thinning), and cornstarch in water, which becomes more solid under the impact (shear-thickening). Toothpaste behaves like a solid until enough pressure is applied to squeeze it out of the tube (Bingham plastic). These fluids are best studied through

[1]Actually, it is called "dynamic viscosity".

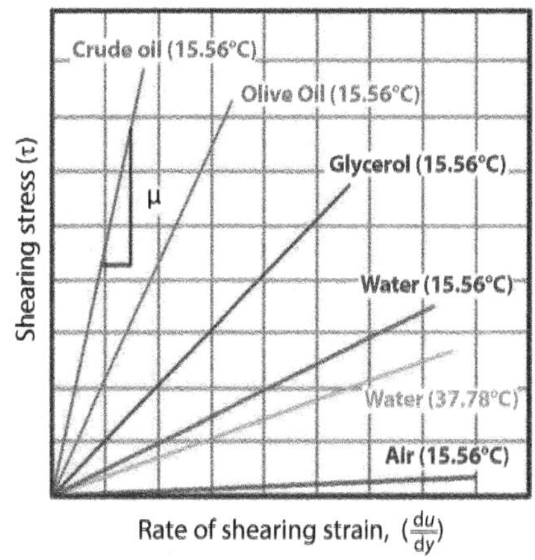

Figure 1.1 The relationship between shearing stress and the rate of shearing strain for typical fluids is linear.

their rheological properties, which describe how they deform and flow under various conditions.

Under normal conditions, most common gases—such as air, nitrogen, and oxygen—are classified as Newtonian fluids, signifying that their viscosity remains constant irrespective of the applied shear rate. This assumption is valid for everyday scenarios and industrial applications. However, under extreme conditions—such as high pressures, elevated temperatures, or in complex mixtures—gases may exhibit non-Newtonian behavior.

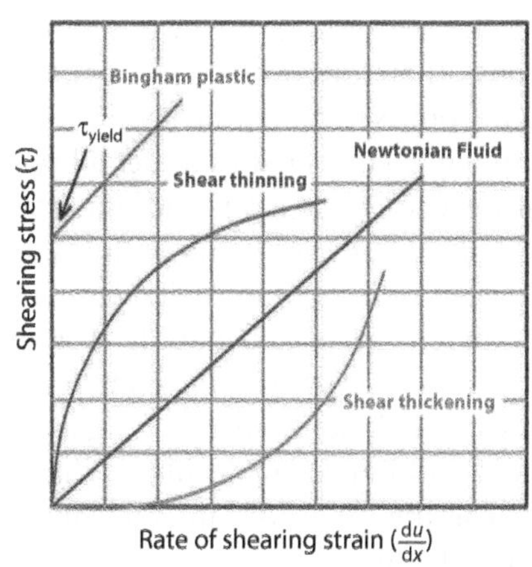

Figure 1.2 The relationship between shearing stress and the rate of shearing strain is examined for various fluid types, including non-Newtonian fluids commonly found.

At high pressures, intermolecular forces become significant, potentially altering the flow behavior of the gas. Elevated temperatures can change the viscosity of a gas, leading to non-Newtonian characteristics.

Additionally, when gases are combined with particulates, aerosols, or other substances, the mixture can display non-Newtonian behavior due to interactions between different components. Consequently, while the assumption of Newtonian fluid behavior is generally valid for gases under standard atmospheric conditions, it is prudent to consider potential deviations in extreme environments or specialized applications. Liquids are inherently more susceptible to exhibiting non-Newtonian behavior than gases. Under typical conditions, gases such as air, nitrogen, and oxygen generally maintain their Newtonian characteristics, meaning their viscosity remains constant irrespective of the applied shear rate. These gases only exhibit non-Newtonian behavior under extreme conditions, such as extremely high pressures, elevated temperatures, or when mixed with particulates or other complex substances.

In contrast, liquids can easily display non-Newtonian behavior even under more typical circumstances. For example, liquids containing suspended particles (like mud or paint) can become shear-thinning, meaning their viscosity decreases with increased shear rate. Emulsions (like mayonnaise) and polymer solutions (like slime or various gels) can also exhibit non-Newtonian properties without extreme conditions. In these cases, the interactions between particles, droplets, or polymer chains lead to significant changes in viscosity in response to shear stress. Furthermore, highly concentrated solutions, such as those with large amounts of dissolved sugar or salt, and many biological fluids, like blood, mucus, and synovial fluid, frequently demonstrate non-Newtonian behavior due to their complex compositions. Therefore, while gases generally require extreme conditions to deviate from Newtonian behavior, liquids are more prone to exhibit non-Newtonian characteristics under a broader range of conditions, including many everyday scenarios and practical applications.

As illustrated in Figure 1.1, the viscosity of fluids, including both gases and liquids, is influenced by temperature. Tables 1.1 and 1.2 display the viscosity value for different gases and liquids at a specific temperature. For gases, dynamic viscosity primarily depends on temperature. As the temperature increases, the kinetic energy of gas molecules increases, causing more frequent and intense molecular collisions. This results in a higher viscosity because the molecules interact more as they move past each other. Unlike liquids, the viscosity of gases increases with temperature. Pressure has a less significant effect on the viscosity of gases at ordinary conditions but can become more relevant at very high pressures.

Table 1.1 The various viscosity of different gases at a room temperature of 20°C and under atmospheric pressure (this table got its data from the website https://roymech.org/Related/Fluids/Fluids_Viscosities.html).

Fluid	Dynamic viscosity
Acetylene (C_2H_2)	0.00935 cP
Ammonia (NH_3)	0.00918 cP
Argon (Ar)	0.021 cP
Carbon dioxide (CO_2)	0.014 cP
Carbon monoxide (CO)	0.0166 cP
Chlorine (Cl)	0.0129 cP
Cyanogen ($(CN)_2$)	0.00928 cP
Ethylene (C_2H_4)	0.0097 cP
Helium (He)	0.0186 cP
Hydrogen (H_2)	0.00835 cP
Hydrogen chloride (HCl)	0.0138 cP
Hydrogen Sulfide (H_2S)	0.0117 cP
Methane (CH_4)	0.0103 cP
Nitric oxide (NO)	0.0178 cP
Nitrous oxide (N_2O)	0.0135 cP
Oxygen (O_2)	0.0192 cP
Sulfur dioxide (SO_2)	0.0117 cP

The increase in viscosity of various gases due to increasing temperature is clearly shown in Table 1.3. For liquids, dynamic viscosity mainly depends on temperature and the intermolecular forces. As temperature rises, the viscosity of a liquid typically decreases because the thermal motion of molecules overcomes the intermolecular attractions, making it easier for them to slide past one another. However, unlike gases, the viscosity of liquids decreases with increasing temperature. The specific composition

Table 1.2 The various viscosity of different liquids at a room temperature of 20°C (this table got its data from the website https://roymech.org/Related/Fluids/Fluids_Viscosities.html).

Fluid	**Dynamic viscosity**
Acetic acid ($C_2H_4O_2$)	1,219 cP
Acetone ($(CH_3)_2CO$)	0,324 cP
Benzene (C_6H_6)	0,647 cP
Bromine (Br)	0.993 cP
Carbon disulfide (CS_2)	0,375 cP
Carbon tetrachloride (CC_{14})	0.972 cP
Chloroform (CHC_{13})	0,569 cP
Glycerol ($C_3H_8O_3$)	1495 cP
Mercury (Hg)	1.552 cP
Methyl alcohol (CH_3OH)	0.594 cP
Nitrobenzene ($C_6H_5NO_2$)	2.03 cP
Olive oil (oleic-oleic-oleic (OOO) triacylglycerol)	84000 cP
Lubricating oil	799,4 cP
Paraffin oil/kerosene ($C_{12}H_{26}C_{15}H_{32}$)	1000 cP
Phenol (C_6H_5OH)	12.74 cP
Toluene ($C_6H_5CH_3$)	0,585 cP
Turpentine ($C_{10}H_{16}$)	1,49 cP
Water (H_2O)	1 cP

of the liquid, including any dissolved substances, also affects its viscosity. External factors such as shear rate can influence the viscosity of non-Newtonian liquids, which display varying viscosities under different flow conditions. The decrease in viscosity of various liquids due to increasing temperature is clearly shown in Table 1.4. According to Tables 1.1 and 1.2, $1\ \text{Pa}\cdot\text{s} = 1\ \text{N}\cdot\text{s}\cdot\text{m}^2 = 10\ \text{poise} = 1000\ \text{milliPa}\cdot\text{s}$.

Table 1.3 The increase of various viscosity of different gases as temperature increases (this table got its data from the website https://roymech.org/Related/Fluids/Fluids_Viscosities.html).

	Temperature			
Fluid	**0°C**	**100°C**	**250°C**	**500°C**
Oxygen (O_2)	19E − 6 Pa·s	24E − 6 Pa·s	30E − 6 Pa·s	39E − 6 Pa·s
Helium (He)	18E − 6 Pa·s	22E − 6 Pa·s	28E − 6 Pa·s	37E − 6 Pa·s
Air (-)	17E − 6 Pa·s	21E − 6 Pa·s	27E − 6 Pa·s	35E − 6 Pa·s
Nitrogen (N_2)	16.5E − 6 Pa·s	21E − 6 Pa·s	26.5E − 6 Pa·s	35E − 6 Pa·s
Carbon monoxide (CO)	16.5E − 6 Pa·s	21E − 6 Pa·s	26E − 6 Pa·s	35E − 6 Pa·s
Carbon dioxide (CO_2)	13.7E − 6 Pa·s	18E − 6 Pa·s	24E − 6 Pa·s	34E − 6 Pa·s
Sulfur dioxide (SO_2)	11.8E − 6 Pa·s	15E − 6 Pa·s	20E − 6 Pa·s	28E − 6 Pa·s
Ammonia (NH_3)	9.5E − 6 Pa·s	13E − 6 Pa·s	18E − 6 Pa·s	27E − 6 Pa·s
Hydrogen (H_2)	8.5E − 6 Pa·s	10.5E − 6 Pa·s	12.5E − 6 Pa·s	16.5E − 6 Pa·s
Ethane (C_2H_6)	8.5E − 6 Pa·s	11.5E − 6 Pa·s	15.5E − 6 Pa·s	22.5E − 6 Pa·s
Propane (C_3H_8)	7.7E − 6 Pa·s	10E − 6 Pa·s	13.2E − 6 Pa·s	18E − 6 Pa·s
Butane (C_4H_{10})	8E − 6 Pa·s	10.5E − 6 Pa·s	14.2E − 6 Pa·s	20E − 6 Pa·s

1.1.2 Importance of the Non-Newtonian Fluids

Non-Newtonian fluids are critically important due to their unique properties and wide-ranging applications across various industries. Unlike Newtonian fluids, their viscosity changes with the applied stress or shear rate, enabling them to perform specialized functions essential in many fields. In the food industry, for example, the shear-thinning behavior of sauces and dressings ensures they are easy to pour but remain thick on the plate. In the medical field, understanding the non-Newtonian nature of blood is vital for developing accurate models for blood flow, which can improve the design of medical devices and treatments for cardiovascular diseases. Additionally, non-Newtonian fluids are used to formulate cosmetics,

Table 1.4 The decrease of various viscosity of different liquids as temperature increases (this table got its data from the website https://roymech.org/Related/Fluids/Fluids_Viscosities.html).

Fluid	Temperature					
	−25°C	0°C	20°C	50°C	100°C	200°C
Water (H_2O)	-	1.7918 cP	1.0028cP	0.5471cP	0.2817cP	0.1346cP
Light machinery oil	-	-	-	28 cP	4.9 cP	-
Spindle oil	-	-	13 cP	5.49 cP	2.0 cP	-
Ethanol (C_2H_5OH)	3.241 cP	1.786 cP	1.201 cP	0.701 cP	0.326 cP	-
Methanol (CH_3OH)	-	0.817 cP	0.584 cP	0.396 cP	-	-
Benzene (C_6H_6)	-	-	0.649 cP	0.436 cP	0.261 cP	0.113 cP
Toluene (C_7H_8)	-	0.773 cP	0.586 cP	0.419 cP	0.269 cP	0.133 cP
Sulfur dioxide (SO_2)	-	0.368 cP	0.304 cP	0.234 cP	-	-
Ammonia (NH_3)	0.215 cP	0.169 cP	0.138 cP	0.103 cP	-	-
Freon 113 ($CClF_2CCl_2F$)	1.4 cP	0.92 cP	0.75 cP	0.5 cP	0.29 cP	-
Fuel oil (sg=0.94)	-	-	1000 cP	100 cP	11 cP	2.5 cP
Crude oil (sg = 0,855)	-	16 cP	7.5 cP	4 cP	2.5 cP	1.5 cP
Paraffin oil/kerosene ($C_{12}H_{26}C_{15}H_{32}$)	-	3.7 cP	2.5 cP	1.4 cP	0.6 cP	-
Sulfuric acid (100%) (H_2SO_4)	90 cP	36 cP	25 cP	10.5 cP	3.4 cP	-
Sulfuric acid (60%) (H_2SO_4)	16 cP	9 cP	6 cP	3.5 cP	1.5 cP	-
Nitric acid (100%) (HNO_3)	2.1 cP	1.5 cP	1.2 cP	0.85 cP	0.50 cP	-
Nitric acid (60%) (HNO_3)	6.0 cP	3.5 cP	2.5 cP	1.4 cP	0.7 cP	-
Mercury (Hg)	2 cP	1.60 cP	1.55 cP	1.5 cP	1.45 cP	1.1 cP

pharmaceuticals, and household products, enhancing their usability and effectiveness. Their unique properties are also leveraged in advanced technologies, such as the development of nanofluids for improved heat transfer in cooling systems. Overall, the study and application of non-Newtonian fluids drive innovation and efficiency in numerous sectors, making them indispensable in everyday products and cutting-edge research.

Non-Newtonian fluids perform various specialized functions across different fields due to their unique properties. In the food industry, they enhance the texture and pourability of products like sauces and dressings. They are crucial for drug delivery systems and prosthetics design in the medical field. Shear-thickening fluids are used in body armor and protective gear, providing flexibility and rigidity upon impact. In the automotive industry, they improve the performance of clutch systems and damping devices. Non-Newtonian fluids are also essential in 3D printing for precise control over the printing process. In environmental engineering, they aid in wastewater treatment and disaster management. Additionally, they are used in cosmetics for better application and stability, in firefighting foams to smother fires effectively, and in construction to improve the flow properties of concrete. Their versatility extends to everyday products like toothpaste shampoo and advanced technologies such as nanofluids for enhanced heat transfer in cooling systems.

1.1.3 Governing Equations for the Newtonian Fluids

1.1.3.1 Vectorial Governing Equations for Newtonian Fluids

Writing the governing equations of continuity, momentum, and energy in vectorial form is crucial because it provides a more comprehensive and generalized representation of these fundamental physical laws. The vectorial form captures the multi-dimensional nature of fluid flow and heat transfer, making the equations applicable to various coordinate systems and geometries. This enhances the accuracy and versatility of the models, allowing for the simultaneous consideration of all directional components. Additionally, vectorial equations offer a compact and elegant notation, simplifying the mathematical manipulation and numerical implementation of these equations. By capturing the full range of physical phenomena, vectorial forms ensure that all relevant forces and interactions are accounted for, leading to more accurate and reliable simulations and analyses across different engineering and scientific applications. Vectorial forms of governing equations for continuity, momentum, and energy are essential when dealing with multi-dimensional problems where the directionality of quantities like velocity, force, and heat flux is crucial. This is particularly important in scenarios involving complex geometries, multiphase flows, anisotropic materials, turbulence modeling, and generalized analyses. Using vectorial form ensures that all directional components are accurately accounted for, providing a unified framework applicable across various coordinate systems and enhancing the precision of simulations and analyses in advanced engineering and scientific applications. Conversely, open form (component form) is typically used for simpler, one-dimensional problems or when specific directional components need isolation for focused analysis.

Therefore, the vectorial forms of constitutive equations of the continuity, momentum, energy, and entropy generation for any arbitrary Newtonian incompressible fluid with non-constant thermal conductivity (i.e., temperature-dependent thermal conductivity) are as follows:

Continuity equation in vectorial form:

$$\nabla \cdot (\mathbf{v}) = 0, \tag{1.3}$$

where ∇ is the gradient operator that measures the rate and direction of change in a field, and $\nabla \cdot$ is the divergence operator. It measures the rate at which a quantity expands or contracts. $\mathbf{v}$ is the velocity vector field with the SI unit of $\mathrm{m/s}$.

Momentum equation (Navier–Stokes) in vectorial form:

$$\rho\left(\frac{D\mathbf{v}}{Dt}\right) = -\nabla p + \mu\nabla^2\mathbf{v} + \mathbf{F}, \tag{1.4}$$

where ρ is the mass density. Every incompressible fluid has its constant mass density, and it has the SI unit of $\mathrm{kg/m^3}$. $D\mathbf{v}/Dt = \partial\mathbf{v}/\partial t + \mathbf{v}\cdot\nabla\mathbf{v}$ is the total derivative or material derivative of the velocity vector field, and it has the SI unit of $\mathrm{m/s^2}$. t is time, and it has the SI unit of s. p is scalar pressure field, and it has the SI unit of Pa. ∇^2 is the Laplacian operator, and it measures the rate at which a quantity diffuses, and $\mathbf{F}$ is the body force per unit volume; in addition, it is a vector as it bolded, and it has the SI unit of $\mathrm{N/m^3}$.

Energy equation in vectorial form:

$$\rho c_p\left(\frac{DT}{Dt}\right) = \nabla \cdot (k\nabla T) + \mu(\nabla\mathbf{v}):(\nabla\mathbf{v}) + \mathbf{F}\cdot\mathbf{v}, \tag{1.5}$$

where c_p is the specific heat at constant pressure, and it has the SI unit of $\mathrm{J/(Kg.\ K)}$. T is the temperature scalar field, and it has the unit of K. $DT/Dt = \partial T/\partial t + \mathbf{v}\cdot\nabla T$ is the total derivative or material derivative of the temperature scalar field, and it has the SI unit of K/s. "." is a dot product, which is the multiplication of vectors producing a scalar quantity.

$k = k(T)$ is the non-constant thermal conductivity, which is a function of the temperature scalar field, and it has the SI unit of $\mathrm{W}/(\mathrm{m}\cdot\mathrm{K})$. μ is the dynamic viscosity, and it has the SI unit of $\mathrm{Pa}\cdot\mathrm{s}$. $(\nabla\mathbf{v}):(\nabla\mathbf{v})$ is viscous dissipation, and it represents the internal frictional energy loss, which has the SI unit of $1/\mathrm{s}$. $\mathbf{F}\cdot\mathbf{v}$ is the work done by forces, which measures the work done by body forces on the fluid and has the SI unit of $\mathrm{N}/(\mathrm{m}^2\cdot\mathrm{s})$. In Equation 1.5, ":" is a double dot product representing the tensor contraction to account for all components of the velocity gradients. This operation between the stress tensor and the velocity gradient tensor calculates the work rate done by viscous forces, essentially measuring the internal frictional energy losses. The vectorial form of the velocity gradient tensor for a three-dimensional velocity field $\mathbf{v} = (u, v, w)$ is expressed as

$$\nabla\mathbf{v} = \begin{pmatrix} \dfrac{\partial u}{\partial x} & \dfrac{\partial u}{\partial y} & \dfrac{\partial u}{\partial z} \\ \dfrac{\partial v}{\partial x} & \dfrac{\partial v}{\partial y} & \dfrac{\partial v}{\partial z} \\ \dfrac{\partial w}{\partial x} & \dfrac{\partial w}{\partial y} & \dfrac{\partial w}{\partial z} \end{pmatrix} \tag{1.6}$$

where u, v, and w are velocity components in the x, y, and z spatial coordinates, respectively, and ∂ are partial derivatives. The double dot product of the velocity gradient tensor with itself is calculated as

$$(\nabla\mathbf{v}):(\nabla\mathbf{v}) = \sum_{i,j}\left(\frac{\partial v_i}{\partial x_j}\right)^2, \tag{1.7}$$

For Newtonian incompressible fluids with non-constant thermal conductivity, the viscous dissipation term in the energy equation is represented as follows:

$$\Phi = \mu(\nabla\mathbf{v}):(\nabla\mathbf{v}) = \mu\left(\sum_{i,j}\left(\frac{\partial v_i}{\partial x_j}\right)^2\right), \tag{1.8}$$

Entropy generation in vectorial form:

$$\sigma = \frac{\nabla \cdot (k\nabla T)}{T^2} + \frac{\mu(\nabla \mathbf{v}):(\nabla \mathbf{v})}{T}, \tag{1.9}$$

where σ is the entropy generation rate, and its SI unit is $\mathrm{W/(m^3 \cdot K)}$. $\nabla T = (\partial T/\partial x, \partial T/\partial y, \partial T/\partial z)$ is the vectorial form of the temperature gradient vector, and its SI unit is K/m.

The entropy generation rate σ in the entropy generation equation can indicate different states based on its value:

- Positive ($\sigma > 0$): Indicates that entropy is being generated within the system. This is typical in real-world processes where irreversibilities such as friction, heat transfer across finite temperature differences, and mixing of different substances are present.
- Zero ($\sigma = 0$): Represents a hypothetical ideal situation where no entropy is generated. This implies a reversible process with no irreversibilities. In reality, this condition is nearly impossible to achieve, but it serves as a benchmark for the most efficient processes.
- Negative ($\sigma < 0$): This situation is physically impossible according to the second law of thermodynamics. Entropy generation cannot be negative as it would imply a process that decreases the total entropy of the system and surroundings, violating the principle that the total entropy of an isolated system can never decrease.

In situations where the applied fluid is Newtonian and incompressible, and thermal conductivity is temperature dependent, the constitutive equations are Equations 1.3–1.9. However, in situations where the utilized fluid is again Newtonian and incompressible, but its thermal conductivity is considered constant, the constitutive Equations 1.3–1.9 are accurate. Still, the energy Equation 1.5 and entropy generation 1.9 should be modified as below:

The vectorial energy equation with constant thermal conductivity:

$$\rho c_p\left(\frac{DT}{Dt}\right) = k\nabla^2 T + \mu(\nabla \mathbf{v}):(\nabla \mathbf{v}) + \mathbf{F}\cdot\mathbf{v}, \tag{1.10}$$

The vectorial entropy generation with constant thermal conductivity:

$$\sigma = \frac{k(\nabla T)\cdot(\nabla T)}{T^2} + \frac{\mu(\nabla \mathbf{v}):(\nabla \mathbf{v})}{T}, \tag{1.11}$$

where Equations 1.3, 1.4, 1.10, and 1.11 can be confidently used for fluids with Newtonian, are incompressible, and have constant temperature-dependent thermal conductivity.

However, there are some situations where Newtonian fluids with non-constant thermal conductivity have to be considered compressible, and the following constitutive equations model those situations:

Continuity equation in vectorial form:

$$\frac{\partial \rho}{\partial t} + \nabla \cdot (\rho \mathbf{v}) = 0, \tag{1.12}$$

Momentum equation (Navier–Stokes) in vectorial form:

$$\rho\left(\frac{D\mathbf{v}}{Dt}\right) = -\nabla p + \nabla \cdot \tau + \mathbf{F}, \tag{1.13}$$

where τ is the stress tensor, and its SI unit is Pa. $\nabla \cdot \tau$ is the divergence of the stress tensor, and its SI unit is $\mathrm{N/m^3}$.

Energy equation in vectorial form:

$$\rho c_v\left(\frac{DT}{Dt}\right) = \nabla \cdot (k\nabla T) + \tau : \nabla \mathbf{v} + \mathbf{F} \cdot \mathbf{v}, \tag{1.14}$$

where c_v is the specific heat at constant volume, and its SI unit is $\mathrm{J/(kg \cdot K)}$. $\tau : \nabla \mathbf{v}$ is viscous dissipation, and its SI unit is 1/s.

Why is the specific heat at constant volume (c_v) used in the energy Equation 1.14 for compressible fluids, whereas the specific heat at constant pressure (c_p) is used in the energy equation for incompressible fluids?

The distinction between specific heat at constant volume (c_v) and specific heat at constant pressure (c_p) in energy equations arises from the behavior of the fluid and the nature of the processes involved. For compressible fluids, changes in both pressure and volume can occur. The energy equation often involves the internal energy, which is associated with c_v, the specific heat at constant volume. This is because internal energy primarily depends on temperature and volume. In the compressible flow, the temperature changes at constant volume are most relevant to the energy transfer within the fluid. The volume remains constant for incompressible fluids, and pressure changes do not affect the volume.

The energy equation for incompressible fluids typically involves the enthalpy, which is associated with c_p, the specific heat at constant pressure. This is because enthalpy accounts for the combined effects of internal energy and the work done due to pressure changes, which are significant in the incompressible flow where the volume does not change, but pressure and temperature variations are crucial. These distinctions ensure that the energy equations accurately reflect the physical processes and energy transformations occurring in the fluid, tailored to whether the fluid is compressible or incompressible.

Entropy generation in vectorial form:

$$\sigma = \frac{\nabla \cdot (k \nabla T)}{T^2} + \frac{\tau : (\nabla \mathbf{v})}{T}, \tag{1.15}$$

In situations where the applied fluid is Newtonian, compressible, and thermal conductivity is temperature dependent, the constitutive equations are Equations 1.12–1.15. However, the situations where the utilized fluid is again Newtonian and compressible, but its thermal conductivity is considered constant, the constitutive Equation 1.12 is true, but the momentum Equation 1.13, the energy Equation 1.14, and entropy generation 1.15 should be modified as below:

The vectorial momentum equation with constant thermal conductivity:

$$\rho\left(\frac{D\boldsymbol{v}}{Dt}\right) = -\nabla p + \mu \nabla^2 v + \mathrm{F}, \tag{1.16}$$

The vectorial energy equation with constant thermal conductivity:

$$\rho c_v\left(\frac{DT}{Dt}\right) = k\nabla^2 T + \mu(\nabla \mathbf{v}) : (\nabla \mathbf{v}) + \mathbf{F} \cdot \mathbf{v}, \tag{1.17}$$

The vectorial entropy generation with constant thermal conductivity:

$$\sigma = \frac{k(\nabla T) \cdot (\nabla T)}{T^2} + \frac{\mu(\nabla \mathbf{v}) : (\nabla \mathbf{v})}{T}, \tag{1.18}$$

Why do the momentum Equation 1.13 feature $\nabla \cdot \tau$ and the energy and entropy Equations 1.14 and 1.15 include $\tau : \nabla \mathbf{v}$ for Newtonian, compressible fluids with non-constant thermal conductivity, whereas for Newtonian, compressible fluids with constant thermal conductivity, the momentum Equation 1.16 contains $\mu\nabla^2\mathbf{v}$ and the energy and entropy Equations 1.17 and 1.18 utilize $\mu(\nabla\mathbf{v}):(\nabla\mathbf{v})$?

This distinction arises because of how these equations handle the variation in fluid properties and their impact on the governing equations.

Non-constant thermal conductivity:

1. Momentum equation ($\nabla\cdot\tau$):
   - For fluids with non-constant thermal conductivity, the stress tensor τ is more complex, and its divergence $\nabla\cdot\tau$ captures the spatial variability of viscosity and its effect on the flow. This approach accounts for the varying resistance to deformation due to temperature changes.
2. Energy equation ($\tau:\nabla\mathbf{v}$):
   - The term $\tau:\nabla\mathbf{v}$ represents the viscous dissipation of energy. In this scenario, because thermal conductivity is temperature-dependent, capturing the detailed interaction between stress and velocity gradients is crucial for accurately modeling the energy conversion into heat.

Constant thermal conductivity:

1. Momentum equation ($\mu\nabla^2\mathbf{v}$):
   - When thermal conductivity is constant, viscosity is also often considered constant. The term $\mu\nabla^2\mathbf{v}$ in the momentum equation simplifies to the Laplacian of the velocity, representing viscous diffusion of momentum. This simplification is valid because it assumes uniform resistance to deformation across the fluid.
2. Energy equation $\left(\mu(\nabla\mathbf{v}):(\nabla\mathbf{v})\right)$:
   - For the energy equation, the term $(\nabla\mathbf{v}):(\nabla\mathbf{v})$ accounts for viscous dissipation. With constant thermal conductivity, the viscous dissipation is modeled by considering the direct interaction of velocity gradients, reflecting a uniform energy conversion rate into heat.

The difference lies in the need to capture spatial variability and complexity when thermal conductivity (and often viscosity) is non-constant, compared to the simplified uniform properties assumption when they are constant. This ensures that the equations accurately represent the physical processes under varying conditions.

1.1.4 Governing Equations for the Non-Newtonian Fluids

1.1.4.1 Vectorial Governing Equations for Non-Newtonian Fluids

The vectorial governing equations for non-Newtonian fluids encompass the conservation of mass, momentum, and energy, incorporating the complex behaviors unique to these fluids. The continuity equation remains the same as Newtonian fluids, ensuring mass conservation. However, the momentum equation is modified to include a stress tensor τ that accounts for the fluid's rheology, which varies with the deformation rate. This tensor captures non-linear relationships between stress and strain rates, often represented by Carreau, Cross, or Herschel–Bulkley models. The energy equation considers internal energy changes and heat conduction, factoring in non-constant thermal conductivity and viscous dissipation effects influenced by the fluid's non-Newtonian properties. These vectorial equations are crucial for accurately modeling and simulating the complex behavior of non-Newtonian fluids in various engineering and scientific applications. Entropy generation, a measure of system irreversibilities, often arises from viscous dissipation, heat conduction, and mass diffusion. When dealing with non-Newtonian fluids, the entropy generation characteristics change significantly compared to Newtonian fluids. Non-Newtonian fluids have complex rheological behaviors leading to non-linear stress–strain relationships, affecting viscous dissipation. For instance, shear-thinning or shear-thickening properties alter internal friction within the fluid, impacting energy conversion to heat. This affects the distribution and rate of entropy generation. Additionally, the non-linear heat conduction properties in non-Newtonian fluids can lead to more complex thermal gradients, further influencing entropy generation. Ultimately, using non-Newtonian fluids introduces additional factors and complexities that must be carefully considered to predict and manage entropy generation in various engineering applications accurately.

Therefore, the vectorial forms of constitutive equations of the continuity, momentum, energy, and entropy generation for any arbitrary non-Newtonian incompressible fluid with non-constant thermal conductivity (i.e., temperature-dependent thermal conductivity) are as follows:

Continuity equation in vectorial form:

$$\nabla \cdot (\mathbf{v}) = 0, \tag{1.19}$$

Momentum equation (Navier–Stokes) in vectorial form:

$$\rho\left(\frac{D\mathbf{v}}{Dt}\right) = -\nabla p + \nabla \cdot \tau + \mathbf{F}, \tag{1.20}$$

Energy equation in vectorial form:

$$\rho c_p\left(\frac{DT}{Dt}\right)=\nabla\cdot(k\nabla T)+\tau:\nabla\mathbf{v}+\mathbf{F}\cdot\mathbf{v}, \tag{1.21}$$

Entropy generation in vectorial form:

$$\sigma=\frac{\nabla\cdot(k\nabla T)}{T^2}+\frac{\tau:(\nabla\mathbf{v})}{T}, \tag{1.22}$$

For a non-Newtonian incompressible fluid with constant thermal conductivity, continuity (Equation 1.19) and momentum (Equation 1.20) equations do not change, but energy (Equation 1.21) and entropy (Equation 1.22) change as follows:

Energy equation in vectorial form with constant thermal conductivity:

$$\rho c_p\left(\frac{DT}{Dt}\right)=k\nabla^2 T+\tau:\nabla\mathbf{v}+\mathbf{F}\cdot\mathbf{v}, \tag{1.23}$$

Entropy generation in vectorial form with constant thermal conductivity:

$$\sigma=\frac{k(\nabla T)\cdot(\nabla T)}{T^2}+\frac{\tau:(\nabla\mathbf{v})}{T}, \tag{1.24}$$

For a non-Newtonian compressible fluid with non-constant thermal conductivity, momentum (Equation 1.20) and entropy generation (Equation 1.22) equations do not change, but continuity (Equation 1.19) and energy (Equation 1.21) equations change as follows:

Continuity equation in vectorial form:

$$\frac{\partial\rho}{\partial t}+\nabla\cdot(\rho\mathbf{v})=0, \tag{1.25}$$

Energy equation in vectorial form:

$$\rho c_v\left(\frac{DT}{Dt}\right)=\nabla\cdot(k\nabla T)+\tau:\nabla\mathbf{v}+\mathbf{F}\cdot\mathbf{v}, \tag{1.26}$$

For a non-Newtonian compressible fluid with constant thermal conductivity, the continuity equation will be the same as Equation 1.25, the momentum equation will be the same as Equation 1.16, the energy equation is the same as Equation 1.17, and entropy generation is the same as Equation 1.18.

1.1.5 Recent Advances in Non-Newtonian Fluids

Modeling non-Newtonian fluids is crucial due to their widespread occurrence and importance in natural and industrial processes. Unlike Newtonian fluids, whose viscosity remains constant, non-Newtonian fluids exhibit complex behavior that varies with shear rate, temperature, and time. Understanding and predicting this behavior is essential for enhancing product design, efficient processing, and maintaining safety and quality control. Non-Newtonian fluids are common in the food, pharmaceutical, and cosmetics industries. Accurate modelling helps design products with desired textures and flow properties, such as creams, gels, and sauces. Efficient processing in manufacturing processes like extrusion, mixing, and coating relies on understanding non-Newtonian fluid characteristics to optimize equipment design and operation [1–4]. Modeling these fluids is equally essential in environmental and biological systems.

Many natural fluids, such as mudflows, lava, and blood, exhibit non-Newtonian behavior. Accurate models contribute to understanding and predicting natural phenomena, aiding disaster management and medical research. For instance, blood, a non-Newtonian fluid, requires precise modelling for developing medical devices like artificial hearts and blood pumps and understanding medical conditions related to blood flow. This aspect of modeling non-Newtonian fluids enhances environmental safety and medical advancements [5–7]. From a scientific and engineering perspective, non-Newtonian fluid modeling provides critical insights. It aids in studying complex materials, leading to new materials with customized properties for specific applications.

Additionally, accurate models allow engineers to simulate and control fluid behavior in complex systems, from pipelines to chemical reactors, enhancing efficiency and safety. Modelling provides the predictive capability to anticipate fluid behavior under various conditions, leading to better design and control of processes, which in turn drives innovation in product development and process engineering [1–3, 8–10]. The benefits of modeling non-Newtonian fluids are significant. By understanding fluid dynamics, industries can optimize processes, reduce waste, and save costs. Modeling these fluids drives product development and process engineering innovation, leading to new and improved products and technologies. Modeling non-Newtonian fluids is foundational for advancing various fields, enhancing product quality, and ensuring efficient and safe operations in numerous applications [1, 2, 8].

Non-Newtonian fluid models are vital tools for understanding and optimizing the behavior of complex fluids across various industrial and scientific applications. The power law model [6, 11–14] is a fundamental non-Newtonian fluid model pivotal for describing fluids that do not follow Newton's law of viscosity. It adeptly characterizes fluids exhibiting shear-thinning or shear-thickening behavior. In simpler terms, this model captures how a fluid's viscosity changes with the rate at which it is sheared. If a fluid becomes less viscous when sheared, it is referred to as shear-thinning, as seen in materials like paint and certain lotions.

Conversely, if a fluid becomes more viscous with shearing, it is termed shear-thickening, exemplified by cornstarch suspensions in water. The power law model is indispensable in various industries, including food processing, cosmetics, and polymer manufacturing. It aids in understanding and controlling the flow characteristics of these complex fluids, ensuring optimal processing conditions and high-quality products.

The Carreau model [8, 15–18] is a widely used non-Newtonian fluid model that describes the shear-thinning behavior of fluids over a broad range of shear rates. It incorporates zero-shear and infinite-shear viscosity to accurately capture the fluid's transition from Newtonian to non-Newtonian behavior as the shear rate increases. This model is particularly effective for polymer solutions and melts, where the viscosity significantly decreases with increasing shear rate. The Carreau model is essential in various industries, including polymer processing, where understanding the viscosity changes under different shear conditions is crucial for optimizing processes like extrusion and injection molding. Its ability to comprehensively describe fluid behavior makes it a valuable tool for theoretical studies and practical applications in fluid dynamics and materials engineering.

The Herschel–Bulkley model [19–21] is a versatile and widely used non-Newtonian fluid model that extends the Bingham plastic model by incorporating shear-thinning or shear-thickening behavior and yield stress. Essentially, this model characterizes fluids that behave as solids under low shear stress until a certain yield stress threshold is surpassed, after which they flow like a viscous fluid. The model further refines this behavior by accounting for changes in viscosity depending on the shear rate. This makes the Herschel–Bulkley model particularly useful in describing the flow of complex materials such as ketchup, toothpaste, and clay suspensions, which exhibit both yield stress and non-linear viscosity. In industrial applications, optimizing processes involving the mixing, pumping, and flow of such materials is crucial to ensuring efficiency and consistency in production. Its ability to model diverse behaviors makes it an invaluable tool in both theoretical research and practical engineering, significantly advancing our understanding and control of non-Newtonian fluids.

The Casson model [6, 22–26] is a significant non-Newtonian fluid model widely used to describe fluids exhibiting yield stress, followed by shear-thinning behavior. Initially developed to describe blood flow characteristics, the model is particularly useful in contexts where the fluid requires a certain amount of stress to begin flowing. This makes it applicable to materials like chocolate, printer inks, and various food products, where precise control over flow properties is essential. The Casson model predicts that below a critical yield stress, the fluid behaves like a solid, and beyond this stress, it flows like a viscous fluid with decreasing viscosity as the shear rate increases. This model has proven invaluable in industries requiring accurate complex fluid modeling to optimize processing conditions and ensure product quality. Its ability to capture the dual behavior of these materials—solid-like under low stress and fluid-like under high stress—makes it a crucial tool for engineers and scientists working with non-Newtonian fluids.

The Bird–Carreau model [8, 9, 17, 27, 28] is a sophisticated and versatile non-Newtonian fluid model extensively utilized to characterize the shear-thinning behavior of complex fluids over a wide spectrum of shear rates. This model augments the foundational Carreau model by incorporating temperature dependency, enhancing its applicability across industries with varied thermal conditions. It effectively captures the fluid's transition from Newtonian to non-Newtonian behavior by accounting for zero-shear and infinite-shear viscosity. The Bird–Carreau model is particularly pertinent in polymer processing, where it plays a crucial role in optimizing operations such as extrusion and injection molding by accurately predicting viscosity changes under different shear conditions. Its comprehensive nature enables precise modelling of polymer solutions and melts, ensuring efficient production processes and superior quality of the final products. By providing a robust theoretical framework, the Bird–Carreau model significantly advances theoretical research and practical applications in fluid dynamics and materials engineering, facilitating a deeper understanding and control of non-Newtonian fluid behaviors.

The Sisko model [29–32] and the Cross model [33, 34] are prominent non-Newtonian fluid models, each offering a nuanced approach to understanding fluid behavior under varying shear rates. The Sisko model is particularly significant for its ability to describe shear-thinning fluids exhibiting a yield stress component. This model is instrumental in applications where accurate viscosity reduction under stress modeling is crucial, such as in lubricants and drilling fluids. On the other hand, the Cross model is widely recognized for its comprehensive depiction of shear-thinning behavior across a broad range of shear rates. By incorporating both a Newtonian plateau at low shear rates and a power-law region at higher shear rates, the Cross model provides a robust framework for understanding and predicting the flow characteristics of polymer solutions and melts. Both models significantly advance our ability to simulate and optimize processes involving complex fluids, enhancing efficiency and control in industrial applications ranging from polymer processing to food production. These models deepen our theoretical understanding of non-Newtonian fluid dynamics and provide practical solutions for engineering challenges.

The Oldroyd models [35–38] and the Jeffreys model [9, 39, 40] are pivotal in non-Newtonian fluid dynamics, each contributing significantly to understanding viscoelastic fluids. The Oldroyd models, including the Oldroyd-A and Oldroyd-B models, extend the Maxwell model by incorporating both elastic and viscous components to describe the complex time-dependent behavior of viscoelastic fluids. These models are instrumental in capturing the intricate interplay between elasticity and viscosity, making them crucial for applications involving polymer solutions and melt, where both instantaneous and delayed stress responses are observed. On the other hand, the Jeffreys model offers a simplified yet robust framework for describing viscoelastic fluids by combining elements of the Maxwell and Kelvin–Voigt models. This dual approach allows the Jeffreys model to effectively represent fluids that exhibit both elastic and viscous characteristics, making it applicable to various materials, from biological tissues to industrial polymers. Both models are essential for advancing the theoretical understanding and practical application of non-Newtonian fluids, providing engineers and scientists with the tools needed to predict and control the behavior of these complex materials under various flow conditions.

Fractional models [8, 41, 42] represent an advanced and nuanced approach to describing non-Newtonian fluids' rheological behavior, incorporating fractional calculus to capture complex time-dependent and frequency-dependent characteristics that conventional models do not adequately address. These models are particularly adept at characterizing fluids with memory effects, where the current stress state depends on the present deformation and the entire history of deformation. This allows for a more accurate representation of materials like biological tissues, polymers, and suspensions, where viscoelastic properties and anomalous diffusion processes play significant roles. By employing fractional derivatives, these models provide a comprehensive framework for understanding the intricate dynamics of non-Newtonian fluids under various conditions, significantly advancing both theoretical research and practical applications. The versatility and precision of fractional models make them invaluable in biomedical engineering, materials science, and complex fluid dynamics, where traditional models fall short of capturing the full spectrum of fluid behaviors.

The Phan–Thien–Tanner (PTT) model [43–46] is a sophisticated non-Newtonian fluid model widely utilized in studying viscoelastic materials, particularly polymer melts and solutions. This model is renowned for accurately capturing the complex rheological behavior of viscoelastic fluids, encompassing both shear-thinning viscosity and normal stress differences. It employs a constitutive equation that effectively describes these materials' stress relaxation and elastic recovery characteristics, which are critical for understanding and predicting their flow behavior under various processing conditions. The PTT model is particularly valuable in the polymer industry, where it informs the design and optimization of processes such as extrusion, injection molding, and fiber spinning. The Phan–Thien–Tanner Model significantly advances both theoretical research and practical applications in materials science and engineering by providing a robust framework for simulating the intricate dynamics of viscoelastic fluids.

The tangent hyperbolic fluid model [47–52] is an advanced constitutive model used to describe the rheological behavior of non-Newtonian fluids, particularly those exhibiting shear-thinning properties. This model is characterized by its ability to capture the complex flow behavior of fluids with a hyperbolic tangent relationship between shear stress and shear rate, making it particularly suitable for materials such as polymer melts and solutions. The tangent hyperbolic model provides a more accurate representation of the viscosity reduction under varying shear rates than traditional models, enhancing the predictive capabilities for industrial processes involving complex fluid flows. Its application extends to various fields, including polymer processing, food engineering, and biomedical applications, where understanding the nuanced flow behavior of non-Newtonian fluids is crucial for optimizing process efficiency and product quality.

The Giesekus model [53–55] is an advanced and widely regarded constitutive model for viscoelastic fluids, primarily used to capture the complex shear-thinning behavior and normal stress differences typical of polymer melts and solutions. This model integrates linear and non-linear viscoelastic elements by incorporating an anisotropic drag term, which is crucial for accurately representing the fluid's response under diverse shear conditions. The Giesekus model is instrumental in the polymer processing industry, enabling precise simulations of extrusion, injection molding, and fiber spinning processes. By accurately predicting the material's rheological behavior, this model aids in optimizing process parameters, enhancing product quality, and ensuring efficiency. Its robust framework and comprehensive approach to modeling viscoelastic properties make the Giesekus model an indispensable tool in both theoretical research and practical applications, significantly advancing complex fluid dynamics.

1.1.6 Summary

The exploration of various non-Newtonian fluid models demonstrates their indispensable roles in accurately capturing and predicting the complex behaviors of non-Newtonian fluids across a range of applications. The power law model effectively describes shear-thinning and shear-thickening behaviors, making it essential for industries like food processing and cosmetics, where understanding viscosity changes with shear rate is crucial. The Carreau model extends this by capturing shear-thinning behavior over a broad range of shear rates, which is particularly useful for polymer solutions and melts and enhances the predictive capabilities for process optimization. The Herschel–Bulkley model combines yield stress with shear-thinning or -thickening behavior, making it invaluable for materials such as paints and drilling muds, where precise control over flow properties is essential. The Casson model addresses fluids with yield stress, such as blood and chocolate, providing critical insights into their flow dynamics, which is vital for medical and food industries. The Bird–Carreau model enhances the Carreau model by incorporating temperature dependence, making it valuable for polymer processing under varied thermal conditions, thereby ensuring efficient and high-quality production processes. The Sisko model describes shear-thinning fluids with a yield stress component, making it useful in applications such as lubricants and drilling fluids, where accurate modelling of viscosity reduction under stress is crucial. The Cross model provides a comprehensive depiction of shear-thinning behavior across a range of shear rates, making it popular in polymer processing, where understanding the full range of viscosity behavior is necessary for process control. The Oldroyd models capture both elastic and viscous components, which are crucial for viscoelastic fluids like polymer solutions, offering a more nuanced understanding of their time-dependent behaviors. The Jeffreys model combines elements of the Maxwell and Kelvin–Voigt models, effectively representing fluids that exhibit both elastic and viscous characteristics, thus broadening its applicability across various materials. Fractional models employ fractional calculus to capture complex time-dependent and frequency-dependent characteristics, which are increasingly used in advanced material studies where traditional models fall short. The Phan–Thien–Tanner (PTT) model accurately captures the viscoelastic behavior of polymer melts, providing insights critical for optimizing processes such as extrusion and injection molding in the polymer industry. The tangent hyperbolic fluid model captures shear-thinning properties using a hyperbolic tangent relationship, making it suitable for accurately describing the flow behavior of complex fluids in various engineering applications. The Giesekus model incorporates an anisotropic drag term to accurately represent the rheological properties of viscoelastic fluids, which is widely used in polymer science to enhance the predictive capabilities of computational fluid dynamics simulations. These models collectively advance our understanding of non-Newtonian fluid dynamics, driving innovation and efficiency across multiple fields by enabling more accurate predictions and optimizing complex fluid behaviors.

- The power law model describes shear-thinning and shear-thickening behaviors widely used in food processing and cosmetics.
- The Carreau model captures shear-thinning behavior over a broad range of shear rates, essential for polymer solutions and melts.
- The Herschel–Bulkley model combines yield stress with shear-thinning/thickening behavior, which is applied in paints and drilling muds.
- The Casson model is used for fluids with yield stress, such as blood and chocolate, and provides critical insights into flow dynamics.
- The Bird–Carreau model enhances the Carreau model with temperature dependence, which is valuable for polymer processing.
- Sisko's model describes shear-thinning fluids with a yield stress component, which is useful in lubricants and drilling fluids.
- A cross-model comprehensive depiction of shear-thinning behavior across a range of shear rates, popular in polymer processing.
- Oldroyd models capture both elastic and viscous components, which is crucial for viscoelastic fluids like polymer solutions.
- Jeffreys's model combines Maxwell and Kelvin–Voigt elements and represents fluids with both elastic and viscous characteristics.

- Fractional models use fractional calculus to capture complex rheological behaviors, which is increasingly employed in advanced material studies.
- The Phan–Thien–Tanner (PTT) model describes the viscoelastic behavior of polymer melts and aids in optimizing industrial processes.
- The tangent hyperbolic fluid model captures shear-thinning properties using a hyperbolic tangent relationship, which is suitable for polymer processing.
- The Giesekus model incorporates an anisotropic drag term that accurately represents viscoelastic fluids, enhancing computational fluid dynamics simulations.

AI disclosure statement: The author acknowledges the use of AI-based tools for minor language editing or grammatical correction. However, all intellectual contributions, analyses, and conclusions presented in this chapter are entirely the author's own.

References

1. Irgens, F., *Rheology and Non-Newtonian Fluids*, Springer International Publishing, Germany, 2013.
2. Farina, A., Fusi, L., Mikelić, A., Saccomandi, G., Sequeira, A., Toro, E.F., Non-Newtonian Fluid Mechanics and Complex Flows Levico Terme, Italy 2016, in: *Lecture Notes in Mathematics*, A. Farina, A. Mikelić, F. Rosso (eds.), Springer International Publishing, 2018.
3. Wu, W.-T. and Massoudi, M., Recent Advances in Mechanics of Non-Newtonian Fluids. *Fluids*, 5, 2020. DOI: 10.3390/fluids5010010.
4. Chhabra, R.P., *Non-Newtonian Fluids: An Introduction, in Rheology of Complex Fluids*, J.M. Krishnan, A.P. Deshpande, P.B.S. Kumar (Eds.), pp. 3–34, Springer New York, New York, NY, 2010.
5. Spagnolie, S.E., Complex fluids in biological systems. *Biol. Med. Phys. Biomed. Eng.*, 2015.
6. Owasit, P. and Sriyab, S., Mathematical modeling of non-Newtonian fluid in arterial blood flow through various stenoses. *Adv. Differ. Equations*, 2021, 1, 340, 2021.
7. Nabwey, H.A., *et al.*, A Comprehensive Review of Non-Newtonian Nanofluid Heat Transfer. *Symmetry*, 15, 2023. DOI: 10.3390/sym15020362.
8. Sun, H., *et al.*, A review of constitutive models for non-Newtonian fluids. *Fract. Calc. Appl. Anal.*, 27, 4, 1483–1526, 2024.

9. Wajihah, S.A. and Sankar, D.S., A review on non-Newtonian fluid models for multi-layered blood rheology in constricted arteries. *Arch. Appl. Mech.*, 93, 5, 1771–1796, 2023.
10. Tedeschi, F., *et al.*, A multi-scale method for complex flows of non-Newtonian fluids. *Math. Eng.*, 4, 6, 1–22, 2022.
11. Thohura, S., Molla, M.M., Sarker, M.M.A., Numerical Simulation of Non-Newtonian Power-Law Fluid Flow in a Lid-Driven Skewed Cavity. *Int. J. Appl. Comput. Math.*, 5, 1, 14, 2019.
12. Chen, Q., *et al.*, Modeling of Newtonian droplet formation in power-law non-Newtonian fluids in a flow-focusing device. *Heat Mass Transfer*, 56, 9, 2711–2723, 2020.
13. Huo, Q. and Wang, X., Mixing mechanism of power-law non-Newtonian fluids in resonant acoustic mixing. *Phys. Fluids*, 36, 2, 2024.
14. Bilal, S., *et al.*, Numerical Analysis of Natural Convection Driven Flow of a Non-Newtonian Power-Law Fluid in a Trapezoidal Enclosure with a U-Shaped Constructal. *Energies*, 14, 2021. DOI: 10.3390/en14175355.
15. Anguiano, M., Bonnivard, M., Suárez-Grau, F.J., Carreau law for non-Newtonian fluid flow through a thin porous media. *Q. J. Mech. Appl. Math.*, 75, 1, 1–27, 2022.
16. Eberhard, U., *et al.*, Determination of the Effective Viscosity of Non-newtonian Fluids Flowing Through Porous Media. *Front. Phys.*, 7, 2019.
17. Zhou, S.-s., *et al.*, Insight into the dynamics of non-Newtonian Carreau fluid when viscous dissipation, entropy generation, convective heating and diffusion are significant. *Appl. Math. J. Chin. Univ.*, 39, 1, 34–46, 2024.
18. Ghorbani, S., *et al.*, Transient heat transfer and electro-osmotic flow of Carreau–Yasuda non-Newtonian fluid through a rectangular microchannel. *Int. J. Numer. Methods Heat Fluid Flow*, 33, 7, 2439–2454, 2023.
19. Ashraf, F. and Park, C.W., Fluid–structure interaction analysis of non-Newtonian Herschel–Bulkley fluid viscosity model for pulsating flow of blood in ω-shaped stenosed arteries. *Korea-Aust. Rheol. J.*, 34, 1, 51–67, 2022.
20. Ashena, R., *et al.*, Stepwise mathematical derivation of the Herschel–Bulkley laminar fluid flow equations—in pipes. *J. Pet. Explor. Prod. Technol.*, 13, 2, 625–643, 2023.
21. Abbas, Z., Shabbir, M.S., Ali, N., Analysis of rheological properties of Herschel-Bulkley fluid for pulsating flow of blood in ω-shaped stenosed artery. *AIP Adv.*, 7, 10, 2017.
22. Srinivasacharya, D. and Kumar, R.S., Artificial neural network modeling of the Casson fluid flow over unsteady radially stretching sheet with Soret and Dufour effects. *J. Therm. Anal. Calorim.*, 147, 24, 14891–14903, 2022.
23. Oke, A.S., *et al.*, Insight into the dynamics of non-Newtonian Casson fluid over a rotating non-uniform surface subject to Coriolis force. *Nonlinear Eng.*, 9, 1, 398–411, 2020.

24. Baitharu, A.P., *et al.*, Dufour Effect on Chemically Reacting Casson Fluid Flow Over a Nonlinearly Stretching Sheet with Radiation Absorption. *Proc. Natl. Acad. Sci., India, Sect. A Phys. Sci.*, 94, 3, 325–333, 2024.
25. Rao, A.S., *et al.*, Mathematical Modelling of Hydromagnetic Casson non-Newtonian Nanofluid Convection Slip Flow from an Isothermal Sphere. 8, 1, 645–660, 2019.
26. Jalili, P., *et al.*, study of non-linear radiative heat transfer with magnetic field for non-Newtonian Casson fluid flow in a porous medium. *Results Phys.*, 48, 106371, 2023.
27. Shende, T., Niasar, V.J., Babaei, M., Effective viscosity and Reynolds number of non-Newtonian fluids using Meter model. *Rheol. Acta*, 60, 1, 11–21, 2021.
28. Hersey, E., Rodriguez, Jr., M., Johnsen, E., Dynamics of an oscillating microbubble in a blood-like Carreau fluid. *J. Acoust. Soc. Am.*, 153, 3, 1836–1845, 2023.
29. Bisht, A. and Sharma, R., Non-similar solution of Sisko nanofluid flow with variable thermal conductivity: a finite difference approach. *Int. J. Numer. Methods Heat Fluid Flow*, 31, 1, 345–366, 2021.
30. Waseem, *et al.*, Exploring hybrid Sisko nanofluid on a stretching sheet: an improved cuckoo search-based machine learning approach. *Eur. Phys. J. Plus*, 139, 8, 760, 2024.
31. Rafiq, T. and Mustafa, M., Steadily revolving flow of Sisko fluid along a stretchable boundary with non-linear radiation effects. *Pramana*, 95, 1–10, 2021.
32. Jayalakshmi, P., *et al.*, Heat transfer analysis of Sisko fluid flow over a stretching sheet in a conducting field with Newtonian heating and constant heat flux. *Energies*, 16, 7, 3183, 2023.
33. Nava, G., *et al.*, Newtonian to non-newtonian fluid transition of a model transient network. *Soft Matter*, 14, 17, 3288–3295, 2018.
34. Hauswirth, S.C., *et al.*, Modeling cross model non-Newtonian fluid flow in porous media. *J. Contam. Hydrol.*, 235, 103708, 2020.
35. Kenea, G. and Ibrahim, W., Non-linear convection stagnation point flow of Oldroyd-B nanofluid with non-Fourier heat and non-Fick's mass flux over a spinning sphere. *Sci. Rep.*, 14, 1, 841, 2024.
36. Deng, C., He, Q., Yan, D., Global Solutions to the 2-Dimensional Incompressible Oldroyd-B Model with Hybrid Dissipation and Partial Damping on Stress Tensor. *Results Math.*, 76, 1–15, 2021.
37. Martin, R., Zinchenko, A., Davis, R., A generalized Oldroyd's model for non-Newtonian liquids with applications to a dilute emulsion of deformable drops. *J. Rheol.*, 58, 3, 759–777, 2014.
38. Tanner, R.I., A modified Oldroyd-B model for non-colloidal suspensions. *Phys. Fluids*, 33, 7, 2021.
39. Prasad, V.R., *et al.*, Numerical study of non-Newtonian Jeffreys fluid from a permeable horizontal isothermal cylinder in non-Darcy porous medium. *J. Braz. Soc. Mech. Sci. Eng.*, 37, 1765–1783, 2015.
40. Asgir, M., *et al.*, Special function form exact solutions for Jeffery fluid: an application of power law kernel. *Adv. Differ. Equations*, 2021, 1–18, 2021.

41. Liu, Y., *et al.*, Finite difference scheme for simulating a generalized two-dimensional multi-term time fractional non-Newtonian fluid model. *Adv. Differ. Equations*, 2018, 1, 442, 2018.
42. Nadeem, S., *et al.*, Fractional Nadeem trigonometric non-Newtonian (NTNN) fluid model based on Caputo-Fabrizio fractional derivative with heated boundaries. *Sci. Rep.*, 13, 1, 21511, 2023.
43. Ferrás, L.L., *et al.*, A generalised Phan–Thien—Tanner model. *J. Non-Newtonian Fluid Mech.*, 269, 88–99, 2019.
44. Latreche, S., *et al.*, Flow Dynamics of PTT and FENE-P Viscoelastic Fluids in Circular and Flat Ducts: An Analytical Study. *Arabian J. Sci. Eng.*, 46, 3, 2783–2792, 2021.
45. Hussain, S., Ali, N., Ullah, K., Peristaltic flow of Phan-Thien-Tanner fluid: effects of peripheral layer and electro-osmotic force. *Rheol. Acta*, 58, 9, 603–618, 2019.
46. Evans, J.D., Palhares Junior, I.L., Oishi, C.M., Stresses of PTT, Giesekus, and Oldroyd-B fluids in a Newtonian velocity field near the stick-slip singularity. *Phys. Fluids*, 29, 12, 2017.
47. Azar, A.A., *et al.*, Analytical solution for MHD nanofluid flow over a porous wedge with melting heat transfer. *Heliyon*, 10, 15, 2024.
48. Abdul Gaffar, S., *et al.*, Free convection flow and heat transfer of non-Newtonian tangent hyperbolic fluid from an isothermal sphere with partial slip. *Arabian J. Sci. Eng.*, 39, 8157–8174, 2014.
49. Gaffar, S.A., Prasad, V.R., Reddy, E.K., Computational study of MHD free convection flow of non-Newtonian tangent hyperbolic fluid from a vertical surface in porous media with Hall/ionslip currents and ohmic dissipation. *Int. J. Appl. Comput. Math.*, 3, 2, 859–890, 2017.
50. Gaffar, S.A., Ramachandra Prasad, V., Bég, O.A., Computational Analysis of Magnetohydrodynamic Free Convection Flow and Heat Transfer of Non-Newtonian Tangent Hyperbolic Fluid from a Horizontal Circular Cylinder with Partial Slip. *Int. J. Appl. Comput. Math.*, 1, 4, 651–675, 2015.
51. Shahzad, F., Sagheer, M., Hussain, S., MHD tangent hyperbolic nanofluid with chemical reaction, viscous dissipation and Joule heating effects. *AIP Adv.*, 9, 2, 2019.
52. Gaffar, S.A., *et al.*, Computation of hydromagnetic tangent hyperbolic non-Newtonian flow from a rotating non-isothermal cone to a non-Darcy porous medium with thermal radiative flux. *Phys. Open*, 19, 100216, 2024.
53. Kudish, I.I., Pashkovski, E., Patterson, R., Line contact lubricated by a fluid described by non-Newtonian Giesekus model. *IMA J. Appl. Math.*, 87, 5, 722–756, 2022.
54. Kudish, I.I. and Volkov, S.S., A General Case of a Line Contact Lubricated by a Non-Newtonian Giesekus Fluid. *Mathematics*, 11, 22, 4679, 2023.
55. Liu, B., *et al.*, Migration of spherical particles in a confined shear flow of Giesekus fluid. *Rheol. Acta*, 58, 639–646, 2019.

2

Viscoelastic Fluid Models

Mukesh Kumar Awasthi[1]*, Atul Kumar Shukla[2] and Dhananjay Yadav[3]

[1]Department of Mathematics, Babasaheb Bhimarao Ambedkar University, Lucknow, India
[2]Department of Mathematics, Ganga Singh College, Chapra, Bihar, India
[3]Department of Mathematical & Physical Sciences, University of Nizwa, Nizwa, Oman

Abstract

This chapter provides a comprehensive exploration of viscoelastic fluids, detailing their fundamental properties, classifications, and mathematical models. It begins by contrasting Newtonian, non-Newtonian, and viscoelastic fluids based on molecular dynamics and rheological behavior. Special emphasis is placed on viscoelastic fluids, which exhibit both viscous and elastic characteristics, making them essential in biomedical, industrial, and environmental applications. The chapter systematically introduces various viscoelastic models—such as Rivlin–Ericksen, Reiner–Rivlin, Maxwell, Oldroyd, power law, Bingham plastic, Ellis, Reiner–Philippoff, Prandtl, Eyring, Power–Eyring, Williamson, and Walters' B fluids—along with their governing equations and relevance in capturing complex fluid behaviors. Real-world applications and emerging trends, including machine-learning integration, multiscale modeling, and smart constitutive laws, are discussed to underscore the growing significance of viscoelastic fluid mechanics in modern science and engineering.

***Keywords*:** Viscoelastic fluids, non-Newtonian fluids, rheological models, maxwell model, oldroyd model, constitutive equations, multiscale modeling

**Corresponding author*: mukeshiitr.kumar@gmail.com; ORCID: 0000-0002-6706-5226
Atul Kumar Shukla: ORCID: 0000-0002-9588-0084
Dhananjay Yadav: ORCID: 0000-0001-8404-2053

Dhananjay Yadav, Mukesh Kumar Awasthi and Harith Mohamed Al-Azri (eds.) Non-Newtonian Fluids for Industrial Applications: Modeling and Simulations, (29–54) © 2026 Scrivener Publishing LLC

2.1 Fluids

Matter in the universe is broadly classified into solids and fluids based on their response to external forces. Solids retain a fixed shape under thermodynamic equilibrium unless subjected to external forces, as their molecular arrangement resists deformation due to strong intermolecular bonds. In contrast, fluids adapt to the shape of their container, as their molecules exhibit relative motion even in the absence of external forces. This fundamental distinction arises from differences in molecular dynamics: in solids, molecules vibrate around fixed positions, while in fluids, molecules slide past one another, enabling flow.

2.1.1 Molecular Perspective

The behavior of solids and fluids can be understood through intermolecular forces. In solids, cohesive forces dominate, creating a rigid lattice structure. When external shear forces exceed a critical threshold (yield stress), some solids, termed "plastic solids", exhibit flow. For example, glaciers flow over geological timescales under sustained stress. Fluids, however, lack a fixed structure. Even minimal forces induce molecular displacement, as seen when water poured into a glass conforms to its shape.

2.1.2 Newtonian Fluids

Newton noted that when two adjacent fluid layers move in straight lines with differing velocities, a tangential force arises between them, directly proportional to their relative speed. Specifically, if one layer travels at velocity u and the neighboring layer at $u+\delta u$, separated by a distance δy, the resulting shear stress τ can be expressed as $\tau = \mu \frac{du}{dy}$ or $\tau = \mu\dot{\gamma}$, where μ (the viscosity coefficient) quantifies the fluid's resistance to deformation, and $\dot{\gamma}$ represents the strain rate. This principle, known as the Newtonian hypothesis, asserts a linear relationship (Figure 2.1) between shear stress and strain rate in fluids. Fluids adhering to this behavior are termed Newtonian fluids [1].

The viscosity coefficient μ has dimensions of $ML^{-1}T^{-1}$ and varies with temperature and pressure but remains unaffected by the strain rate. When plotted, the shear stress versus strain rate for such fluids forms a straight line (termed a "flow curve") with a slope equal to μ. Additionally, the ratio of dynamic viscosity μ to fluid density ρ defines the kinematic viscosity

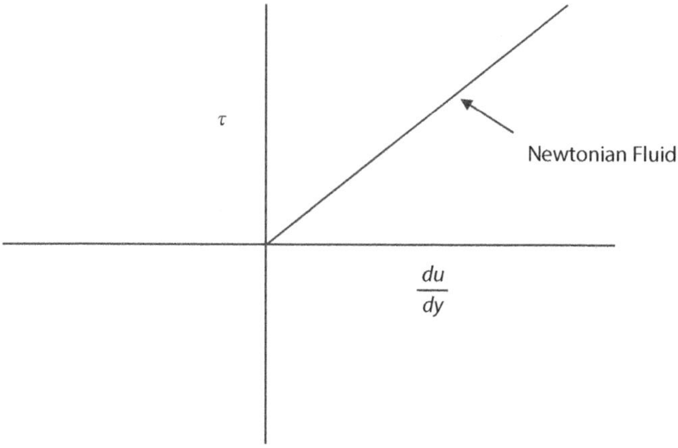

Figure 2.1 The relationship between shear stress τ and rate of strain $\frac{du}{dy}$.

ν, expressed as $\nu = \mu/\rho$, which has dimensions of L^2T^{-1}. This parameter helps characterize a fluid's resistance to flow under gravitational forces.

2.1.3 Non-Newtonian Fluids

Non-Newtonian fluids are defined by their deviation from the linear stress–strain rate relationship observed in Newtonian fluids. Unlike their Newtonian counterparts, the viscosity of non-Newtonian fluids is not fixed under constant temperature and pressure but instead varies with factors such as shear rate, container geometry, and even the fluid's mechanical history. Common Newtonian examples include water, air, and mercury, whereas paints, coal tar, and polymer solutions exemplify non-Newtonian behavior. Viscous effects in fluid flow arise from two primary mechanisms:

Cohesive molecular forces: Intermolecular forces resist internal relative motion. Elevated temperatures increase molecular separation, reducing these cohesive forces.
Molecular vibrations: Enhanced thermal agitation at higher temperatures amplifies molecular motion, influencing viscosity.

Newtonian fluid dynamics are governed by the Navier–Stokes equations. In contrast, Stokes [2] theory of couple-stress fluids introduces additional stress components arising from microscopic torque interactions, particularly notable in polymer solutions. Experimental studies reveal deviations from Newtonian behavior, necessitating specific material constants. The constitutive equations for couple-stress fluids are expressed as follows:

$$T_{(ij)} = \left(-p + \lambda D_{kk}\right)\delta_{ij} + 2\mu D_{ij}, T_{[ij]} = -2\eta \mathbf{W}_{ij,kk} - \frac{\rho}{2}\varepsilon_{ijs}G_s$$

and $$M_{ij} = 4\eta\omega_{j,i} + 4\eta'\omega_{i,j}$$

where $$D_{ij} = \frac{1}{2}\left(V_{i,j} + V_{j,i}\right), \mathbf{W}_{ij} = -\frac{1}{2}\left(V_{i,j} - V_{j,i}\right) \text{ and } \omega_i = \frac{1}{2}\varepsilon_{ijk}V_{k,j}.$$

Here, $T_{ij}, T_{(ij)}, T_{[ij]}, \mathbf{M}_{ij}, D_{ij}, \mathbf{W}_{ij}, \omega_i, G_s, \varepsilon_{ijs}, V, \rho, \lambda, \mu, \eta$, and η' are the stress tensor, the symmetric part of T_{ij}, the anti-symmetric part of T_{ij}, the couple-stress tensor, the deformation tensor, the vorticity tensor, the vorticity vector, the body couple, the alternating unit tensor, the velocity field, the density, and material constants, respectively. The dimensions of λ and μ are those of viscosity, whereas the dimensions of η and η' are those of momentum.

2.2 Viscoelastic Fluids

Viscoelastic fluids are a unique class of materials that exhibit both viscous (liquid-like) and elastic (solid-like) properties when subjected to mechanical stress. Unlike simple liquids or solids, these fluids "remember" their past deformations to some extent, leading to complex behaviors such as stress relaxation, creep, and recoil. This duality arises from their molecular or microstructural composition, which often includes long-chain polymers, colloidal particles, or biological macromolecules that entangle or interact dynamically. The study of viscoelastic fluids bridges the gap between classical fluid mechanics and solid mechanics, offering insights into phenomena that cannot be explained by Newtonian fluid models alone.

The physical significance of viscoelastic fluids lies in their ubiquity across natural and industrial systems. For instance, biological fluids like blood and mucus, synthetic materials such as polymer melts, and everyday substances like shampoo or toothpaste all display viscoelastic traits. Understanding these fluids is critical for optimizing processes in manufacturing, healthcare, and environmental engineering. Their ability to store and dissipate energy under deformation makes them essential in applications requiring shock absorption, lubrication, or precise flow control.

2.2.1 Differences between Newtonian, Non-Newtonian, and Viscoelastic Fluids

Newtonian fluids, such as water or air, obey a linear relationship between shear stress and shear rate, as described by Newton's law of viscosity. Their viscosity remains constant regardless of the applied force, making their behavior predictable under most conditions. In contrast, non-Newtonian fluids defy this linearity, with viscosity varying depending on the shear rate or stress. This category includes shear-thinning fluids (e.g., ketchup), shear-thickening fluids (e.g., cornstarch–water mixtures), and viscoelastic fluids.

Viscoelastic fluids are a specialized subset of non-Newtonian fluids distinguished by their elastic response. While all non-Newtonian fluids exhibit nonlinear viscosity, viscoelastic fluids uniquely combine this with elasticity, enabling them to partially recover their shape after deformation. For example, a Newtonian fluid like honey flows steadily under gravity, whereas a viscoelastic fluid like polymer solution might form a rotating rod-climbing effect (the "Weissenberg effect") due to elastic normal stresses. This elasticity introduces time-dependent behaviors, such as delayed deformation under constant stress (creep) or gradual stress reduction under fixed strain (stress relaxation), which are absent in purely viscous non-Newtonian fluids.

2.2.2 Real-World Examples and Applications

Viscoelastic fluids play vital roles in industries and everyday life. In biomedical engineering, synovial fluid in joints acts as a viscoelastic lubricant, reducing friction during movement while cushioning impacts. Similarly, mucus in the respiratory tract traps pathogens and is expelled through coordinated elastic recoil during coughing. In consumer products, shampoos and conditioners leverage viscoelasticity to balance flowability during pouring with adherence to hair strands, enhancing texture and effectiveness.

Industrial applications are equally profound. In enhanced oil recovery, viscoelastic polymer solutions are injected into reservoirs to improve sweep efficiency by reducing water mobility and altering flow patterns in porous rock. The food industry relies on viscoelasticity in products like melted cheese or gelatin desserts, where texture and mouth feel depend on controlled elasticity. Even manufacturing processes such as 3D printing of soft materials or extrusion of plastics require precise management of viscoelastic properties to avoid defects like warping or die swell.

Environmental and technological applications further highlight their importance. Viscoelastic surfactants are used in fracturing fluids for hydraulic fracking, minimizing environmental impact by reducing water usage. In microfluidics, viscoelastic fluids enable particle focusing and sorting without external forces, revolutionizing lab-on-a-chip diagnostics. As research advances, these fluids continue to unlock innovations in soft robotics, smart materials, and sustainable technologies, underscoring their indispensable role in modern science and engineering.

2.3 Viscoelastic Fluid Models

Elastic materials exhibit a direct relationship between stress and strain, meaning the internal force generated is solely a function of deformation. When stress is applied, these materials deform but regain their original shape upon stress removal, demonstrating a near-perfect "memory" of their initial configuration. In contrast, fluids respond to stress through deformation rates rather than deformation itself. Removing stress halts ongoing deformation, yet any prior distortion remains permanent, indicating fluids inherently lack memory. However, certain substances like polymer solutions or liquid soap blend fluidic and elastic traits, earning the classification of viscoelastic fluids.

At rest, viscoelastic fluids are isotropic and homogeneous. Under shear stress, their behavior is primarily governed by shear rate, yet diverse responses to stress have led to multiple rheological models. Notable examples include the following.

2.3.1 Rivlin–Ericksen Fluids

Rivlin–Ericksen fluids are a class of non-Newtonian fluids that fall under the category of differential type fluids, particularly used to model viscoelastic behavior. These fluids are characterized by a constitutive relationship that incorporates not only the rate of deformation tensor but also its time derivatives, thereby accounting for the fluid's memory effects.

They are named after Ronald S. Rivlin and James L. Ericksen [3], who formulated a general theory for the stress–deformation relationship in complex fluids. These models are especially useful in describing polymeric liquids, biological fluids, and other complex materials where the classical Newtonian assumption fails.

The general form of the constitutive equation for a Rivlin–Ericksen fluid of order n is given as

$$\tau = -p\boldsymbol{I} + \sum_{i=1}^{n} \alpha_i \boldsymbol{A}_i \tag{2.1}$$

where τ denotes the Cauchy stress tensor, p is the hydrostatic pressure, $\boldsymbol{I}$ is the identity tensor, α_i represent the material constants, and $\boldsymbol{A}_i$ are the Rivlin–Ericksen tensors of order i.

The first Rivlin–Ericksen tensor, $\boldsymbol{A}_1$, is defined as

$$\boldsymbol{A}_1 = (\nabla \boldsymbol{q}) + (\nabla \boldsymbol{q})^T \tag{2.2}$$

where $\boldsymbol{q}$ is the velocity field.

The second Rivlin–Ericksen tensor, $\boldsymbol{A}_2$, is

$$\boldsymbol{A}_2 = \frac{D}{Dt}\boldsymbol{A}_1 + \boldsymbol{A}_1 \cdot \nabla \boldsymbol{q} + (\nabla \boldsymbol{q})^T \cdot \boldsymbol{A}_1 \tag{2.3}$$

Higher-order tensors are similarly defined recursively.

In geophysical fluid dynamics, Rivlin–Ericksen (R-E) fluids serve as a framework for analyzing viscous flow behavior in channels under time-varying pressure gradients. Dharamendra *et al.* [4] explored interfacial dynamics between a Newtonian viscous liquid and an R-E viscoelastic fluid, employing the normal mode method to assess instability patterns at their boundary. Subsequent studies expanded on this foundation: Awasthi *et al.* [5] evaluated how heat and mass transfer influence cylindrical interfaces involving R-E and viscous fluids, while Srija *et al.* [6] analyzed rotational effects on such interfaces during mass transport. In scenarios involving heat transfer, Awasthi *et al.* [7] demonstrated that applying an axial electric field to the R-E-viscous fluid interface suppresses instability development. Additional investigations [8–10] have further explored heat transfer's role in fluid dynamics across diverse geometries.

2.3.2 Reiner–Rivlin Fluids

Reiner–Rivlin fluids belong to a class of non-Newtonian fluids introduced by Mark Reiner [11] and Ronald Rivlin [12] to model complex fluid behaviors that cannot be captured by Newtonian models. Unlike Newtonian fluids (where the stress is linearly proportional to the rate of strain), Reiner–Rivlin fluids incorporate nonlinear dependencies, making them suitable for describing materials like polymers, suspensions, and biological fluids.

The constitutive equation for a Reiner–Rivlin fluid is a generalization of the Newtonian fluid model and is based on the Rivlin–Ericksen theory. It expresses the Cauchy stress tensor τ in terms of the rate-of-deformation tensor and other kinematic quantities.

The general form of the Reiner–Rivlin constitutive equation is

$$\tau = -p\boldsymbol{I} + \alpha_1 \boldsymbol{A}_1 + \alpha_2 \boldsymbol{A}_2 + \alpha_3 \boldsymbol{A}_1^2 + \dots\dots\dots \tag{2.4}$$

where τ denotes the Cauchy stress tensor; p is the hydrostatic pressure; $\boldsymbol{I}$ is the identity tensor; α_i represent the material constants. The Rivlin–Ericksen tensors $\boldsymbol{A}_1, \boldsymbol{A}_2$ are defined in Equations 2.2 and 2.3.

If $\alpha_1 = \mu$ (viscosity) and all other $\alpha_i = 0$, the Reiner–Rivlin model reduces to Newtonian fluid behavior. Keeping only α_1 and α_2 nonzero, one obtains a second-order fluid, a subset often used for weakly viscoelastic flows. If $\alpha_3 \neq 0$, it introduces nonlinear effects like shear-thickening or thinning behavior.

Kosterin [13] critiqued the applicability of the Reiner–Rivlin rheological model in accurately characterizing normal stress effects. Subsequently, Gao and Hartnett [14] numerically investigated Reiner–Rivlin fluid flow through rectangular ducts, with their analysis revealing a pronounced enhancement in heat transfer under the influence of secondary normal stresses. In separate work, Attia [15] explored unsteady Reiner–Rivlin fluid dynamics near a rotating permeable disk, incorporating suction and injection effects. The study concluded that suction significantly amplifies heat transfer compared to injection. Expanding on this, Attia [16] examined Reiner–Rivlin fluid behavior with ion slip and Hall current effects in a rotating disk system, highlighting that ion slip exerts a more marked influence on axial velocity in Reiner–Rivlin fluids than in Newtonian counterparts. Meanwhile, Tabassum and Mustafa [17] analyzed partial slip effects in Reiner–Rivlin fluid flow driven by a rotating disk, demonstrating that sustaining steady disk rotation necessitates elevated torque magnitudes and stronger slip parameters.

2.3.3 Maxwell Fluids

Maxwell fluids are a class of viscoelastic fluids that exhibit both elastic and viscous behavior. The model was originally proposed by James Clerk Maxwell [18] to describe the flow behavior of materials that do not fit neatly into purely viscous (Newtonian) or purely elastic (Hookean solid) categories. Maxwell fluids are particularly important in the study of polymer melts, biological fluids, and complex industrial materials.

The Maxwell model combines a Newtonian dashpot (viscous element) and a Hookean spring (elastic element) in series. In tensor notation for incompressible Maxwell fluids, the constitutive equation is given as

$$\tau + \lambda \frac{D\tau}{Dt} = 2\mu \boldsymbol{D} \tag{2.5}$$

where τ denotes the stress tensor, and $\boldsymbol{D} = \frac{1}{2}\left((\nabla \boldsymbol{q}) + (\nabla \boldsymbol{q})^T\right)$ is the rate of deformation tensor. The term $\frac{D\tau}{Dt}$ denotes the upper-convected derivative of stress, accounting for material rotation and deformation defined as

$$\frac{D\tau}{Dt} = \frac{\partial \tau}{\partial t} + \boldsymbol{q}\cdot\nabla\tau - \tau\cdot\nabla\boldsymbol{q} - (\nabla\boldsymbol{q})^T\cdot\tau \tag{2.6}$$

2.3.4 Oldroyd Fluids

Oldroyd fluids [19] are a class of non-Newtonian viscoelastic fluids that model the behavior of materials exhibiting both viscous and elastic characteristics. These fluids are particularly useful in describing the flow behavior of polymeric solutions and melts, which do not conform to Newton's law of viscosity. The most common forms of Oldroyd fluids are (i) Oldroyd-A model, (ii) Oldroyd-B model, and (iii) Oldroyd-B' (or modified Oldroyd-B). Among these, the Oldroyd-B model is widely studied and applied due to its mathematical tractability and ability to capture essential features of viscoelasticity.

The Oldroyd-B model combines elements of the upper-convected Maxwell model with a Newtonian solvent, resulting in the following constitutive equation:

$$\tau + \lambda_1 \frac{D\tau}{Dt} = \mu\left(\boldsymbol{D} + \lambda_2 \frac{D\boldsymbol{D}}{Dt}\right) \tag{2.7}$$

where τ denotes the stress tensor; λ_1, λ_2 are the relaxation and retardation time, respectively; and $D = \frac{1}{2}\left((\nabla q) + (\nabla q)^T\right)$ is the rate of deformation tensor. The term $\frac{D\tau}{Dt}$ denotes the upper-convected derivative of stress, accounting for material rotation and deformation defined as in equation (2.6).

This model incorporates both elastic memory (through the convected derivative and relaxation time) and viscous damping (through the Newtonian term), making it suitable for simulating time-dependent stress–strain relationships.

The interplay of simultaneous mass and heat transfer at the interface between a viscous gas and a viscoelastic liquid has been extensively investigated. Research by Fu *et al.* [20] explored the influence of these coupled phenomena on Kelvin–Helmholtz instability at such interfaces. Building on this, Awasthi [21] investigated how heat transport modifies capillary instability in similar viscous gas–viscoelastic liquid systems. Further studies by Awasthi *et al.* [22–24] expanded the scope to include interfaces involving Oldroyd fluids. Notably, all these works employed Oldroyd-B-type viscoelastic fluid models and leveraged potential flow theory to analyze interfacial dynamics, emphasizing the role of viscoelasticity and thermal effects in destabilizing or stabilizing the interfaces.

2.3.5 Power Law Fluids

Power law fluids (Ostwald–de Waele model [25, 26]) are a class of viscoelastic fluids whose viscosity changes with the rate of shear strain. Unlike Newtonian fluids, whose viscosity remains constant regardless of the applied shear rate, power law fluids exhibit a shear-dependent viscosity that can either decrease (shear-thinning) or increase (shear-thickening) with the shear rate.

These fluids are commonly found in many industrial and biological applications, such as in the processing of polymer solutions, paints, slurries, and blood flow in narrow vessels.

The constitutive equation for a power law fluid relates the shear stress to the shear rate through a nonlinear relationship, given by

$$\tau = K\left(\frac{du}{dy}\right)^n \tag{2.8}$$

where τ = shear stress; $\frac{du}{dy}$ = shear rate or velocity gradient;
K = consistency index (analogous to viscosity); and
n = flow behavior index.

- When $n = 1$, the fluid behaves like a **Newtonian fluid** with viscosity $\mu = K$.
- When $n < 1$, the fluid is **shear-thinning** or **pseudoplastic**, meaning the apparent viscosity decreases with increasing shear rate. Examples include ketchup, paint, and blood.
- When $n > 1$, the fluid is **shear-thickening** or **dilatant**, meaning the apparent viscosity increases with increasing shear rate. Examples include cornstarch–water suspensions.

Power law fluids are part of the **generalized Newtonian fluid** family, where the stress tensor τ is given by

$$\tau = 2\eta(\dot{\gamma})\boldsymbol{D} \tag{2.9}$$

Here, $\mathbf{D}$ = rate of strain tensor;
$\dot{\gamma} = \sqrt{2\mathbf{D}:\mathbf{D}}$ = magnitude of the shear rate; and
$\eta(\dot{\gamma}) = K(\dot{\gamma})^{n-1}$ = apparent viscosity.

Research by Wang *et al.* [27] investigated the influence of heat and mass transfer on power-law viscoelastic jets in gaseous environments, revealing that these phenomena induce destabilization and alter jet dynamics. Parallel studies by Awasthi *et al.* [28] and Shukla and Awasthi [29] further examined how thermal and mass transfer processes affect the stability of viscoelastic fluid interfaces. Extending this work, Awasthi *et al.* [30, 31] analyzed the role of heat transfer in modulating interfacial stability across multiphase systems, emphasizing its critical impact on phase interactions.

2.3.6 Bingham Plastic Fluids

A Bingham plastic fluid [32] is a type of non-Newtonian fluid that behaves like a rigid body at low stresses but flows as a viscous fluid when a critical stress (yield stress) is exceeded. These fluids do not flow until the applied shear stress surpasses this yield stress (τ_0).

This behavior is commonly observed in materials such as toothpaste, mud, paint, ketchup, and drilling fluids, where some initial force is required to initiate flow.

The constitutive relation (shear stress–shear rate relationship) for Bingham plastic fluids is given by

$$\begin{aligned} \tau &= \tau_0 + \mu_p \dot{\gamma} && \text{for} \quad |\tau| > \tau_0 \\ \dot{\gamma} &= 0 && \text{for} \quad |\tau| \le \tau_0 \end{aligned} \tag{2.10}$$

where τ denotes the shear stress, τ_0 represents the yield stress, μ_p is the plastic viscosity (viscosity after yield stress is exceeded), and $\dot{\gamma}$ is the shear rate (rate of strain).

- When $|\tau| \le \tau_0$, the material behaves like a solid, showing no deformation (i.e., $\dot{\gamma} = 0$).
- When $|\tau| > \tau_0$, the material begins to flow, and the relationship between stress and strain rate becomes linear, similar to a Newtonian fluid but with an offset due to the yield stress.

2.3.7 Ellis Fluids

Ellis fluids [33] are a class of non-Newtonian fluids that exhibit shear-thinning behavior—that is, their viscosity decreases with increasing shear rate. Unlike Newtonian fluids (where viscosity remains constant), Ellis fluids are useful for modeling materials such as polymer solutions, paints, and some biological fluids, which do not follow a linear stress–strain relationship.

They are particularly useful for describing fluids that transition from Newtonian behavior at low shear rates to power-law (non-Newtonian) behavior at higher shear rates, but without a yield stress.

The constitutive equation for an Ellis fluid defines the relationship between the shear stress (τ) and shear rate ($\dot{\gamma}$). It is given by

$$\mu_{eff} = \mu_0 \left[1 + \left(\frac{|\tau|}{\tau_{1/2}} \right)^{\alpha - 1} \right]^{-1} \tag{2.11}$$

Here, μ_{eff} represents the effective or apparent viscosity, μ_0 denotes the zero-shear viscosity (viscosity at very low shear rates), τ is the shear stress, $\tau_{1/2}$ is the shear stress at which the viscosity is half of μ_0, and α denotes a dimensionless flow behavior index (typically $\alpha > 1$).

This equation can also be rearranged to express the shear rate as a function of shear stress:

$$\dot{\gamma} = \frac{\tau}{\mu_0}\left[1 + \left(\frac{|\tau|}{\tau_{1/2}}\right)^{\alpha-1}\right] \tag{2.12}$$

The viscosity of Ellis fluid decreases as shear rate increases. For large shear stresses, the fluid behaves like a power-law fluid. Unlike Bingham plastics or Herschel–Bulkley fluids, Ellis fluids begin to flow immediately when a shear stress is applied. At very low shear rates ($\tau \ll \tau_{1/2}$), the fluid behaves as Newtonian with viscosity μ_0.

2.3.8 Reiner–Philippoff Fluids

Reiner–Philippoff fluids [34] are a type of non-Newtonian fluid that exhibit shear-thinning behavior, meaning their viscosity decreases with increasing shear rate. These fluids are particularly useful in modeling biological fluids, polymer solutions, and lubricants where the flow behavior cannot be accurately captured by Newtonian assumptions.

The constitutive equation for a Reiner–Philippoff fluid can be written as

$$\tau = \frac{\mu_0}{1 + k\dot{\gamma}^2}\dot{\gamma} \tag{2.13}$$

where τ denotes the stress tensor, μ_0 is the limiting viscosity at low shear rates, and k represents the material parameter (related to fluid structure and flow behavior).

Alternatively, in terms of the viscosity function $\mu(\dot{\gamma})$, the model becomes

$$\mu(\dot{\gamma}) = \frac{\mu_0}{1 + k\dot{\gamma}^2} \tag{2.14}$$

As $\dot{\gamma} \to 0$, the viscosity approaches μ_0, behaving like a Newtonian fluid. As $\dot{\gamma} \to \infty$, the viscosity diminishes, highlighting the shear-thinning nature.

2.3.9 Prandtl Fluids

Prandtl fluids [35] are a class of non-Newtonian fluids that do not follow Newton's linear relationship between shear stress and shear rate. The concept originates from Ludwig Prandtl's work on boundary layer theory and has since been applied in modeling materials with a yield stress—such as toothpaste, mud, blood, ketchup, and drilling fluids—which behave like solids until a certain stress threshold is exceeded.

These fluids exhibit Bingham plastic-like behavior at low shear rates and power-law behavior at higher shear rates. Therefore, Prandtl fluids are ideal for modeling viscoplastic materials where the transition from solid-like to fluid-like behavior is smooth.

The constitutive equation for a Prandtl fluid is nonlinear and represents the relationship between shear stress τ and shear rate $\dot{\gamma}$ (or strain rate). It is commonly expressed as

$$\begin{aligned} \tau = \tau_0 + k\dot{\gamma}^n \quad & \text{for} \quad \tau > \tau_0 \\ \dot{\gamma} = 0 \quad & \text{for} \quad \tau \le \tau_0 \end{aligned} \tag{2.15}$$

where τ denotes the shear stress, τ_0 represents the yield stress, k is the consistency index, n is the flow behavior index, and $\dot{\gamma}$ is the shear rate (rate of strain).

This model smoothly combines yield-stress behavior with shear-dependent viscosity, and it is considered a generalization of both Bingham and power-law models.

2.3.10 Eyring Fluids

Eyring fluids [36] are a class of non-Newtonian fluids that exhibit shear-thinning behavior, meaning their viscosity decreases with increasing shear rate. These fluids are particularly important in engineering and biological applications, such as polymer processing, lubrication, and blood flow modeling.

The Eyring fluid model is based on the Eyring theory of absolute reaction rates, originally developed for chemical kinetics, and later adapted to describe the flow behavior of complex fluids. It assumes that molecular rearrangements in a flowing fluid occur *via* a thermally activated process over an energy barrier, and the rate of shear is related nonlinearly to the applied shear stress.

The constitutive equation for an Eyring fluid relates the shear rate $\dot{\gamma}$ to the shear stress τ as

$$\dot{\gamma} = \dot{\gamma}_0 \sinh\left(\frac{\tau}{\tau_0}\right) \tag{2.16}$$

where τ denotes the shear stress, τ_0 is a characteristic shear stress that represents the stress scale for molecular rearrangements, $\dot{\gamma}_0$ is a reference shear rate (depends on temperature and material properties), and $\dot{\gamma}$ is the shear rate (rate of strain).

For small values of shear stress ($\tau \ll \tau_0$), the Eyring model behaves linearly, and the fluid shows Newtonian behavior as

$$\dot{\gamma} \approx \dot{\gamma}_0\left(\frac{\tau}{\tau_0}\right) \Rightarrow \tau = \mu\dot{\gamma} \tag{2.17}$$

where $\mu = \dfrac{\tau_0}{\dot{\gamma}_0}$ is the viscosity.

For large shear stresses ($\tau \gg \tau_0$), the fluid exhibits nonlinear shear-thinning behavior as

$$\dot{\gamma} \prec \exp\left(\frac{\tau}{\tau_0}\right) \tag{2.18}$$

2.3.11 Power–Eyring Fluids

Power–Eyring fluids [37] are a class of non-Newtonian fluids that exhibit complex rheological behavior, combining features of power-law (shear-thinning/thickening) and Eyring (activation energy-based) models. These fluids are particularly useful for modeling the behavior of materials like polymer melts, biological fluids, lubricants, and slurries, where viscosity is not constant and varies with shear rate or stress.

The constitutive relation for Power–Eyring fluids typically blends the Eyring model with a power-law form of the stress–strain rate relationship. A general expression is

$$\tau = \tau_0 \sinh^{-1}\left(\frac{\mu\dot{\gamma}}{\tau_0}\right)^n \tag{2.19}$$

where τ denotes the shear stress, τ_0 is a characteristic shear stress that represents the stress scale for molecular rearrangements, μ is the dynamic viscosity coefficient, and $\dot{\gamma}$ is the shear rate (rate of strain).

In a more compact form, the model may be described using the combined Eyring–power-law relation:

$$\dot{\gamma} = \left(\frac{\tau_0}{\mu}\right)\sinh\left(\frac{\tau}{\tau_0}\right)^{1/n} \tag{2.20}$$

For $n = 1$, the model reduces to the Eyring fluid. When $\tau_0 \to \infty$, the model approximates the power-law fluid.

2.3.12 Williamson Fluids

Williamson fluid [38] is a type of non-Newtonian fluid that exhibits shear-thinning (pseudoplastic) behavior. Unlike Newtonian fluids (where viscosity is constant), the viscosity of Williamson fluids decreases with increasing shear rate. This fluid model is often used to describe the flow behavior of certain biological fluids, polymer solutions, and industrial suspensions. The Williamson model is particularly useful because it captures shear-thinning characteristics using a relatively simple formulation, while still remaining mathematically tractable.

The constitutive relation (i.e., the relationship between stress and rate of deformation) for a Williamson fluid is given by

$$\mu(\dot{\gamma}) = \mu_\infty + \frac{(\mu_0 - \mu_\infty)}{1 + \lambda\dot{\gamma}} \tag{2.21}$$

where $\mu(\dot{\gamma})$ represents the effective (apparent) viscosity as a function of shear rate $\dot{\gamma}, \mu_0$ denotes the zero-shear-rate viscosity (viscosity at very low shear), μ_∞ is the infinite-shear-rate viscosity (viscosity at very high shear), λ is the time constant (relaxation parameter), and $\dot{\gamma}$ is the shear rate.

2.3.13 Walters' B Fluids

Walters' B fluid [39] is a type of non-Newtonian fluid that belongs to the category of viscoelastic fluids, specifically those that exhibit both elastic and viscous behavior. This fluid model was introduced by K. Walters to describe materials such as polymer solutions, which display memory effects and stress relaxation properties not captured by classical Newtonian models.

The Walters' B model is a simplified viscoelastic model derived from more general theories like the Oldroyd-B fluid but focuses on small deformations and short memory effects. It is particularly useful in theoretical and computational studies where moderate viscoelastic behavior is observed.

The constitutive relation of a Walters' B fluid expresses the extra stress tensor τ as a function of the rate of strain and the history of deformation. The equation is given by

$$\tau = -p\boldsymbol{I} + 2\mu\boldsymbol{D} + k_0 \frac{D\boldsymbol{D}}{Dt} \tag{2.22}$$

where τ denotes the Cauchy stress tensor, p is the hydrostatic pressure, $\boldsymbol{I}$ is the identity tensor, μ represents the dynamic viscosity, $\boldsymbol{D} = \frac{1}{2}\left((\nabla\boldsymbol{q}) + (\nabla\boldsymbol{q})^T\right)$ is the rate of deformation tensor, and k_0 is the material modulus characterizing viscoelastic effects (known as elastic viscosity or relaxation parameter). Awasthi *et al.* [40–42] investigated the behavior of Walter's B fluid under thermal transport conditions, revealing that heat transfer enhances the stability of the interfacial region.

2.4 Applications of Viscoelastic Fluids in Industry and Nature

2.4.1 Biomedical Engineering: Blood Flow and Circulatory Dynamics

Viscoelastic fluids play a critical role in understanding blood flow, particularly in the human circulatory system. Blood, a complex suspension of cells in plasma, exhibits non-Newtonian behavior, where its viscosity changes under shear stress. This viscoelasticity ensures efficient flow through narrow vessels, prevents clot formation, and influences conditions like atherosclerosis. In microcirculation, red blood cells' deformability—a viscoelastic trait—enables them to navigate capillaries smaller than their diameter. Pathological states, such as sickle cell anemia, disrupt this behavior, altering blood's rheology and impairing oxygen delivery. Computational models incorporating viscoelastic principles help design medical devices like stents and artificial heart valves, optimizing their interaction with dynamic blood flow.

2.4.2 Biomedical Innovations: Targeted Drug Delivery

Viscoelastic hydrogels and polymer-based carriers revolutionize drug delivery by enabling controlled release of therapeutics. These materials respond to stimuli (e.g., pH, temperature) to release drugs at specific sites, minimizing systemic side effects. For instance, shear-thinning hydrogels—used in injectable therapies—flow easily through syringes but solidify in tissues, forming localized drug reservoirs. In cancer treatment, viscoelastic nanoparticles navigate the bloodstream, adhere to tumor vasculature, and penetrate dense tumor matrices due to their adaptable rheology. Such systems mimic biological fluids' behavior, ensuring precise, sustained therapeutic action while accommodating the body's mechanical stresses.

2.4.3 Polymer Processing: Manufacturing and Material Design

The polymer industry relies heavily on viscoelastic properties to shape materials into films, fibers, and molded products. During extrusion or injection molding, polymers undergo shear thinning, where viscosity decreases under high stress, allowing efficient processing. Viscoelastic memory effects, however, can cause defects like die swell (post-extrusion expansion) or residual stresses in cooled products. Additives like plasticizers or fillers modulate these behaviors to achieve desired mechanical properties. Advanced simulations of viscoelastic flow guide the design of industrial equipment, ensuring uniformity in products ranging from plastic bottles to automotive components.

2.4.4 Food Industry: Texture and Stability

In food science, viscoelasticity dictates texture, mouthfeel, and shelf life. Products like ketchup, mayonnaise, and yogurt exhibit yield stress—a Bingham plastic-like behavior—where they remain semi-solid until force is applied. This prevents settling of particles in suspensions and ensures easy dispensing. Dairy proteins and polysaccharides (e.g., xanthan gum) impart viscoelasticity to sauces and gels, creating structures that resist syneresis (water separation). In baking, dough's viscoelastic nature allows it to trap gases during fermentation, enabling bread to rise while maintaining structural integrity during baking.

2.4.5 Geophysical Flows: Lava, Glaciers, and Mudslides

Natural viscoelastic materials shape the Earth's surface through phenomena like volcanic eruptions and glacial movement. Lava, a molten silicate suspension, displays shear-dependent viscosity: it flows rapidly on steep slopes but solidifies into rugged terrain upon cooling. Glaciers, behaving as Bingham plastics, deform under their weight over millennia, carving valleys. Mudslides—mixtures of water, soil, and debris—transition from solid-like to fluid-like states under stress, posing hazards during heavy rains. Understanding these flows aids in predicting disasters and mitigating risks in vulnerable regions.

2.4.6 Environmental Engineering: Oil Spills and Sediment Transport

Viscoelastic surfactants are deployed in oil spill remediation, forming gel-like barriers that encapsulate oil slicks and prevent coastal contamination. These fluids retain structural coherence under waves, enhancing cleanup efficiency. Similarly, in river systems, sediment-laden flows exhibit complex rheology: clay-rich slurries behave as viscoelastic fluids, resisting erosion and affecting delta formation. Managing such flows is critical for maintaining navigation channels and preventing reservoir siltation.

2.4.7 Energy and Industrial Fluids: Hydraulic Fracturing and Drilling

In oil and gas extraction, viscoelastic fracturing fluids are pumped into shale formations to create fractures, releasing trapped hydrocarbons. These fluids suspend proppants (sand or ceramics) to keep fractures open, relying on their shear-thinning properties to minimize pumping pressure while maximizing proppant transport. Drilling muds, another viscoelastic application, lubricate drill bits, stabilize boreholes, and carry rock cuttings to the surface. Their ability to transition between solid-like and fluid-like states under varying shear rates ensures efficient, safe drilling operations.

2.4.8 Ecological Adaptations: Biological Fluids and Mucus

In nature, viscoelasticity aids survival strategies. For example, snail mucus—a viscoelastic gel—allows adhesion to surfaces while enabling locomotion. Spider silk, with its unique blend of elasticity and strength, dissipates energy when capturing prey. Biofilms, colonies of microbes embedded in viscoelastic matrices, resist mechanical disruption and antibiotics, posing challenges in medical and environmental contexts. Studying these biological systems inspires synthetic materials with adaptive rheology for applications in robotics, adhesives, and antimicrobial coatings.

2.5 Recent Advances and Emerging Trends in Viscoelastic Fluid Flow

2.5.1 Machine Learning in Viscoelastic Flow Modeling

Recent years have witnessed a paradigm shift in viscoelastic flow modeling through the integration of machine learning (ML). Traditional constitutive equations often struggle to capture the nonlinear, history-dependent behavior of viscoelastic fluids, especially in complex geometries. ML algorithms, such as neural networks and reinforcement learning, are now being trained on high-fidelity simulation datasets to predict stress–strain relationships, transient flow patterns, and instability thresholds. For instance, physics-informed neural networks (PINNs) combine governing equations with experimental data, enabling accurate predictions even in data-sparse regimes. Emerging trends include the use of generative adversarial networks (GANs) to simulate turbulent viscoelastic flows and hybrid models that embed rheological principles into ML architectures, ensuring both interpretability and predictive power. These tools are accelerating the design of polymer processing systems and microfluidic devices while reducing reliance on computationally expensive simulations.

2.5.2 Data-Driven Constitutive Model Discovery

A groundbreaking application of ML lies in autonomously discovering constitutive laws from experimental or numerical data. By leveraging symbolic regression and sparse identification techniques, researchers are identifying compact mathematical expressions that describe viscoelastic stress responses without prior assumptions about material class (e.g., Oldroyd-B vs. Giesekus models). This approach is particularly valuable for complex fluids like bio-polymer solutions or nanocomposites, where existing models fail to generalize. Transfer learning is also gaining traction, allowing models trained on one fluid type to adapt to others with minimal retraining. Challenges remain in ensuring robustness across varying deformation rates and temperatures, but the fusion of ML with rheological physics promises a new era of "smart" constitutive models.

2.5.3 Multiscale Modeling: Bridging Molecular and Continuum Scales

Multiscale modeling has emerged as a critical tool for viscoelastic fluids, where macroscopic flow behavior is intrinsically linked to microscale polymer dynamics. Coarse-grained molecular dynamics (CGMD) and kinetic theory-based approaches are now coupled with continuum-level simulations to resolve phenomena like shear banding or elastic turbulence. For example, hybrid frameworks combine Brownian dynamics for polymer chains with finite-element methods for bulk flow, enabling precise predictions of entanglement effects in polymer melts. Recent advances include adaptive resolution schemes that dynamically switch between scales, optimizing computational efficiency. These methods are vital for designing advanced materials, such as self-healing hydrogels or drag-reducing additives, where molecular architecture dictates macroscopic performance.

2.5.4 Hybrid Approaches for Complex Flow Regimes

Hybrid modeling integrates viscoelastic constitutive equations with other physical phenomena, such as heat transfer, phase change, or multiphase interactions. For instance, coupled electrohydrodynamic models are being developed to study viscoelastic droplets in electric fields, relevant to inkjet printing and lab-on-a-chip systems. Similarly, reactive flow models now account for viscoelasticity-driven mixing enhancements in chemical reactors. Emerging trends focus on GPU-accelerated simulations and reduced-order modeling (ROM) techniques to handle real-world geometries, such as porous media or biological vasculature. These approaches are unlocking insights into flow instabilities and energy dissipation mechanisms previously deemed intractable.

2.5.5 Cutting-Edge Experimental Techniques for Validation

Validating theoretical and computational models demands advanced experimental tools. Microparticle image velocimetry (μPIV) and dynamic light scattering (DLS) now achieve sub-micron resolution, mapping velocity and stress fields in confined viscoelastic flows. Rheo-optical setups, combining rheometry with high-speed imaging, reveal transient microstructure changes during shear thinning or yielding. Additionally, *in situ* X-ray scattering at synchrotron facilities provides nanoscale insights into polymer alignment under flow. Emerging techniques like 3D-printed microfluidic devices with embedded sensors enable precise control of deformation history, bridging the gap between idealized lab conditions and industrial applications.

2.5.6 Machine Vision and Real-Time Feedback Loops

Machine vision algorithms are revolutionizing experimental data analysis by automating the extraction of flow features from high-speed videos or laser scans. Deep-learning-based image segmentation can quantify filament stretching, droplet breakup, or elastic recoil in real time. Coupled with adaptive control systems, these tools enable closed-loop experiments where flow conditions are dynamically adjusted based on instantaneous feedback. For example, stress-controlled rheometers now use ML to optimize test protocols, minimizing experimental runtime while maximizing data quality. This synergy between experimentation and AI is critical for validating multiscale models under non-equilibrium conditions.

2.5.7 Interdisciplinary Fusion and Future Directions

The convergence of ML, multiscale modeling, and advanced experimentation is driving viscoelastic fluid research into uncharted territories. Emerging trends include digital twins for industrial processes, where real-time sensor data updates simulations to predict equipment wear or product defects. Bio-inspired studies are also flourishing, with viscoelastic models applied to cellular mechanics and mucus dynamics. However, challenges persist, such as standardizing datasets for ML training and improving the portability of multiscale codes. Future breakthroughs will likely hinge on interdisciplinary collaboration, leveraging expertise in rheology, data science, and soft matter physics to decode the intricate "memory" of viscoelastic systems.

2.5.8 Sustainability and Industry 4.0 Applications

Finally, advances in viscoelastic fluid research are aligning with sustainability goals. ML-optimized polymer processing reduces energy consumption, while multiscale models guide the development of biodegradable viscoelastic materials. In Industry 4.0, predictive models integrated with IoT-enabled manufacturing systems enable real-time quality control for products ranging from adhesives to personal care formulations. As these trends mature, viscoelastic fluid mechanics will play a pivotal role in shaping next-generation technologies across energy, healthcare, and environmental engineering.

2.6 Conclusion

This chapter has provided a comprehensive overview of viscoelastic fluid models, emphasizing their unique ability to exhibit both viscous and elastic behaviors. Starting from the fundamental differences between Newtonian, non-Newtonian, and viscoelastic fluids, it delved into the molecular underpinnings and the broad spectrum of rheological models developed to describe viscoelastic behavior, such as Rivlin–Ericksen, Oldroyd, Maxwell, and others. Each model captures specific aspects of fluid memory, stress relaxation, and nonlinear stress–strain relationships observed in real-world materials. The chapter also highlighted their critical role in natural phenomena and industrial applications—from biomedical engineering and polymer processing to geophysical and environmental systems—showcasing how understanding and accurately modeling viscoelastic fluids are vital for innovation in science and engineering.

AI disclosure statement: The author acknowledges the use of AI-based tools for minor language editing or grammatical correction. However, all intellectual contributions, analyses, and conclusions presented in this chapter are entirely the author's own.

References

1. Raisinghania, M.D., *Fluid Dynamics*, S Chand & Co Ltd, India, 2003.
2. Stokes, V.K., Couple stresses in fluids. *Phys. Fluids*, 9, 1709–1715, 1966.
3. Rivlin, R.S. and Ericksen, J.L., Stress-deformation relaxations for isotropic materials. *J. Ration. Mech. Anal.*, 4, 323–329, 1955.

4. Dharamendra, Awasthi, M.K., Kumar, V., Chauhan, A., Dhiman, N., Study of instability of Rivlin-Ericksen viscoelastic fluid film, in: *AIP Conference Proceedings*, vol. 2481, p. 040005, 2022.
5. Mostafa, D.M., Viscoelastic potential flow instability theory of Rivlin-Ericksen electrified fluids of cylindrical interface. *J. Ocean. Eng. Sci.*, 2022. (In Press).
6. Awasthi, M.K., Dharamendra, Yadav, D., Instability of Rivlin-Ericksen fluid film with heat transport. *Int. Comm. Heat Mass Trans.*, 135, 106085, 2022.
7. Srija, R., Singh, A.K., Awasthi, M.K., Yadav, D., Nair, S.B., Instability analysis of swirling cylindrical Rivlin-Ericksen viscoelastic fluid-viscous fluid interface with heat and mass transfer. *Int. J. Mod. Phys. B.*, 38, 31, 2450416, 2024.
8. Awasthi, M.K., Dutt, N., Kumar, A., Kumar, S., Electrohydrodynamic capillary instability of Rivlin-Ericksen viscoelastic fluid film with mass and heat transfer. *Heat Transfer*, 53, 115–133, 2024.
9. Awasthi, M.K., Dutt, N., Kumar, A., Kumar, S., Electrohydrodynamic capillary instability of Rivlin–Ericksen viscoelastic fluid film with mass and heat transfer. *Heat Transfer*, 1–19, 2023. doi:10.1002/htj.22944.
10. Shukla, A.K. and Awasthi, M.K., Stability characteristics of planar Rivlin–Ericksen fluid interface with mass and heat transfer. *J. Fluids Eng.*, 145, 031302, 2023. (6 pages).
11. Reiner, M., A mathematical theory of dilatancy. *Am. J. Math.*, 67, 3, 350–362, 1945.
12. Rivlin, R.S., The hydrodynamics of non-Newtonian fluids. *I. Proc. R. Soc. Lond. Ser. A Math. Phys. Sci.*, 193, 1033, 260–281, 1948.
13. Kosterin, A.F., Some rheological properties of Reiner–Rivlin fluids. *J. Eng. Phys.*, 35, 3, 1121–1122, 1978.
14. Gao, S.X. and Hartnett, J.P., Heat transfer behavior of Reiner–Rivlin fluids in rectangular ducts. *Int. Commun. Heat Mass Transf.*, 39, 6, 1317–1324, 1996.
15. Attia, H.A., Numerical study of the flow and heat transfer in a Renier–Rivlin fluid on a rotating porous disk. *J. Appl. Mech. Tech.*, 46, 1, 68–76, 2005.
16. Attia, H.A., The effect of ion slip on the flow of Reiner–Rivlin fluid due a rotating disk with heat transfer. *J. Appl. Mech. Tech.*, 21, 1, 174–183, 2007.
17. Tabassum, M. and Mustafa, M., A numerical treatment for partial slip flow and heat transfer of non-Newtonian Reiner–Rivlin fluid due to rotating disk. *Int. Commun. Heat Mass Transf.*, 123, 979–987, 2018.
18. Maxwell, J.C., On the Dynamical Theory of Gases. *Philos. Trans. R. Soc. London*, 157, 49–88, 1867.
19. Oldroyd, J.G., On the formulation of rheological equations of state. *Proc. R. Soc. London Ser. A. Math. Phys. Sci.*, 200, 1063, 523–541, 1950.
20. Fu, Q.F., Deng, X.D., Yang, L.J., Kelvin-Helmholtz instability analysis of confined Oldroyd-B liquid film with heat and mass transfer. *J. Non-Newtonian Fluid Mech.*, 267, 28–34, 2019.
21. Awasthi, M.K., Capillary instability of viscoelastic liquid film with heat and mass transfer. *ASME J. Heat Transfer*, 142, 022108, 2020. (5 pages).

22. Shukla, A.K., Awasthi, M.K., Agarwal, S., Stability analysis of spherical viscous fluid-Oldroyd B viscoelastic fluid interface: A Viscous Potential Flow approach. *Chin. J. Phys.*, 86, 148–159, 2023.
23. Awasthi, M.K. and Hoshoudy, G.A., Study of heat and mass transfer on the instability of a swirling viscoelastic liquid film. *Eur. Phys. J. E*, 44, 1-9, 2021.
24. Awasthi, M.K., Kelvin-Helmholtz instability of viscoelastic liquid -viscous gas interface with heat and mass transfer. *Int. J. Therm. Sci.*, 161, 106710, 2021.
25. Ostwald, W., Ueber die rechnerische Darstellung des Strukturgebietes der Viskosität. *Kolloid-Zeitschrift*, 47, 176–187, 1929.
26. De Waele, A., Viscometry and Plastometry. *Oil Colour Chemist's Assoc. J.*, 6, 33–36, 19231923.
27. Wang, X.T., Ning, Z., Lu, M., Temporal instability analysis of a confined non-Newtonian liquid jet with heat and mass transfer. *Eur. J. Mech. B. Fluids*, 84, 350–356, 2020.
28. Awasthi, M.K., Shukla, A.K., Yadav, D., Rayleigh Instability of Power-Law Viscoelastic Liquid With Heat and Mass Transfer. *Int. Commun. 388 Heat Mass Transfer*, 129, 105657, 2021. 389.
29. Shukla, A.K. and Awasthi, M.K., Interfacial Characteristics of Power-Law Viscoelastic Fluid With Heat and Mass Transfer in Planar Configuration. *390 ASME J. Fluids Eng.*, 144, 10, 101303, 2022.
30. Awasthi, M.K., Dharamendra, Yadav, D., Stability characteristics of the cylindrical power-law viscoelastic fluid interface with heat transfer, in: *Waves in Random and Complex Media*, pp. 1–15, 2023.
31. Awasthi, M.K. and Shukla, A.K., Instability of power-law viscoelastic-viscous fluid interface with heat and mass transfer. *Chin. J. Phys.*, 89, 1441–1455, 2024.
32. Bingham, E.C., *Fluidity and Plasticity*, McGraw-Hill Book Company, New York, 1922.
33. Weidner, D.E. and Schwartz, L.W., Contact-line motion of shear-thinning liquids. *Phys. Fluids*, 6, 3535–3538, 1994.
34. Na, T.-Y., Boundary layer flow of Reiner-Philippoff fluids. *Int. J. Non-Linear Mech.*, 29, 6, 871–877, 1994.
35. Prandtl, L., *Motion of fluids with very little viscosity*, No. NACA-TM-452, 1928.
36. Eyring, H., Viscosity, plasticity, and diffusion as examples of absolute reaction rates. *J. Chem. Phys.*, 4, 4, 283–291, 1936.
37. Bird, R.B., Calvin Armstrong, R., Hassager, O., Dynamics of polymeric liquids. Vol. 1: Fluid mechanics, 1987.
38. Williamson, R.V., The flow of pseudoplastic materials. *Ind. Eng. Chem.*, 21, 11, 1108–1111, 1929.
39. Walters, K., Non-Newtonian effects in some elastic-viscous liquids whose behavior at small rates of shear is characterized by a general linear equation of state. *Quart. J. Mech. App. Math*, 15, 63–76, 1962.

40. Awasthi, M.K., Dharamendra, Yadav, D., Stability characteristics of Walter's B viscoelastic fluid in a cylindrical configuration with heat transfer. *Proc. IMechE Part C J. Mech. Eng. Sci.*, 236, 10370–10377, 2022.
41. Shukla, A.K., Awasthi, M.K., Singh, S., Impact of heat and mass transport on Rayleigh–Taylor instability of Walter's B viscoelastic fluid layer. *Microgravity Sci. Technol.*, 35, 1, 2023.
42. Srija, R., Singh, A.K., Awasthi, M.K., Yadav, D., Impact of swirling on capillary instability of Walter's B viscoelastic fluid-viscous fluid interface with heat and mass transfer. *Proc. Inst. Mech. Engg. Part C J. Mech. Engg. Sci.*, 238, 11, 5359–5368, 2024.

3

Computational Fluid Dynamics (CFD) for Non-Newtonian Fluids

K. Jyothi[1], Yeddula Rameswara Reddy[2], Ramachandra Reddy Vaddemani[3], Raghunath Kodi[4]* and Dhananjay Yadav[5]

[1]Department of Mathematics, G. Pulla Reddy Engineering College, Kurnool, Andhra Pradesh, India
[2]Department of Mechanical Engineering, JNTUA College of Engineering Pulivendula, Constitute College of Jawaharlal Nehru Technological University Anantapur, Ananthapuramu, Andhra Pradesh, India
[3]Department of Humanities and Sciences, K.S.R.M College of Engineering, Kadapa, Andhra Pradesh, India
[4]Department of Humanities and Sciences, St. Johns College of Engineering and Technology, Yemmiganur, Andhra Pradesh, India
[5]Department of Mathematical & Physical Sciences, University of Nizwa, Nizwa, Oman

Abstract

This study investigates the impact of nonlinear thermal radiation on Carreau nanofluid within the context of Marangoni boundary conditions and wedge geometry. By applying similarity transformations, the governing equations are transformed into ordinary differential equations. Key parameters such as the local Nusselt number (Nux), the local Sherwood number (Shx), and the skin friction coefficient (Cfx) are analyzed. The results demonstrate favorable profiles for velocity, temperature, and concentration, indicating that Marangoni boundary conditions significantly enhance heat transfer. The equations are solved using the finite-element method (FEM) implemented in Mathematica 10.4, with the findings presented through graphs and tables. A notable observation is that the velocity profiles vary with changes in the magnetic field parameter (M), illustrating the magnetic field's influence on the fluid dynamics.

**Corresponding author*: kraghunath25@gmail.com

Dhananjay Yadav, Mukesh Kumar Awasthi and Harith Mohamed Al-Azri (eds.) Non-Newtonian Fluids for Industrial Applications: Modeling and Simulations, (55–84) © 2026 Scrivener Publishing LLC

Keywords: Carreau nanofluid, Marangoni boundary conditions, thermal radiation, Brownian motion and thermophoresis

3.1 Introduction

In the present era, nanofluids have captivated the interest of researchers globally. Nanofluids find applications in a wide array of fields such as coolant technologies, heat transfer, mass transfer, and even in medical applications, impacting various aspects of our daily lives. Recent studies have introduced highly conductive nano ribbons for the advancement of next-generation batteries. Another promising option for heat storage and release involves phase change materials (PCMs) and sugar alcohols in relation to nanomaterials. The primary focus of nanofluid research revolves around exploring their heat transfer capabilities, advantages, applications, and innovations. These applications extend beyond industrial settings to include agricultural fields as well. Countless innovative ideas continue to capture the attention of researchers in this field. The term "nanofluids," coined by Choi *et al.* [1], refers to a suspension of nanotubes in oils that has significantly expanded the research possibilities for scientists. This innovative concept not only enhances thermal conductivity but also paves the way for a wide range of applications involving nanotubes. In a study by Eastman *et al.* [2], nanofluids with ethylene glycol have 40% higher thermal conductivity than pure ethylene glycol. However, the study did not yield any significant findings regarding particle size. Chakraborty *et al.* [3] focused on the initial stages of producing nanofluids and their practical applications. Special attention was paid to the stability of nanofluids. Abbas *et al.* [4] studied a comparative analysis of the distinctive characteristics of nanofluids in contrast to conventional fluids. Phase change materials (PCMs) have emerged as promising contenders for various thermal exchange applications, with Munyalo *et al.* [5] delving into their energy storage capabilities.

Ebaid *et al.* [6] empirically unveiled the efficacy of photovoltaic cooling panels in harnessing two distinct nanofluids as the cooling agent, such as H_2O. Tawfik *et al.* [7] delved into a study focusing on the transient outlook and prospective applications of nanofluids in augmenting the thermal efficiency of traditional fluids. Kim *et al.* [8] observed that altering pressure fluctuations resulted in changes in heat transfer, yielding favorable outcomes through computational fluid dynamics (CFD). Sheremet *et al.* [9] utilized Marangoni effects within a cavity-cubical shaped, filled with nanofluid, with water as the base fluid, to analyze mass and heat transfer processes.

Song *et al.* [10] assessed the thermal conductivity and melting characteristics of Sutterby nanofluid, noting improvements in velocity profiles in conjunction with Marangoni numbers. Sheikholeslami *et al.* [11] employed a model of two-phase to investigate magnetohydrodynamics (MHD) heat transfer under conventional Marangoni conditions. The significance of enhancing heat transfer with hybrid Marangoni nanofluids was discussed by Wahid *et al.* [12]. Abdullah *et al.* [13] conducted a stability analysis on conventional Marangoni convection in a heated layer with nanofluids, demonstrating that the study of the state yielded linear and exponential temperature distributions for different volume fractions, which exhibited good agreement.

Jawad *et al.* [14] provided a theoretical basis for nanofluid behavior using the Maxwell power-law model, considering unsteady flow, Welan gum solution, analyzing Joule heating effects with a magnetic field. Gowda *et al.* [15] explored potential of using blood as a conventional fluid and gold nanoparticles for treating cancer cells. The applied techniques yielded realistic results and achieved significant outcomes. Gowda *et al.* [16] studied chemical reactions on 2nd grade steady Marangoni liquid models to understand the impact of activation energy. Li *et al.* [17] conducted a study on Marangoni phenomena in Al_2O_3–Cu–H_2O Casson hybrid nanofluids. Ellahi *et al.* [18] primarily focused on the impact of particle size on Marangoni convection, specifically discussing the effects of needle- and disc-shaped particles. Sheikholeslami *et al.* [19] illustrated impact of magnetohydrodynamics on CuO–H_2O nanofluids under Marangoni convection, demonstrating a substantial level of concurrence. Chaudhary *et al.* [20] employed the method Galerkin finite element to scrutinize the MHD Marangoni boundary layer flow of electrically conductive copper–water nanofluids. Hashim *et al.* [21] executed a numerical investigation on magnetohydrodynamic (MHD) Williamson nanofluid flow over wedge geometry. The ramifications of thermophoresis and Brownian motion were investigated considering wedge geometry by Tlili *et al.* [22]. Furthermore, Hashim *et al.* [23] explored the H & T transfer characteristics of various nanofluids over wedge geometry using MATLAB solver bvp4c. Kameswaran *et al.* [24] have utilized heat and mass transfer from an isothermal wedge in nanofluids with Soret effect. Xiu *et al.* [25] have discussed dyamics of ternary-hybrid nanofluids due to dual stretching on wedge surfaces when volume of nanoparticles is small and large: forced convection of water at different temperatures. Rizwan *et al.* [26] have possessed flow characteristics of heat and mass for nanofluid under different operating temperatures over wedge and plate. Dharmaiah *et al.* [27] have reviewed nonNewtonian nanofluid characteristics over a melting wedge.

Babu *et al.* [28] discussed mixed convective flow of heat and mass transfer of nanofluids over a static wedge with convective boundary conditions. Yacob *et al.* [29] reviewed Falkner–Skan problem for a static and moving wedge with prescribed surface heat flux in a nanofluid. De *et al.* [30] studied influence of nanofluids on magnetohydrodynamic heat and mass transfer over a non-isothermal wedge with variable viscosity and thermal radiation. Dharamaiah *et al.* [27], Hussain *et al.* [31], Sheikholeslami *et al.* [32], Sreedevi *et al.* [33] incorporated studied the wedge geometry of various nanofluids to enhance H & T transfer. Babu *et al.* [28] delved the phenomenon of steady blending of traditional fluid H & T transfer. Yacob *et al.* [29] concentrated on utilizing the Falkner–Skan power law to analyze the continuous-flow, H & T transfer over a mobile wedge. Soret–Dufour displayed a profound interest in exploring nanofluid convective flow under the influence of magnetohydrodynamics, as elaborated by De *et al.* [30]. Hussain *et al.* [31] used the FEM to analyze heat and mass transfer characteristics of diverse nanofluids. Sheikholeslami *et al.* [32] employed the finite-element method in conjunction with adaptive meshing techniques. The introduction of (NEPCM) in lieu of pure phase change material (PCM) exhibited heightened energy transfer efficiencies. Additionally, Sreedevi *et al.* [33] applied FEM to determine the values of the homogeneous–heterogeneous parameter and Maxwell parameter reactions.

Upon meticulous examination of the literature, it becomes apparent that a significant research gap exists concerning the investigation of Carreau nanofluid under Marangoni boundary conditions as it flows through a wedge geometry, while considering the impact of Brownian motion and thermophoresis. Consequently, we have embarked upon the endeavor of addressing this issue in the present manuscript by integrating the influences of Brownian motion and thermophoresis into our model.

3.2 Mathematical Formulation of the Problem

We analyze a permeable wedge structure moving in Carreau nanofluid at velocity $u_w(x)$. Heat and mass transfer are examined, along with magnetic field and thermal radiation, considering suction v0 at the wall. Carreau nanofluid is free from stream velocity and follows a power-law form. Velocity $u_w(x)$ is positive on the stretching surface and negative on the shrinking surface of the wedge. Temperature T of the Carreau nanofluid is observed at the stretching surface, while nanoparticles do not accumulate at the wall due to consistent concentration levels. Concentration of the

Carreau nanofluid is denoted as C within the boundary layer. The angle of the wedge is represented by beta phi, with beta defined as the Hartree pressure gradient.

$$\frac{\partial u}{\partial x}+\frac{\partial v}{\partial y}=0 \tag{3.1}$$

$$u\frac{\partial u}{\partial x}+v\frac{\partial u}{\partial y}=u_e\frac{du_e}{dx}+v\frac{\partial^2 u}{\partial y^2}\left(1+\Gamma^2\left(\frac{\partial u}{\partial y}\right)^2\right)^{\frac{n-1}{2}}$$
$$+v(n-1)\Gamma^2\frac{\partial^2 u}{\partial y^2}\left(\frac{\partial u}{\partial y}\right)^2\left(1+\Gamma^2\left(\frac{\partial u}{\partial y}\right)^2\right)^{\frac{n-3}{2}}-\frac{\sigma_{nf}}{\rho_{nf}}B_0^{\,2}(u-u_e) \tag{3.2}$$

$$u\frac{\partial T}{\partial x}+v\frac{\partial T}{\partial y}=\frac{k}{\rho_{cp}}\frac{\partial^2 T}{\partial y^2}+\tau\left[D_B\frac{\partial T}{\partial y}\frac{\partial C}{\partial y}+\frac{D_T}{T_\infty}\left(\frac{\partial T}{\partial y}\right)^2\right]+\frac{1}{\rho_{cp}}\frac{\partial q_r}{\partial y} \tag{3.3}$$

$$u\frac{\partial C}{\partial x}+v\frac{\partial C}{\partial y}=D_B\frac{\partial^2 C}{\partial y^2}+\frac{D_T}{T_\infty}\frac{\partial^2 T}{\partial y^2}-K_r(c-c_\infty) \tag{3.4}$$

The related boundary situations are

$$u=L\frac{\partial u}{\partial y},v=v_w,w=0,\left(\mu\frac{\partial u}{\partial y}\right)=-\frac{\partial\sigma}{\partial y}=\sigma\left(\gamma_c\frac{\partial C}{\partial z}+\gamma_T\frac{\partial T}{\partial z}\right),$$
$$C=C_w+k_2\frac{\partial C}{\partial z}\quad \text{at}\quad y=0 \tag{3.5}$$

$$u=u_e=bx^m,T\to T_\infty,C\to C_\infty,\quad \text{at}\quad y\to\infty \tag{3.6}$$

The stream position can be restricted

$$u=\frac{\partial\psi}{\partial y},\quad v=-\frac{\partial\psi}{\partial x} \tag{3.7}$$

The adaptations simplify the mathematical investigation

$$\psi = \left(\frac{2\nu b}{m+1}\right)^{1/2} x^{\frac{m+1}{2}} f(\eta), \quad \eta = \sqrt{\frac{b(m+1)}{2\nu}} x^{\frac{m-1}{2}} y,$$

$$\theta(\eta) = \frac{T - T_\infty}{T_w - T_\infty}, \quad \varnothing(\eta) = \frac{C - C_\infty}{C_w - C_\infty} \tag{3.8}$$

Radiative heat flux,

$$q_r = -\frac{4\sigma^*}{3K^*}\frac{\partial T^4}{\partial z} \tag{3.9}$$

and

$$T^4 = T_\infty{}^4 + 4T_\infty{}^3(T - T_\infty) + 6T_\infty{}^2(T - T_\infty)^2 + \ldots \tag{3.10}$$

Radiative heat flux is defined using Rosseland approximation for radiation.

$$q_r = -\frac{4\sigma^*}{3K^*}\frac{\partial T^4}{\partial z} = -\frac{16\sigma^* T^3}{3K^*}\frac{\partial T}{\partial z}$$

The nondimensional temperature $\theta(\eta) = \frac{T - T_\infty}{T_w - T_\infty}$ can be simplified as

$$T = T_\infty\left(1 + (\theta_w - 1)\theta\right) \tag{3.11}$$

where $\theta_w = \frac{T_w}{T_\infty}$ is the temperature parameter.

$$f'''\left[1 + We^2 f''^2\right]^{\frac{n-3}{2}}\left(1 + n\, We^2 f''^2\right) + f\, f'' + \gamma\left[1 - f'^2\right] - M(f' - 1) = 0 \tag{3.12}$$

$$\theta'' + \Pr f\,\theta' + \Pr \mathrm{Nt}\theta'^2 + \Pr \mathrm{Nb}\theta'\phi' + \frac{4}{3\,R}\left[1 + \theta(\theta_w - 1)\right]^3 \theta''$$
$$+ \frac{4}{R}\left[1 + \theta(\theta_w - 1)\right]^2 (\theta_w - 1)\theta'^2 = 0, \tag{3.13}$$

$$S'' + Le\ f\ S' + \frac{Nt}{Nb}\theta'' - C_1\ S = 0 \tag{3.14}$$

The associate transformed conditions are

$$f(0)=0,\quad f''(0)=-2(1+r),\quad \theta(0)=1,\quad S(0)=1 \tag{3.15}$$

$$f'(\infty)\to 0,\qquad \theta(\infty)\to 0,\qquad S(\infty)\to 0 \tag{3.16}$$

The nondimensional parameters are defined.

$$\begin{aligned}
&\Pr = \frac{\mu_{cf}}{k},\ Nt = \frac{\rho_{cp}D_T(T_w - T_\infty)}{T_\infty \nu \rho_{cf}},\ Nb = \frac{\rho_{cp}D_B(C_w - C_\infty)}{\nu \rho_{cf}},\ \theta_w = \frac{T_w}{T_\infty},\\
&Le = \frac{\nu}{D_B},\ \gamma = \frac{2m}{m+1},\\
&C_1 = \frac{2\nu k_r}{D_B b(m+1)x^{m-1}},\quad M = \frac{\sigma B_0^2}{2\omega_1 \rho_f},\quad R = \frac{16T_\infty{}^3\sigma^*}{3K^* k_f},\\
&We = \left(\frac{b^3\Gamma^2(m+1)x^{3m-1}}{2\nu}\right)^{1/2}
\end{aligned} \tag{3.17}$$

The objective is to calculate the skin-friction coefficient, the Nusselt number, and the Sherwood number.

$$C_{fx} = \frac{\tau_w}{\rho u_e^2/2},\quad Nu_x = \frac{xq_w}{k(T_w - T_\infty)},\quad Sh_x = \frac{xJ_w}{D_m(C_w - C_\infty)}$$

where

$$\begin{aligned}
&\tau_w = \mu_0 \frac{\partial u}{\partial y}\left[\left(1+\Gamma^2\left(\frac{\partial u}{\partial y}\right)^2\right)^{\frac{n-1}{2}}\right]_{y=0},\ q_w = -k\left(\frac{\partial T}{\partial y}\right)_{y=0} + (q_r)_{y=0},\\
&J_w = -D_m\left(\frac{\partial C}{\partial y}\right)_{y=0},
\end{aligned} \tag{3.18}$$

Using similarity variables equations take form.

$$Re^{\frac{1}{2}}C_{fx} = \frac{2}{\sqrt{2-\beta}} f''(0)\left[1+We^2(f''(0))^2\right]^{\frac{n-1}{2}}$$

$$Re^{-\frac{1}{2}}Nu_x = -\frac{2}{\sqrt{2-\beta}}\theta'(0)\left[1+\frac{4}{3N_R}[1+(\theta_w-1)(\theta(0))^3\right]^{\frac{n-1}{2}}$$

$$Re^{-\frac{1}{2}}Sh_x = -\frac{2}{\sqrt{2-\beta}}S'(0), \text{ where } Re=\frac{x\,u_e}{\nu} \text{ is the Reynolds number.}$$

3.3 Numerical Method of Solution

3.3.1 The Finite-Element Method

The finite-element method (FEM) solves differential equations by dividing the domain into smaller elements. The FEM is used in engineering for various analyses. Steps in the FEM include domain division and equation solving.

- Finite-element discretization divides the domain into elements to form the mesh.
- Generate element equations by isolating a typical element. Construct a variation formulation over the element. Assume an approximate solution and substitute it into the system. Stiffness matrix constructed with interpolation functions.
- Elemental equations are derived, and algebraic equations are combined by enforcing inter-element continuity conditions, resulting in a global finite-element model governing the entire domain.
- Application of boundary conditions
    - o Essential and natural boundary conditions enforced on combined equations.

Resolution of combined equations

These equations can be solved using numerical techniques like Gauss elimination and LU decomposition. Shape functions are important for approximating real functions.

To solve nonlinear ODE systems 3.12–3.14 with BCs 3.15 and 3.16, we first make an assumption.

$$\frac{df}{d\eta} = h \tag{3.19}$$

Equations 3.12 to 3.14 then reduce to

$$h''\left[1+We^2(h')^2\right]^{\frac{n-3}{2}}\left(1+nWe^2(h')^2\right)+f\,h'+\gamma\left[1-h^2\right]+M(1-h)=0 \tag{3.20}$$

$$\theta''+\Pr f\,\theta'+\Pr Nt\,\theta'^2+\Pr Nb\theta'\phi'+\frac{4}{3\,R}\left[1+\theta(\theta_w-1)\right]^3$$
$$\theta''+\frac{4}{R}\left[1+\theta(\theta_w-1)\right]^2\theta''(\theta_w-1)\theta'^2=0 \tag{3.21}$$

$$\phi''+Lef\,\phi'+\frac{Nb}{Nt}\theta''-C_1\,\phi=0 \tag{3.22}$$

The boundary conditions take form.

$$\eta=0,\ f=0,\ -2(1+r)=h',\ \theta=1,\ \phi=1 \tag{3.23}$$

$$\eta\to\infty,\quad h=0,\quad \theta=0,\quad \phi=0 \tag{3.24}$$

3.3.2 Variational Formulation

The variational form for Equations 3.20 to 3.24 over a linear element (η_e,η_{e+1}) is given by

$$\int_{\eta_e}^{\eta_{e+1}} w_1\left(\frac{df}{d\eta}-h\right)d\eta=0 \tag{3.25}$$

$$\int_{\eta_e}^{\eta_{e+1}} w_2\left(h''\left[1+We^2(h')^2\right]^{\frac{n-3}{2}}\left(1+nWe^2(h')^2\right)+f\,h'+\gamma\left[1-h^2\right]+M(1-h)\right)\mathrm{d}\eta=0 \tag{3.26}$$

$$\int_{\eta_e}^{\eta_{e+1}} w_3\left(\phi''+Lef\,\phi'+\frac{Nb}{Nt}\theta''-C_1\phi\right)\mathrm{d}\eta=0 \tag{3.27}$$

where w_1, w_2, and w_3 are arbitrary test functions and may be viewed as the variations in f, h, θ, and ϕ, respectively.

3.3.3 Finite-Element Formulation

The finite-element model is obtained by substituting approximations of the form

$$f=\sum_{j=1}^{2} f_j\psi_j,\quad h=\sum_{j=1}^{2} h_j\psi_j,\quad \theta=\sum_{j=1}^{2}\theta_j\psi_j,\quad \phi=\sum_{j=1}^{2}\phi_j\psi_j. \tag{3.28}$$

with $w_1=w_2=w_3=\psi_i,\ (i=1,2,3)$.
where ψ_i are the shape functions for a typical element $(\eta_e,\ \eta_{e+1})$ and are defined as

$$\psi_1^e=\frac{(\eta_{e+1}+\eta_e-2\eta)(\eta_{e+1}-\eta)}{(\eta_{e+1}-\eta_e)^2},\quad \psi_2^e=\frac{4(\eta-\eta_e)(\eta_{e+1}-\eta)}{(\eta_{e+1}-\eta_e)^2}$$
$$\psi_3^e=\frac{(\eta_{e+1}+\eta_e-2\eta)(\eta-\eta_e)}{(\eta_{e+1}-\eta_e)^2},\quad \eta_e\le\eta\le\eta_{e+1}. \tag{3.29}$$

The finite-element model of the equations thus formed is given by

$$\begin{bmatrix} [K^{11}] & [K^{12}][K^{13}] & [K^{14}] \\ [K^{21}] & [K^{22}][K^{23}] & [K^{24}] \\ [K^{31}] & [K^{32}][K^{33}] & [K^{34}] \\ \multicolumn{3}{c}{[K^{41}][K^{42}][K^{43}][K^{44}]} \end{bmatrix}\begin{bmatrix} f \\ h \\ \theta \\ \phi \end{bmatrix}=\begin{bmatrix} \{r^1\} \\ \{r^2\} \\ \{r^3\} \\ \{r^4\} \end{bmatrix}$$

where $[K^{mn}]$ and $[r^{m}]$ ($m, n = 1, 2, 3, 4$) are defined as

$$K_{ij}^{11} = \int_{\eta_e}^{\eta_{e+1}} \psi_i \frac{\partial \psi_j}{\partial \eta} \mathrm{d}\eta, \quad K_{ij}^{12} = -\int_{\eta_e}^{\eta_{e+1}} \psi_i \psi_j \, \mathrm{d}\eta, \quad K_{ij}^{13} = K_{ij}^{14} = 0.$$

$$K_{ij}^{21} = 0,$$

$$K_{ij}^{22} = \int_{\eta_e}^{\eta_{e+1}} \left[1 + We^2 (\frac{\partial \psi_i}{\partial \eta})^2\right]^{\frac{n-3}{2}} \left(1 + nWe^2 (\frac{\partial \psi_i}{\partial \eta})^2\right) \frac{\partial \psi_i}{\partial \eta} \frac{\partial \psi_j}{\partial \eta} \mathrm{d}\eta$$
$$+ \bar{f_1} \int_{\eta_e}^{\eta_{e+1}} \psi_i \psi_1 \frac{\partial \psi_j}{\partial \eta} \mathrm{d}\eta + \bar{f_2} \int_{\eta_e}^{\eta_{e+1}} \psi_i \psi_2 \frac{\partial \psi_j}{\partial \eta} \mathrm{d}\eta - \frac{1}{2} \gamma \bar{h_1} \int_{\eta_e}^{\eta_{e+1}} \psi_i \psi_1 \psi_j \mathrm{d}\eta$$
$$- \frac{1}{2} \gamma \bar{h_2} \int_{\eta_e}^{\eta_{e+1}} \psi_i \psi_j \mathrm{d}\eta - M \int_{\eta_e}^{\eta_{e+1}} \psi_i \psi_j \mathrm{d}\eta.$$

$$K_{ij}^{23} = 0, K_{ij}^{24} = 0, \quad K_{ij}^{31} = 0, \; K_{ij}^{32} = 0,$$

$$K_{ij}^{33} = \int_{\eta_e}^{\eta_{e+1}} \frac{\partial \psi_i}{\partial \eta} \frac{\partial \psi_j}{\partial \eta} \mathrm{d}\eta + Pr \, \bar{f_1} \int_{\eta_e}^{\eta_{e+1}} \psi_i \psi_1 \frac{\partial \psi_j}{\partial \eta} \mathrm{d}\eta + Pr \, \bar{f_2} \int_{\eta_e}^{\eta_{e+1}} \psi_i \psi_j \frac{\partial \psi_j}{\partial \eta} \mathrm{d}\eta$$
$$+ Pr \bar{\theta_1} Nt \int_{\eta_e}^{\eta_{e+1}} \psi_i \frac{\partial \psi_1}{\partial \eta} \frac{\partial \psi_j}{\partial \eta} \mathrm{d}\eta + Pr \bar{\theta_2} Nt \int_{\eta_e}^{\eta_{e+1}} \psi_i \frac{\partial \psi_2}{\partial \eta} \frac{\partial \psi_j}{\partial \eta} \mathrm{d}\eta$$
$$+ Pr \bar{\phi_1} Nb \int_{\eta_e}^{\eta_{e+1}} \psi_i \psi_1 \frac{\partial \psi_j}{\partial \eta} \mathrm{d}\eta + Pr \bar{\phi_2} Nb \int_{\eta_e}^{\eta_{e+1}} \psi_i \psi_2 \frac{\partial \psi_j}{\partial \eta} \mathrm{d}\eta$$
$$+ \frac{4}{3R} \int_{\eta_e}^{\eta_{e+1}} \left[1 + \psi_i (\theta_w - 1)\right]^3 \frac{\partial \psi_i}{\partial \eta} \frac{\partial \psi_j}{\partial \eta} \mathrm{d}\eta$$
$$+ \frac{4}{R} \bar{\theta_1} \int_{\eta_e}^{\eta_{e+1}} \left(\left[1 + \psi_i (\theta_w - 1)\right]^2 (\theta_w - 1)\right) \frac{\partial \psi_i}{\partial \eta} \frac{\partial \psi_j}{\partial \eta} \mathrm{d}\eta$$
$$+ \frac{4}{R} \bar{\theta_2} \int_{\eta_e}^{\eta_{e+1}} \left(\left[1 + \psi_i (\theta_w - 1)\right]^2 (\theta_w - 1)\right) \frac{\partial \psi_i}{\partial \eta} \frac{\partial \psi_j}{\partial \eta} \mathrm{d}\eta, \; K_{ij}^{34} = 0.$$

$$K_{ij}^{41} = 0,\ K_{ij}^{42} = 0, K_{ij}^{43} = \frac{Nt}{Nb}\int_{\eta_e}^{\eta_{e+1}} \frac{\partial \psi_i}{\partial \eta}\frac{\partial \psi_j}{\partial \eta} d\eta,$$

$$K_{ij}^{44} = \int_{\eta_e}^{\eta_{e+1}} \frac{\partial \psi_i}{\partial \eta}\frac{\partial \psi_j}{\partial \eta} d\eta + Le\,\overline{f_1} \int_{\eta_e}^{\eta_{e+1}} \psi_i \psi_1 \frac{\partial \psi_j}{\partial \eta} d\eta$$
$$+ Le\,\overline{f_2} \int_{\eta_e}^{\eta_{e+1}} \psi_i \psi_2 \frac{\partial \psi_j}{\partial \eta} d\eta - C_1 \int_{\eta_e}^{\eta_{e+1}} \psi_i \psi_j d\eta$$

$$r_i^2 = 0, r_i^2 = -\left(\psi_i \frac{d\psi_i}{d\eta}\right)_{\eta_e}^{\eta_{e+1}}, r_i^3 = -\left(\psi_i \frac{d\psi_i}{d\eta}\right)_{\eta_e}^{\eta_{e+1}}, r_i^4 = -\left(\psi_i \frac{d\psi_i}{d\eta}\right)_{\eta_e}^{\eta_{e+1}}.$$

3.4 Results and Discussions

Examine Carreau nanofluid flow over wedge with Marangoni effect, analyze numerically, and display results graphically and in tables. Illustrate the problem with graphs showing concentration, temperature, and velocity for different parameters.

Figure 3.1 shows how magnetic field parameter (M) affects velocity distribution. Higher M values lead to overall decrease. Thermal energy is

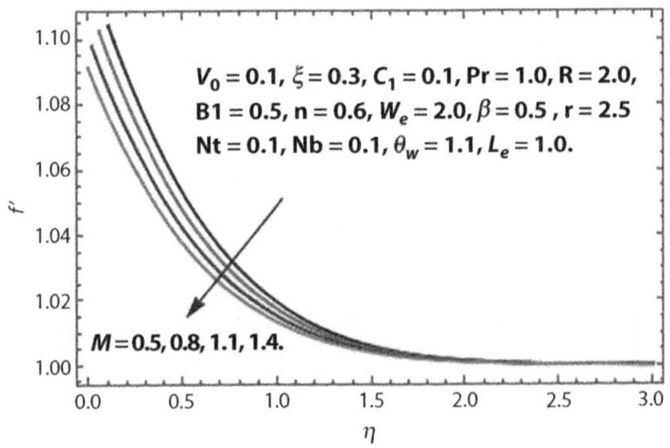

Figure 3.1 Velocity distributions incorporating (M).

Lorentz force pulling nanofluid away from magnetic field, increasing thermal boundary layer thickness. Figures 3.2 to 3.4 show the characteristics of the wedge angle parameter (B1) (γ in the geometry), with velocity profiles decreasing while temperature and concentration profiles increase. Figures 3.5 and 3.6 show temperature and concentration profiles decreasing with increasing ξ in Carreau nanofluid flow. Figure 3.7 illustrates the impact of C1 on Carreau nanofluid flow, with lower values for higher C1 values. The radiation parameter increases, reducing temperature and enhancing concentration profiles of Carreau nanofluid as shown in Figures 3.8 and 3.9.

As β values decrease, boundary layer thickness is affected (Figure 3.10). Figure 3.11 shows higher velocity profiles with increased n values. Figures 3.12–3.14 display weakening profiles with higher We values. Figures 3.15–3.17

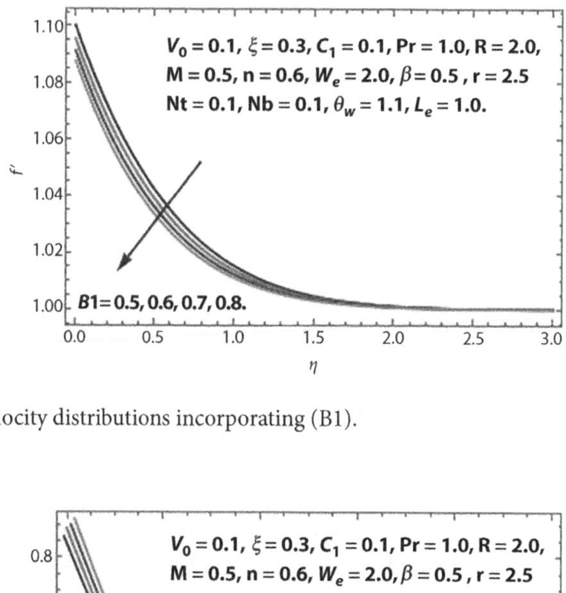

Figure 3.2 Velocity distributions incorporating (B1).

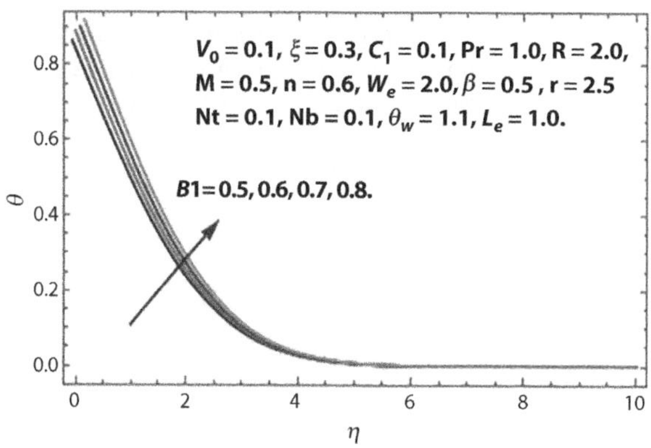

Figure 3.3 Temperature distributions incorporating (B1).

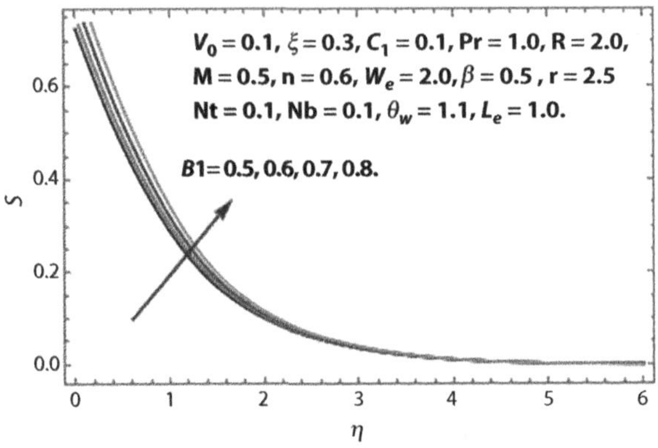

Figure 3.4 Concentration distributions incorporating (B1).

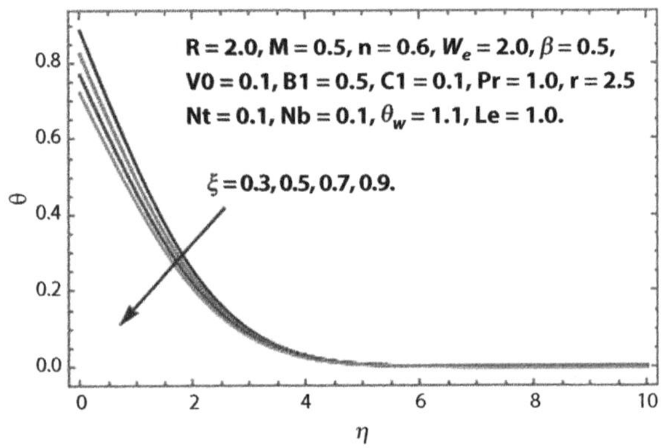

Figure 3.5 Temperature distributions incorporating (ξ).

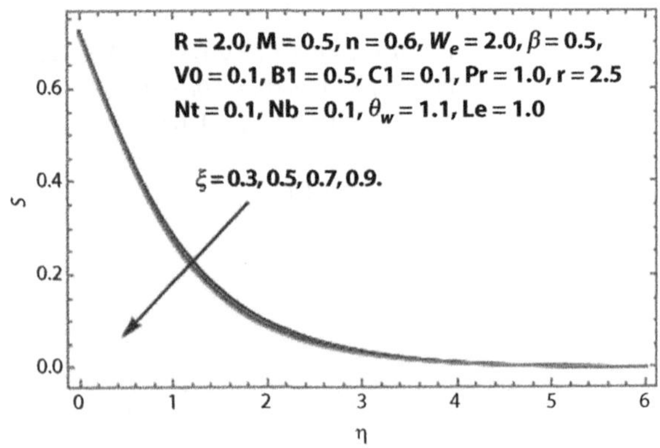

Figure 3.6 Concentration distributions incorporating (ξ).

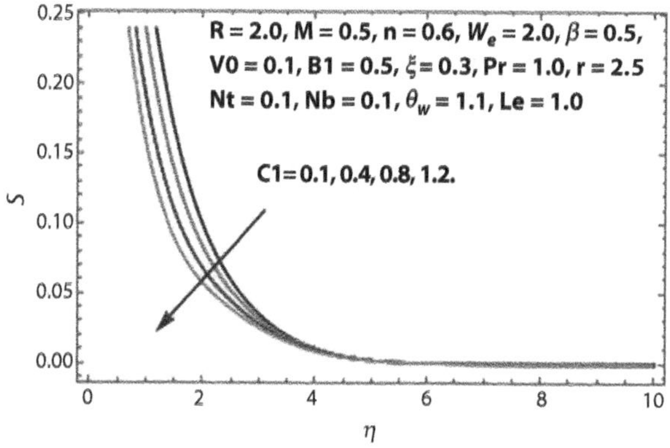

Figure 3.7 Concentration distributions incorporating (C1).

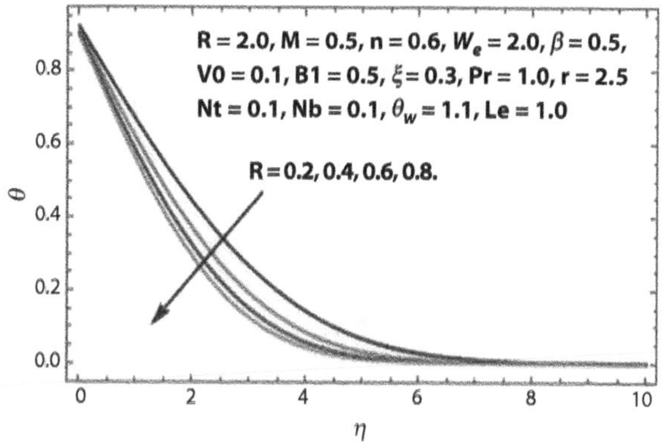

Figure 3.8 Temperature distributions incorporating (R).

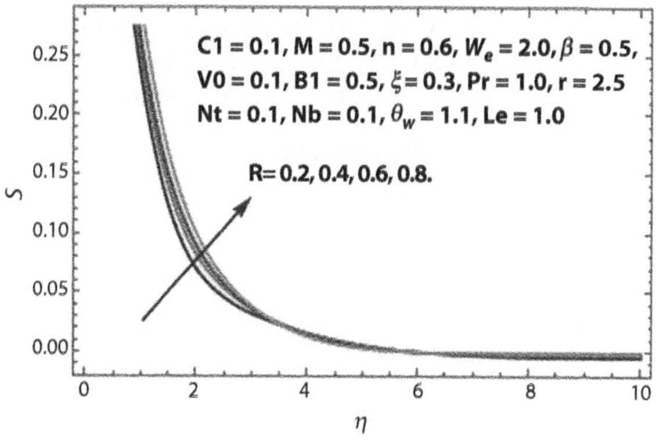

Figure 3.9 Concentration distributions incorporating (R).

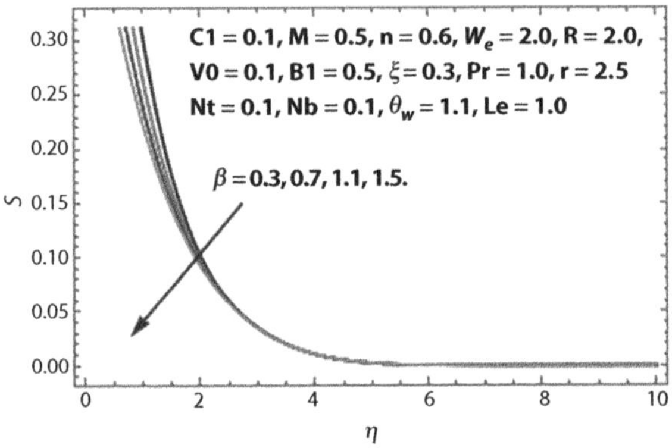

Figure 3.10 Concentration distributions incorporating (β).

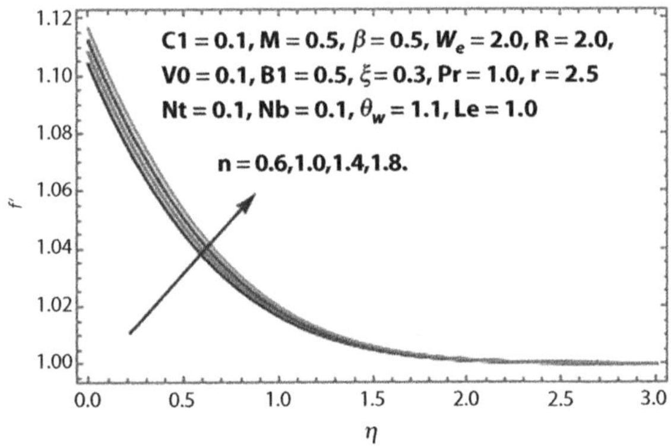

Figure 3.11 Velocity distributions incorporating (n).

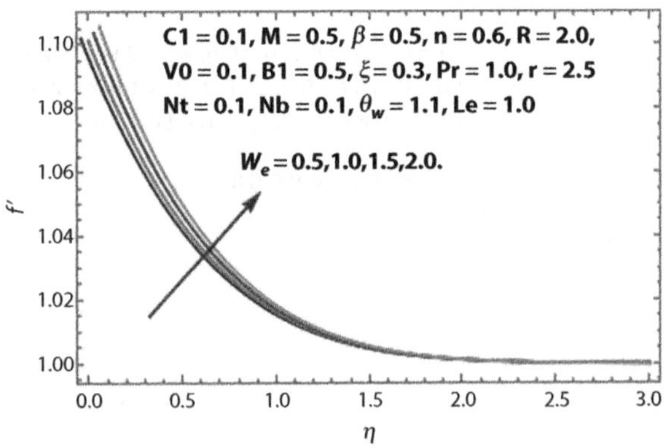

Figure 3.12 Velocity distributions incorporating (We).

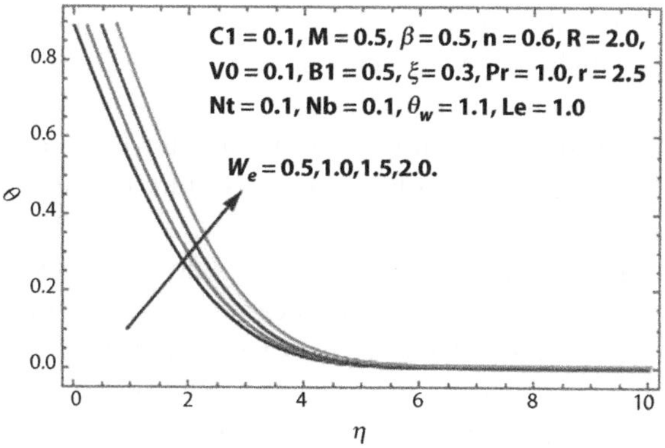

Figure 3.13 Temperature distributions incorporating (We).

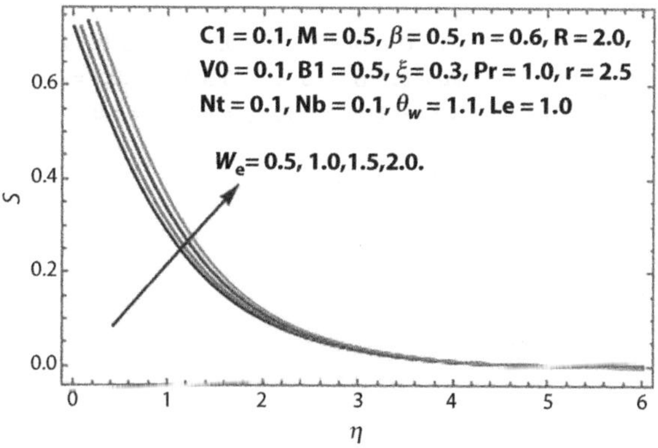

Figure 3.14 Concentration distributions incorporating (We).

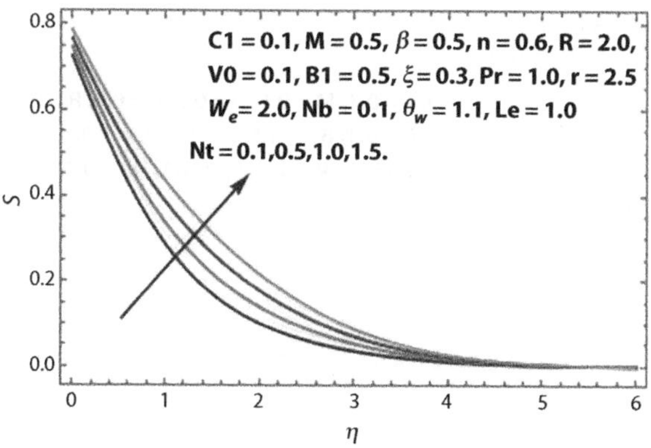

Figure 3.15 Concentration distributions incorporating (Nt).

highlight Nt, Nb, and Le effects on Carreau nanofluid flow. Concentration field decreases with higher Nt values (Figure 3.15). Discrepancy in Nb is seen in Figure 3.16. Concentration profiles decrease with Le (Figure 3.17).

Moving on to the temperature ratio parameter (θw), as θw increases, velocity and temperature profiles intensify, but concentration profiles decline. Figures 3.18–3.20 show this. Figures 3.21 and 3.22 show temperature decreasing and concentration increasing along the wedge with higher Prandtl number (Pr). The Prandtl number affects momentum and thermal boundary layer thickness. Heat diffuses faster than velocity for small Pr. In liquid metals, the thermal boundary layer is thicker than the velocity boundary layer. Regarding the suction parameter (V0), velocity decreases for higher V0 in Figure 3.23. Temperature and concentration decrease for higher V0 along the wedge in Figures 3.24 and 3.25. The Marangoni

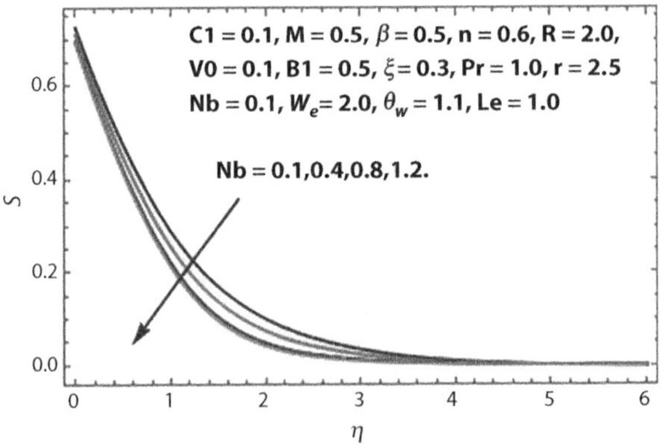

Figure 3.16 Concentration distributions incorporating (Nb).

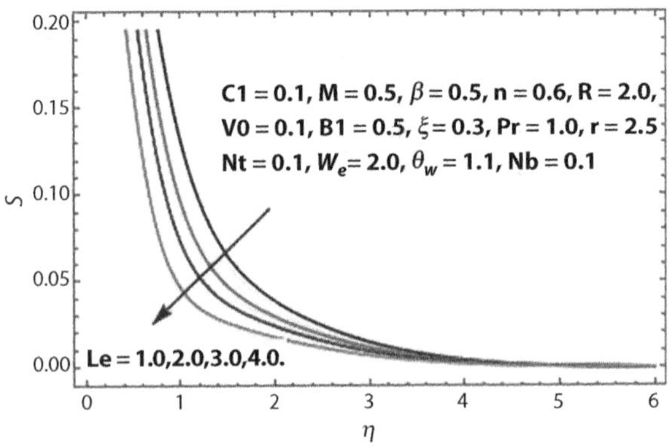

Figure 3.17 Concentration distributions incorporating (Le).

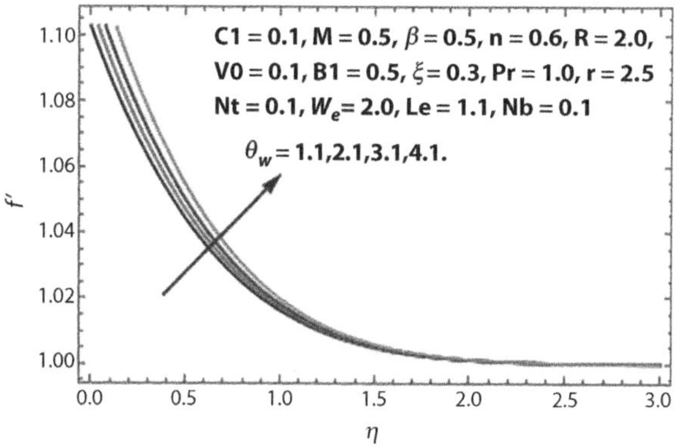

Figure 3.18 Velocity distributions incorporating (θ_w).

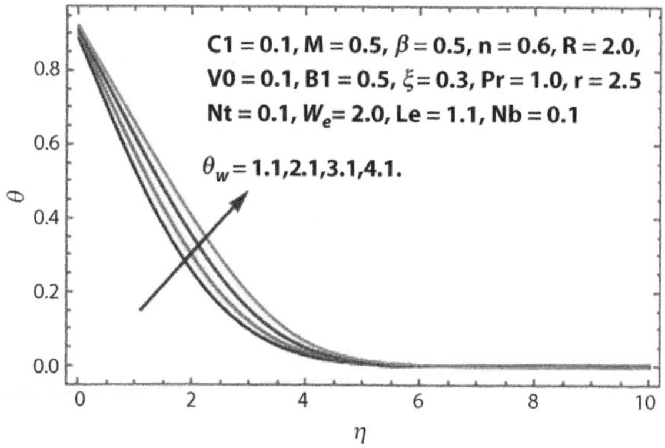

Figure 3.19 Temperature distributions incorporating (θ_w).

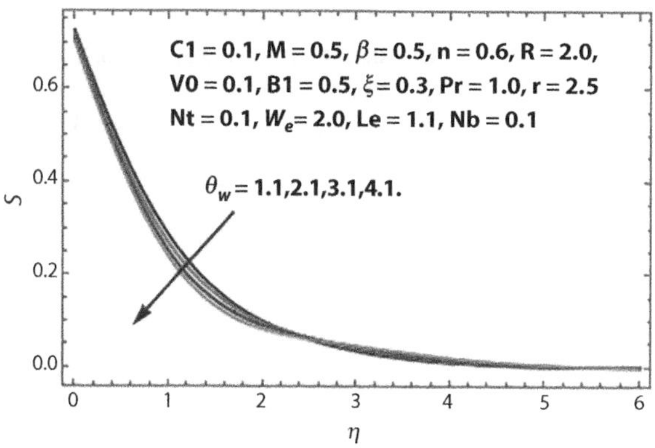

Figure 3.20 Concentration distributions incorporating (θ_w).

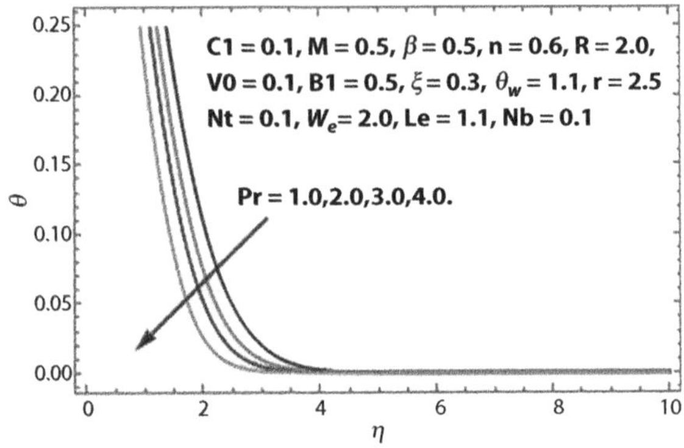

Figure 3.21 Temperature distributions incorporating (Pr).

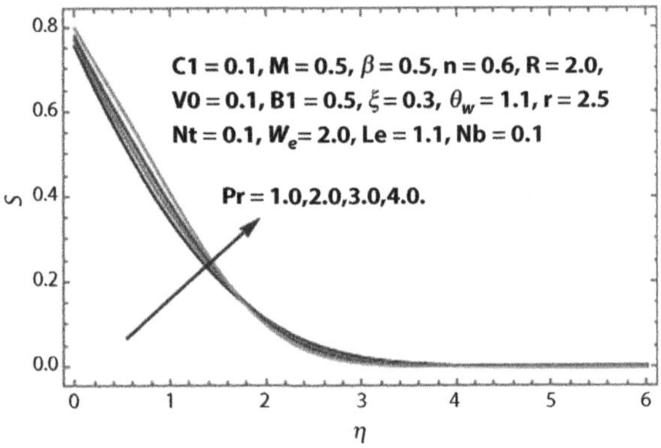

Figure 3.22 Concentration distributions incorporating (Pr).

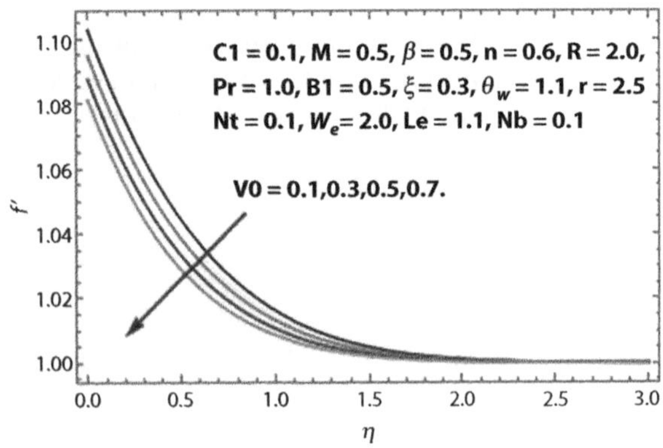

Figure 3.23 Velocity distributions incorporating (V0).

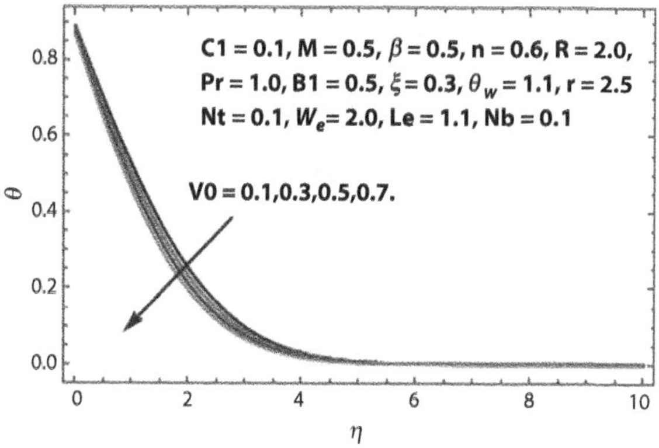

Figure 3.24 Temperature distributions incorporating (V0).

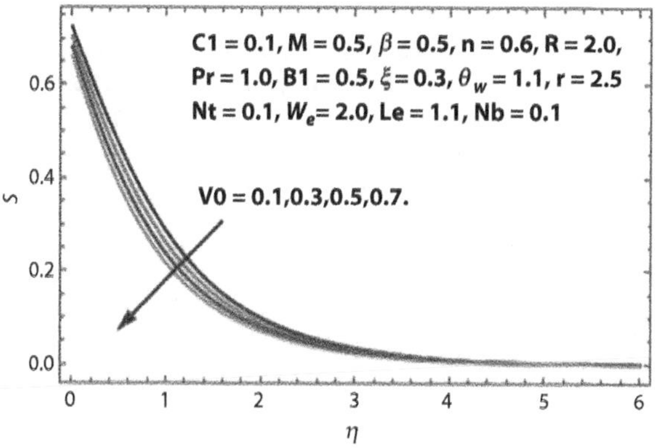

Figure 3.25 Concentration distributions incorporating (V0).

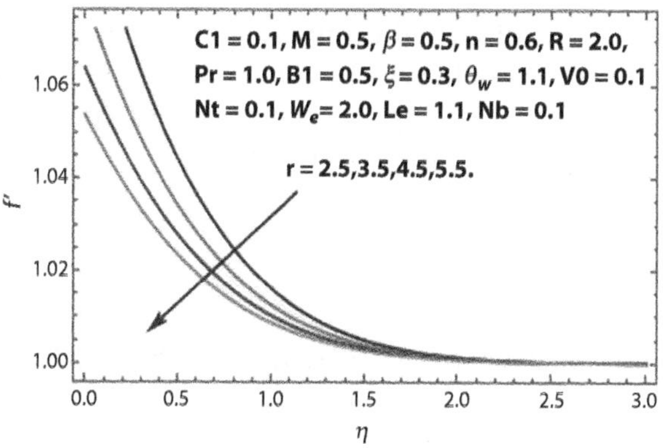

Figure 3.26 Velocity distributions incorporating (r).

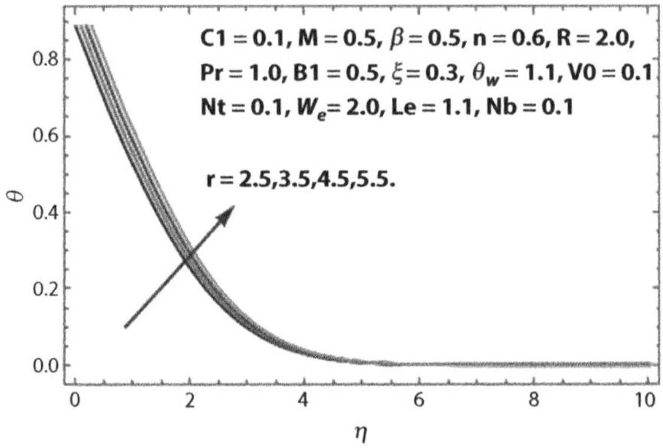

Figure 3.27 Temperature distributions incorporating (r).

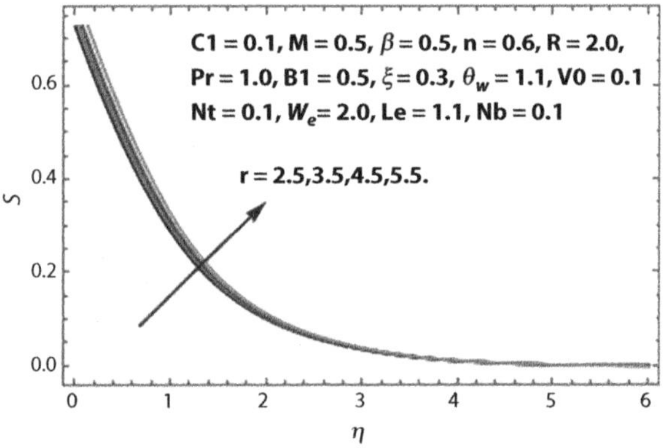

Figure 3.28 Concentration distributions incorporating (r).

ratio parameter (r) is discussed through Figures 3.26–3.28. The increase in r results in the decrease in velocity profiles in Figure 3.26. Temperature profiles decrease with increasing r values, a similar trend in concentration profiles over a wedge with Carreau nanofluid.

3.5 Table Discussions

Table 3.1 shows data on the skin-friction coefficient, Nu_x and Sh_x for parameters M, B1, ξ, R, and β, and We increasing M leads to higher values of these coefficients, as shown in the table. With higher wedge angle (B1), thermal boundary layer thickness, Nu_x, and concentration boundary

Table 3.1 The numerical values of skin-friction coefficient, Nusselt number, and Sherwood number for various values of parameters.

Parameters						Skin-friction coefficient	Nusselt number	Sherwood number
M	B1	ξ	R	β	We			
0.5	0.5	0.3	2.0	0.5	2.0	0.25237	1.14770	0.89977
0.8	0.5	0.3	2.0	0.5	2.0	0.25186	1.14714	0.89940
1.1	0.5	0.3	2.0	0.5	2.0	0.25137	1.14662	0.89905
1.4	0.5	0.3	2.0	0.5	2.0	0.25091	1.14614	0.89783
0.5	0.5	0.3	2.0	0.5	2.0	0.25184	1.14712	0.89939
0.5	0.6	0.3	2.0	0.5	2.0	0.25078	1.14600	0.89864
0.5	0.7	0.3	2.0	0.5	2.0	0.24985	1.14504	0.89799
0.5	0.8	0.3	2.0	0.5	2.0	0.24904	1.14419	0.89742
0.5	0.5	0.3	2.0	0.5	2.0	0.25329	1.14720	0.89944
0.5	0.5	0.5	2.0	0.5	2.0	0.25242	1.06008	0.90928
0.5	0.5	0.7	2.0	0.5	2.0	0.25122	0.98468	0.91762
0.5	0.5	0.9	2.0	0.5	2.0	0.25035	0.91916	0.92497
0.5	0.5	0.3	2.0	0.5	2.0	0.21764	1.88501	0.97396
0.5	0.5	0.3	0.2	0.5	2.0	0.21701	2.18170	0.95015
0.5	0.5	0.3	0.4	0.5	2.0	0.21688	2.38691	0.93261
0.5	0.5	0.3	0.6	0.5	2.0	0.21656	2.53312	0.91959
0.5	0.5	0.3	0.8	0.3	2.0	0.23713	1.98509	1.78717
0.5	0.5	0.3	2.0	0.7	2.0	0.23703	1.99002	1.38164
0.5	0.5	0.3	2.0	1.1	2.0	0.23685	1.99313	1.12610
0.5	0.5	0.3	2.0	1.5	2.0	0.23646	1.99527	0.95338
0.5	0.5	0.3	2.0	0.5	0.5	0.25536	1.14021	0.89706
0.5	0.5	0.3	2.0	0.5	1.0	0.25528	1.14013	0.89432
0.5	0.5	0.3	2.0	0.5	1.5	0.25483	1.14099	0.89363
0.5	0.5	0.3	2.0	0.5	2.0	0.25461	1.14095	0.89231

layer thickness decrease. Cf_x coefficient and Nu_x number decrease with higher temperature slip parameter (ξ), while Sh_x number shows a different trend. There is a downward trend in skin-friction coefficient and Sh_x for larger R values, while the opposite is true for Nu_x. Increasing β leads to decrease in Cf_x coefficient and Sh_x, with Nu_x showing an opposing trend. All three boundary layers exhibit a diminishing pattern with increasing We at the wedge.

Table 3.2 delves into the parameters Nt, Nb, Le, θ_w, Pr, and r concerning Cf_x coefficient, Nu_x, and Sh_x in this section. Initially, Nt affects boundary layer thickness. Increasing Nt reduces thickness. Cf_x and Sh_x decrease with higher Nb. Nu_x decreases with higher Nb. Larger Le leads to lower Cf_x coefficient and higher Sh_x and Nu_x. With increasing θ_w, skin friction

Table 3.2 The numerical values of skin-friction coefficient, Nusselt number, and Sherwood number for various values of parameters.

Parameters						Skin-friction coefficient	Nusselt number	Sherwood number
Nt	Nb	Le	θ_w	Pr	r			
0.1	0.1	1.0	1.1	1.0	2.5	0.25165	1.14471	0.93148
0.5	0.1	1.0	1.1	1.0	2.5	0.25143	1.14225	0.89977
1.0	0.1	1.0	1.1	1.0	2.5	0.25140	1.13683	0.76458
1.5	0.1	1.0	1.1	1.0	2.5	0.25138	1.13145	0.69937
0.1	0.1	1.0	1.1	1.0	2.5	0.25623	1.14222	0.87749
0.1	0.4	1.0	1.1	1.0	2.5	0.25691	1.14136	0.94816
0.1	0.8	1.0	1.1	1.0	2.5	0.25725	1.12248	0.99652
0.1	1.2	1.0	1.1	1.0	2.5	0.25734	1.11001	1.00861
0.1	0.1	1.0	1.1	1.0	2.5	0.25332	1.14765	1.22828
0.1	0.1	2.0	1.1	1.0	2.5	0.25179	1.14787	1.33888
0.1	0.1	3.0	1.1	1.0	2.5	0.25125	1.14813	1.43037
0.1	0.1	4.0	1.1	1.0	2.5	0.25034	1.14865	1.57595
0.1	0.1	1.0	1.1	1.0	2.5	0.25166	22.2838	0.91774
0.1	0.1	1.0	2.1	1.0	2.5	0.25188	20.3941	0.93128
0.1	0.1	1.0	3.1	1.0	2.5	0.25196	18.5488	0.95703
0.1	0.1	1.0	4.1	1.0	2.5	0.25244	17.0228	0.97490
0.1	0.1	1.0	1.1	1.0	2.5	0.23795	1.54408	0.80286
0.1	0.1	1.0	1.1	2.0	2.5	0.23882	1.69161	0.76259
0.1	0.1	1.0	1.1	3.0	2.5	0.23904	1.81937	0.72602
0.1	0.1	1.0	1.1	4.0	2.5	0.23993	2.03391	0.66134
0.1	0.1	1.0	1.1	1.0	2.5	0.23579	1.14626	0.89747
0.1	0.1	1.0	1.1	1.0	3.5	1.93558	1.14395	0.89709
0.1	0.1	1.0	1.1	1.0	4.5	0.15678	1.14158	0.89540
0.1	0.1	1.0	1.1	1.0	5.5	0.13167	1.13995	0.89424

coefficient and Sh_x increase while Nu_x decreases. For high Pr values, there is a change in all three parameters leading to increase in skin friction coefficient and Nu_x and decrease in Sh_x. As Marangoni ratio parameter r increases, all three boundary layer thicknesses decrease.

3.6 Conclusions

In this analysis, we explore the impact of Marangoni forces and slip effects on H & M transfer in Buongiorno's model for flow of a Carreau nanofluid over a wedge. The FEM is used to solve nonlinear ODEs for momentum, temperature, and concentration. Results are presented and discussed.

By considering Marangoni forces and slip effects, we endeavor to comprehend their impact on nanofluid flow over a wedge in a chemically reactive milieu.

- Through graph analysis, we observed escalating temperature and concentration profiles corresponding to elevated Marangoni ratio parameter values.
- The skin friction, Nu_x number, and Sh_x number values decrease with increasing Weber number.
- As the wedge angle parameter increases, the Nu_x and Sh_x numbers decrease.
- The trends of Cf_x, Nu_x number, and Sh_x number decrease as Marangoni ratio parameter (r) increases.

AI disclosure statement: The author confirms that no artificial intelligence (AI)-generated content or AI-assisted writing tools were used in the preparation of this chapter (including summaries and the abstract). All text, analysis, and conclusions are the result of the author's original work.

References

1. Choi, S.U.S., Zhang, Z.G., Yu, W., Lockwood, F.E., Grulke, E.A., Anomalous thermal conductivity enhancement in nano-tube suspensions. *Appl. Phys.*, 79, 2252–2254, 2001.
2. Eastman, J.A., Choi, S.U.S., Li, S., Yu, W., Thompson, L.J., Anomalously increased effective thermal conductivities of ethylene glycol-based nanofluids containing copper nano-particles. *Appl. Phys.*, 78, 718–720, 2001.
3. Chakraborty, S. and Panigrahi, P.K., Stability of nanofluid: A review. *Appl. Therm. Eng.*, 174, 115259, 2020.
4. Abbas, F., Ali, H.M., Shah, T.R., Babar, H., Janjua, M.M., Sajjad, U., Amer, M., Nanofluid: Potential evaluation in automotive radiator. *J. Mol. Liq.*, 297, 112014, 2020.
5. Munyalo, J.M. and Zhang, X., Particle size effect on thermophysical properties of nanofluid and nanofluid based phase change materials: A review. *J. Mol. Liq.*, 265, 77–87, 2018.
6. Ebaid, M.S.Y., Ghrair, A.M., Al-Busoul, M., Experimental investigation of cooling photovoltaic (PV) panels using (TiO_2) nanofluid in water-polyethylene glycol mixture and (Al_2O_3) nanofluid in water-cetyltrimethylammonium bromide mixture. *Energy Convers. Manage.*, 155, 324–343, 2018.
7. Tawfik, M.M., Experimental studies of nanofluid thermal conductivity enhancement and applications: A review. *Renewable Sustainable Energy Rev.*, 75, 1239–1253, 2017. https://doi.org/10.1016/j.rser.2016.11.111.

8. Kim, S., Song, H., Yu, K., Tserengombo, B., Choi, S.-H., Chung, H., Kim, J., Jeong, H., Comparison of CFD simulations to experiment for heat transfer characteristics with aqueous Al2O3 nanofluid in heat exchanger tube. *Int. Commun. Heat Mass Transfer*, 95, 123–131, 2018.
9. Sheremet, M.A. and Pop, I., Marangoni natural convection in a cubical cavity filled with a nanofluid. *J. Therm. Anal. Calorim.*, 135, 1, 357–369, 2018.
10. Song, Y.-Q., Waqas, H., Al-Khaled, K., Farooq, U., Ullah Khan, S., Ijaz Khan, M., Chu, Y.-M., Qayyum, S., Bioconvection analysis for Sutterby nanofluid over an axially stretched cylinder with melting heat transfer and variable thermal features: A Marangoni and solutal model. *Alexandria Eng. J.*, 60, 5, 4663–4675, 2021.
11. Sheikholeslami, M. and Chamkha, A.J., Influence of Lorentz forces on nanofluid forced convection considering Marangoni convection. *J. Mol. Liq.*, 225, 750–757, 2017.
12. Wahid, N.S., Arifin, N.M., Khashi'ie, N.S., Pop, I., Marangoni hybrid nanofluid flow over a permeable infinite disk embedded in a porous medium. *Int. Commun. Heat Mass Transfer*, 126, 105421, 2021.
13. Abdullah, A.A., Alraiqib, N.M., Lindsay, K.A., Modelling the stability of Marangoni convection in a layer of nanofluid. *Int. J. Therm. Sci.*, 151, 106228, 2020.
14. Jawad, M., Saeed, A., Gul, T., Shah, Z., Kumam, P., Unsteady thermal Maxwell power law nanofluid flow subject to forced thermal Marangoni Convection. *Sci. Rep.*, 11, 1, 2021.
15. Gowda, R.J.P., Mallikarjuna, H.B., Prasannakumara, B.C., Kumar, R.N., Manjunatha, P.T., Dynamics of thermal Marangoni stagnation point flow in dusty Casson nanofluid. *Int. J. Model. Simul.*, 42, 5, 707–715, 2021.
16. Punith Gowda, R.J., Naveen Kumar, R., Jyothi, A.M., Prasannakumara, B.C., Sarris, I.E., Impact of Binary Chemical Reaction and Activation Energy on Heat and Mass Transfer of Marangoni Driven Boundary Layer Flow of a Non-Newtonian Nanofluid. *Processes*, 9, 4, 702, 2021.
17. Li, Y.-X., Khan, M.I., Gowda, R.J.P., Ali, A., Farooq, S., Chu, Y.-M., Khan, S.U., Dynamics of aluminum oxide and copper hybrid nanofluid in nonlinear mixed Marangoni convective flow with entropy generation: Applications to renewable energy. *Chin. J. Phys.*, 73, 275–287, 2021.
18. Ellahi, R., Zeeshan, A., Hassan, M., Particle shape effects on Marangoni convection boundary layer flow of a nanofluid. *Int. J. Numer. Methods Heat Fluid Flow*, 26, 7, 2160–2174, 2016.
19. Sheikholeslami, M. and Ganji, D.D., Influence of magnetic field on CuO–H2O nanofluid flow considering Marangoni boundary layer. *Int. J. Hydrogen Energy*, 42, 5, 2748–2755, 2017.
20. Chaudhary, S. and Kanika, K., Viscous dissipation and Joule heating in MHD Marangoni boundary layer flow and radiation heat transfer of Cu–water nanofluid along particle shapes over an exponential temperature. *Int. J. Comput. Math.*, 97, 5, 943–958, 2019.

21. Hashim, Khan, M., Hamid, A., Numerical investigation on time-dependent flow of Williamson nanofluid along with heat and mass transfer characteristics past a wedge geometry. *Int. J. Heat Mass Transfer*, 118, 480–491, 2018.
22. Tlili, I., Hamadneh, N.N., Khan, W.A., Thermodynamic Analysis of MHD Heat and Mass Transfer of Nanofluids Past a Static Wedge with Navier Slip and Convective Boundary Conditions. *Arabian J. Sci. Eng.*, 44, 2, 1255–1267, 2018.
23. Hashim, Khan, M., Ul Huda, N., Hamid, A., Non-linear radiative heat transfer analysis during the flow of Carreau nanofluid due to wedge-geometry: A revised model. *Int. J. Heat Mass Transfer*, 131, 1022–1031, 2019.
24. Kameswaran, P.K., Narayana, M., Shaw, S., Sibanda, P., Heat and mass transfer from an isothermal wedge in nanofluids with Soret effect. *Eur. Phys. J.*, Plus, 129(7), 2014.
25. Xiu, W., Animasaun, I.L., Al-Mdallal, Q.M., Alzahrani, A.K., Muhammad, T., Dynamics of ternary-hybrid nanofluids due to dual stretching on wedge surfaces when volume of nanoparticles is small and large: forced convection of water at different temperatures. *Int. Commun. Heat Mass Transfer*, 137, 106241, 2022.
26. Rizwan, M., Hassan, M., Asjad, M.I., Tag-ElDin, E.M., Flow Characteristics of Heat and Mass for Nanofluid under Different Operating Temperatures over Wedge and Plate. *Micromachines*, 13, 12, 2080, 2022.
27. Dharmaiah, G., Sridhar, W., AL-Farhany, K., Balamurugan, K.S., Ali, F., Non-Newtonian nanofluid characteristics over a melting wedge: A numerical study. *Heat Transfer*, 51, 5, 4620–4640, 2022. Portico.
28. Babu, M.S., Ramana, V.V., Shankar, G.R., Raju, C.s.k., Mixed convective flow of heat and mass transfer of nanofluids over a static wedge with convective boundary conditions. *J. Therm. Eng.*, 7, 14, 1958–1969, 2021.
29. Yacob, N.A., Ishak, A., Nazar, R., Pop, I., Falkner–Skan problem for a static and moving wedge with prescribed surface heat flux in a nanofluid. *Int. Commun. Heat Mass Transfer*, 38, 2, 149–153, 2011.
30. De, P., Mondal, H., Bera, U.K., Influence of Nanofluids on Magnetohydrodynamic Heat and Mass Transfer Over a Non-Isothermal Wedge with Variable Viscosity and Thermal Radiation. *J. Nanofluids*, 3, 4, 391–398, 2014.
31. Hussain, S., Finite Element Solution for MHD Flow of Nanofluids with Heat and Mass Transfer through a Porous Media with Thermal Radiation, Viscous Dissipation and Chemical Reaction Effects. *Adv. Appl. Math. Mech.*, 9, 4, 904–923, 2017.
32. Sheikholeslami, M. and Ghasemi, A., Solidification heat transfer of nanofluid in existence of thermal radiation by means of FEM. *Int. J. Heat Mass Transfer*, 123, 418–431, 2018.
33. Sreedevi, P., Sudarsana Reddy, P., Sheremet, M.A., Impact of homogeneous–heterogeneous reactions on heat and mass transfer flow of Au–Eg and Ag–Eg Maxwell nanofluid past a horizontal stretched cylinder. *J. Therm. Anal. Calorim.*, 141, 1, 533–546, 2020.

Nomenclature

C_f	Skin-friction coefficient	C_p	Specific heat at constant pressure
k_f	Thermal conductivity of basefluid	Nu_x	Nusselt number
Re_x	Local Reynolds number	u_∞	Velocity of mainstream
T_w	Wall constant temperature	T_∞	Ambient temperature
T	Fluid temperature	C	Fluid concentration
θ_w	Temperature ratio parameter	n	Power-law index
q_w	Wall heat flux	J_w	Wall mass flux
f	Dimensionless stream function	u_w	Velocity of the wall
u_e	Velocity of the far flow	k	Thermal conductivity [W/m K]
K^*	Mean absorption coefficient	σ^*	Stephan–Boltzmann constan
Sh_x	Sherwood number	Pr	Prandtl number
(u,v)	Velocity components in x- and y-axis	N_R	Radiation parameter
τ_w	Shear stress	Sc	Schmidt number
We	Local Weissenberg number	M	Magnetic field parameter
D_m	Diffusion coefficient	Sc	Schmidth number
t	Time	U	Composite velocity
(x,y)	Direction along and perpendicular to the wedge	C_1	Chemical reaction parameter

V_0	Suction parameter	C_w	Nanoparticle volume fraction at wall
Greek symbols			
α	Thermal diffusivity of base fluid	ν	Kinematic viscosity
μ	Fluid viscosity	ρ_p	Nanoparticle mass density
ψ	Stream function	γ	Eigen value
S	Dimensionless nanoparticle volume fraction	η	Similarity variable
θ	Dimensionless temperature	λ	Velocity slip parameter
σ	Electrical conductivity	γ	Wedge angle
ξ	Thermal slip parameter	Γ	Material parameter
Subscripts			
∞	Condition far away from cone surface		
f	Base fluid	nf	Nanofluid

4

Exploring Heat and Mass Diffusion in Non-Newtonian Fluid Flow over a Stretching Surface in a Non-Darcy Variable Porous Medium: An Analysis by Finite Difference Scheme

Sahin Ahmed*, Bikash Das and Anil Nangkar

Department of Mathematics, Rajiv Gandhi University, Rono Hills, Doimukh, Arunachal Pradesh, India

Abstract

The steady two-dimensional motion of Casson fluid over a vertical stretching sheet embedded in a porous regime of variable porosity and permeability has been analysed in this investigation. The current study's novelty is to examine the impact of non-uniform porous medium on heat transmission by considering the ohmic heating and viscous dissipation. In this regard, the first-order velocity term and inertial term have been encompassed in the heat diffusion equation. The nonlinear governing equations has been transformed into a system of ODEs by employing suitable similarity transformations. Finally, the solutions are evaluated by using the bvp4c MATLAB routine. The impression of nondimensional parameters on the fluid velocity and temperature has been illustrated with the help of graphs. It is reported that the Forchheimer parameter and Casson fluid parameter act as a resistive force and reduce the motion of the fluid, while the temperature fluid rises for higher values of Fr and M. The decreased permeability and elevated values of the porosity parameter led to a higher drag at the surface. Due to the limitations of Newtonian fluid, the study of non-Newtonian fluid in this investigation can be beneficial in the areas of magneto-aerodynamics, MHD propulsion, thermo-fluid dynamics, and material processing.

**Corresponding author*: nanofluid.sahin@gmail.com

Dhananjay Yadav, Mukesh Kumar Awasthi and Harith Mohamed Al-Azri (eds.) Non-Newtonian Fluids for Industrial Applications: Modeling and Simulations, (85–102) © 2026 Scrivener Publishing LLC

Keywords: Variable porosity, variable permeability, Darcy–Forchheimer, vertical stretching sheet

4.1 Introduction

The study of fluid flow and heat transfer in porous media has received extensive attention due to its significance in thermal insulation, energy storage, electronic cooling, filtration, and industrial applications. Many of these studies are based on Darcy's law, which governs creeping flow through porous materials at low Reynolds numbers. However, at higher flow velocities, the conventional Darcy model fails to capture inertia effects, necessitating an extension of the model. Forchheimer [1] introduced a correction term incorporating a quadratic velocity component, which was later formalized by Muskat [2] as the Forchheimer term. This modification provides a more accurate representation of high-speed flows in porous structures where inertial forces are significant.

Numerous studies have progressively explored heat and mass transport within porous media across varying flow scenarios. Hussain and Ahmed [3] examined the behaviour of a micropolar fluid over a stretching surface immersed in a Darcy porous medium, concluding that fluid velocity decreases as permeability reduces. Similarly, Ahmed and Ahmed [4] investigated biofluid transport under chemical reactions and porous medium resistance, emphasizing the role of porosity variations in heat and mass transfer. Other studies [5, 6] have explored the impact of governing parameters on fluid motion in porous structures, particularly in non-Newtonian fluids, where viscosity varies nonlinearly.

A significant development in non-Newtonian porous media flow was presented in the work of Das and Ahmed [7], where they numerically analysed bioconvection and heat transmission in Prandtl nanofluids across an inclined stretching sheet using a finite difference approach. Although their study focused on nanofluids, the findings emphasized the influence of inclination angles and stretching dynamics on convective transport, which is also relevant within porous domains with varying permeability. Similarly, another investigation on Darcy–Forchheimer flow of Williamson fluid over a stretching surface with varying thermal conductivity [8] highlighted the complex interaction between non-Newtonian rheology, heat transfer, and porosity effects. These findings contribute to the growing body of research demonstrating that heat transfer characteristics in porous media are highly sensitive to fluid type, boundary conditions, and permeability variations.

A common assumption in porous media research is that permeability remains uniform across the domain. However, experimental and numerical studies suggest that permeability and porosity often vary spatially. Research [7, 8] has demonstrated that permeability is highest near boundaries and decreases further into the domain. To capture these variations, Vafai [9] introduced spatially dependent permeability models, significantly improving the understanding of fluid behaviour in non-uniform porous media. Chandrashekar and Vortmeyer [10] extended this concept by analysing how variable porosity influences convective heat transfer, showing that permeability variations strongly affect thermal transport.

Magnetohydrodynamics (MHD) refers to the study of how electrically conductive fluids behave when subjected to magnetic fields. Typical examples include plasmas, liquid metals, electrolytic solutions, and biological fluids such as blood. MHD flow is relevant to various applications, including electromagnetic pumps, crude oil processing, cooling of polymer sheets, and biomedical engineering. The application of magnetic fields allows researchers to regulate flow behaviour and enhance heat transfer in porous and nonporous environments.

The foundations of MHD theory were established by Hannes Alfvén, whose pioneering research provided insights into the effects of magnetic fields on fluid motion. Since then, numerous studies [11–13] have examined how Lorentz force affects the transfer of heat and mass within the fluid. Anuar *et al.* [14] examined MHD flow within porous domains and found that Lorentz forces enhance skin friction while improving heat diffusion. Similarly, Khan *et al.* [15] investigated MHD flow in microchannels, reporting that fluid velocity decreases under an applied magnetic field, while temperature increases due to Joule heating. Additionally, Beg *et al.* [16] performed numerical simulations on MHD flow through porous structures, whereas Khan *et al.* [17] assessed the effect of magnetic fields on the flow dynamics and heat transfer characteristics of a non-Newtonian Williamson fluid.

Despite significant progress in understanding Newtonian and non-Newtonian fluid dynamics within permeable media, further research is required to examine the integrated influence of Darcy–Forchheimer drag, permeability variations, and MHD interactions on heat transfer and flow behaviour. While previous studies have contributed valuable insights, many have either neglected inertia effects in porous media or assumed uniform permeability, leading to an incomplete understanding of real-world applications.

This study seeks to address the existing research gap by analysing the heat and fluid flow behaviour over a vertically stretching sheet embedded in a porous domain, influenced by the magnetohydrodynamic (MHD) effects. This investigation encompasses both Newtonian and non-Newtonian fluids, incorporating Darcy–Forchheimer flow conditions to capture the impact of inertial and porous drag forces. By accounting for variable permeability and thermal conductivity, this research offers a comprehensive evaluation of convective transport mechanisms in complex porous structures, providing deeper perceptions into the relationship between fluid rheology, thermal variations, and magnetic field influences.

4.1.1 Research Questions

- How does permeability of a porous medium (K) affect velocity distribution in the presence of Forchheimer drag, and what is the interaction between inertial resistance (Fr) and permeability in altering flow properties?
- What is the role of magnetic field strength (M) in damping fluid velocity, and how does it change with various Prandtl numbers (Pr) in an MHD boundary layer?
- How does the parameter of Casson fluid (β) affect velocity variation, and how do wall injection and suction (F_w) change the boundary layer structure in non-Newtonian flow?
- How does the magnetic parameter (M) and Prandtl number (Pr) influence heat transfer, and how does the thermal boundary layer react to changes in these parameters under a magnetic field application?
- What are the effects of raising the Forchheimer number (Fr) on thermal profile. Also, what is the role of inertial resistance in relation to thermal energy retention in flow through porous media?
- To what degree does the Eckert number (Ec) contribute to heat rise due to viscous dissipation, and how does it vary with respect to different values of inertial resistance (Fr)?
- How are the cumulative impacts of magnetic fields, permeability, and non-Newtonian fluid behaviour affecting the total heat and momentum transfer properties in porous structure flows?

4.2 Mathematical Formulation

Consider a two-dimensional steady Casson fluid flow across a permeable vertical sheet embedded in a non-Darcy porous regime. The heat diffusion equation includes the combined effects of viscous dissipative, ohmic heating, and Darcian and non-Darcian term. The physical representation of the flow model has been presented in Figure 4.1. The velocity along (x, y) directions are (u, v), repectively. A uniform magnetic field of strength B_0 is incorporporated horizontally, and wall suction $(v_w < 0)$ or injection $(v_w > 0)$ has been imposed at the wall of the vertical stretching sheet. The stretching surface's velocity is taken to be $u_w = bx$, the wall temperature of the stretching sheet is taken to be T_w, while T_∞ is the temperature at free stream.

To simplify the analysis, the following assumptions have been made:

- Porous material is considered as a non-Darcy medium with Forchheimer.
- Porosity and permeability are assumed to be non-uniform throughout the domain.
- Electrical conduction within the fluid is taken into account.
- Due to the small magnetic Reynolds number, the influence of the induced magnetic field is considered negligible.
- The fluid is considered non-Newtonian with the flow presumed to be laminar two-dimensional, where the stretching sheet is assumed to be infinitely long.
- The wall temperature is maintained above the ambient temperature $(T_w > T_\infty)$.
- Surface is assumed to be permeable.

Taking into account the preceding assumptions, the governing equations are expressed as follows [18]:

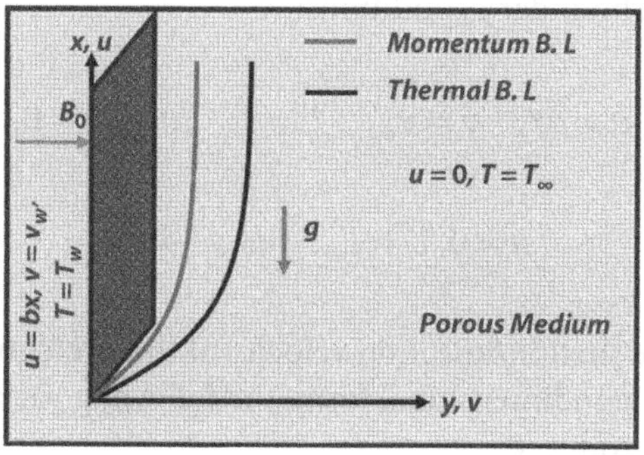

Figure 4.1 Flow domain representation.

$$\frac{\partial u}{\partial x}+\frac{\partial v}{\partial y}=0, \tag{4.1}$$

$$u\frac{\partial u}{\partial x}+v\frac{\partial u}{\partial y}=\nu\left(1+\frac{1}{\beta}\right)\left(\frac{\partial^2 u}{\partial y^2}\right)+g\beta_T\left(T-T_\infty\right)$$
$$-\frac{\sigma B_0^2}{\rho}u-\frac{\nu\epsilon\left(y\right)}{k\left(y\right)}u-\frac{C_b\epsilon^2\left(y\right)}{\sqrt{k\left(y\right)}}u^2, \tag{4.2}$$

$$\left(u\frac{\partial T}{\partial x}+v\frac{\partial T}{\partial y}\right)=\left\{\begin{array}{c}\dfrac{\kappa_{eff}}{\left(\rho C_p\right)}\dfrac{\partial^2 T}{\partial y^2}+\dfrac{\mu}{\rho C_p}\left(1+\dfrac{1}{\beta}\right)\left(\dfrac{\partial u}{\partial y}\right)^2+\\ \dfrac{\mu}{\rho C_p}\dfrac{\epsilon^2\left(y\right)}{k\left(y\right)}u^2+\dfrac{1}{\rho C_p}\dfrac{C_b\epsilon^2\left(y\right)}{\sqrt{k\left(y\right)}}u^3+\dfrac{\sigma B_0^2}{\rho C_p}u^2\end{array}\right\}, \tag{4.3}$$

where $\rho,\mu,C_p,\kappa,$ and σ are the density, dynamic viscosity, specific heat capacity, thermal conductivity and electrical conductivity, respectively, and $\epsilon\left(y\right)=\epsilon_0\left(1+d^*e^{-\frac{y}{a}}\right)$ and $k\left(y\right)=k_0\left(1+de^{-\frac{y}{a}}\right)$ are the variable porosity and permeability, respectively [19], where d^* and d are constants. The value of a is such that the dimension of y and a is the same. The effective thermal conductivity κ_{eff} is given by the relation $\kappa_{eff}=\epsilon\left(y\right)\kappa_f+\left[1-\epsilon\left(y\right)\right]\kappa_s$, where κ_f and κ_s are the fluid's and solid's thermal conductivity, respectively.

Subject to the boundary conditions:

$$u=bx,\quad v=v_w,\quad T=T_w\quad at\quad y=0, \tag{4.4}$$

$$u=0, T=T_\infty,\quad as\ y\to\infty, \tag{4.5}$$

v_w is the surface suction/injection velocity of the stretching sheet.

To obtain the nondimensional form, the following variable transformations [18] are introduced:

$$\left\{\begin{array}{c}\psi=(vb)^{\frac{1}{2}}xf(\eta),\quad \eta=\left(\frac{b}{v}\right)^{\frac{1}{2}}y\\ u=bxf'(\eta),v=-(bv)^{\frac{1}{2}}f(\eta),\\ \theta=\dfrac{T-T_{\infty}}{T_w-T_{\infty}}\end{array}\right\}. \tag{4.6}$$

Employing Equation 4.6, eq. 4.2 – 4.3 reduces to

$$\left(1+\frac{1}{\beta}\right)f'''(\eta)+\frac{Gr_T}{Re}\theta-M^2f'(\eta)-K\left[\frac{\epsilon_0\left(1+d^*e^{-\eta}\right)}{\left(1+d^*e^{-\eta}\right)}\right]f'(\eta)$$
$$-Fr\left[\frac{\epsilon_0\left(1+d^*e^{-\eta}\right)^2}{\left(1+d^*e^{-\eta}\right)^{\frac{1}{2}}}\right]f'^2(\eta)-f'^2(\eta)+f(\eta)f''(\eta)=0, \tag{4.7}$$

$$\frac{\theta''(\eta)}{Pr}\left[\epsilon_0\left(1+d^*e^{-\eta}\right)+\lambda\left\{1-\epsilon_0\left(1+d^*e^{-\eta}\right)\right\}\right]+Ec\left(1+\frac{1}{\beta}\right)f''^2(\eta)$$
$$+\left\{\begin{array}{c}KEc\dfrac{\epsilon_0\left(1+d^*e^{-\eta}\right)^2}{\left(1+de^{-\eta}\right)}f'(\eta)+EcFr\left[\dfrac{\epsilon_0\left(1+d^*e^{-\eta}\right)^2}{\left(1+de^{-\eta}\right)^{\frac{1}{2}}}\right]f'^3(\eta)+\\ M^2Ecf'^2(\eta)-f'(\eta)\theta(\eta)\end{array}\right\}=0. \tag{4.8}$$

Subject to nondimensional boundary conditions:

$$\left\{\begin{array}{c}f(0)=-F_w,f'(0)=1,\theta(0)=1,\\ f'(\infty)=0,\theta(\infty)=0\end{array}\right\}, \tag{4.9}$$

where a prime $(')$ signifies differentiation with respect to η, and the associated dimensionless parameters are defined as follows:

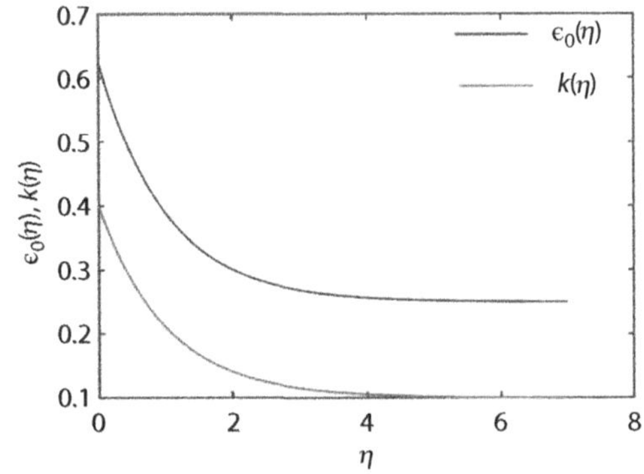

Figure 4.2 Variation of porosity and permeability.

$$\left\{\begin{array}{c} M^2=\dfrac{\sigma B_0^2}{\rho b}, Fr=\dfrac{C_b\epsilon_0 x}{\sqrt{k_0}}, K=\dfrac{\nu}{k_0 b}, Pr=\dfrac{\nu}{\alpha} \\ Gr_T=\dfrac{g\beta_T\nu\left(T_w-T_\infty\right)}{(bx)^3}, \\ Re=\dfrac{u_w x}{\nu}, \lambda=\dfrac{\kappa_s}{\kappa_f}, \quad Ec=\dfrac{u_w^2}{C_p\left(T_w-T_\infty\right)}, \\ F_w=\dfrac{v_w}{(\nu b)^{\frac{1}{2}}} \end{array}\right\}. \tag{4.10}$$

The nondimensional porosity and permeability are $\epsilon_0(\eta)=\epsilon_0\left(1+d^*e^{-\eta}\right)$ and $k_0(\eta)=k_0\left(1+de^{-\eta}\right)$. The variation of porosity and permeability in the region of interest is in Figure 4.2. The permeability and the porosity are maximum near the surface, while the permeability and porosity gradually degrade, as the horizontal distance from the vertical stretching sheet increases. Thus, it can be seen that the resistance is maximum near the surface, while it is minimum at the free stream.

4.3 Research Methodology

To solve the nondimensional system of ODEs given by Equations 4.7 and 4.8 along with boundary conditions given by Equation 4.9, bvp4c MATLAB routine was employed. In order to employ the methods, one has to renovate the higher-order ODEs into first-order ODEs as discussed below:

Assume that

$$\left\{\begin{matrix} f = y_1, f' = y_2, f'' = y_3; \\ \theta = y_4, \theta' = y_5; \end{matrix}\right\}. \tag{4.11}$$

Equations 4.7 and 4.8 transform to

$$y_3' = \left(\frac{1}{\left(1+\frac{1}{\beta}\right)}\right)\left\{\begin{matrix} -\frac{Gr_T}{Re}y_4 + M^2 y_2 + K\left[\frac{\epsilon_0\left(1+d^*e^{-\eta}\right)}{\left(1+d^*e^{-\eta}\right)}\right]y_2 \\ +Fr\left[\frac{\epsilon_0\left(1+d^*e^{-\eta}\right)^2}{\left(1+d^*e^{-\eta}\right)^{\frac{1}{2}}}\right]y_2^2 + y_2^2 - y_1 y_3 \end{matrix}\right\}, \tag{4.12}$$

$$y_5' = -Ec\left(1+\frac{1}{\beta}\right)y_3^2 - \left\{\begin{matrix} KEc\frac{\epsilon_0\left(1+d^*e^{-\eta}\right)^2}{\left(1+de^{-\eta}\right)}y_2 + M^2 Ecy_2^2 - y_2 y_4 \\ +EcFr\left[\frac{\epsilon_0\left(1+d^*e^{-\eta}\right)^2}{\left(1+de^{-\eta}\right)^{\frac{1}{2}}}\right]y_2^3 + \end{matrix}\right\}, \tag{4.13}$$

4.4 Stability and Validation

The stability and convergency of the numerical schemes are illustrated in Table 4.1. The grid point analysis has been carried out for two separate step sizes $h = 0.01$ and $h = 0.0001$. The asymptotic behaviour of velocity and temperature profile can be observed from the numerical values of $f'(\eta)$ and $\theta(\eta)$ as $\eta \rightarrow 5$. The current analysis has been validated with the results already published by Rasheed *et al.* [18] for the values of $-\theta'(0)$ and demonstrated to be in good consistency as deliberated in Table 4.2.

Table 4.1 Numerical values of $f'(\eta)$ at different grid points.

	$h = 0.01$		$h = 0.0001$	
η	$f'(\eta)$	$\theta(\eta)$	$f'(\eta)$	$\theta(\eta)$
0	1	1	1	1
1	0.4441	0.4338	0.4465	0.4375
2	0.2006	0.0742	0.2029	0.0766
3	0.0883	0.0045	0.0887	0.0048
4	0.0323	0.0009	0.0324	0.0000
5	0	0	0	0

Table 4.2 Validation of current investigation for $-\theta'(0)$ with previously published works.

Pr	**Current**	**Rasheed *et al.* [18]**
0.3	0.2648	0.2568
0.5	0.3342	0.4446
0.7	0.3952	0.6344

4.5 Results and Discussion

This section of the study investigates the significance of multiple dimensionless factors that appeared in the governing nondimensional equations over nondimensional velocity and temperature distributions are illustrated and discussed in detail. The solution of the current physical problem is obtained with the help of bvp4c MATLAB routine.

The influence of the porous parameter K and the Forchheimer number Fr over the velocity of the Casson fluid is presented in Figure 4.3. The augmented values of K and Fr both suppress the fluid velocity. This is attributed to the fact that the nondimensional porous parameter K is inversely proportional to the permeability of the porous media. Hence, increased values of K result from reduced permeability of the porous domain due to which fluid experiences more resistance to flow through

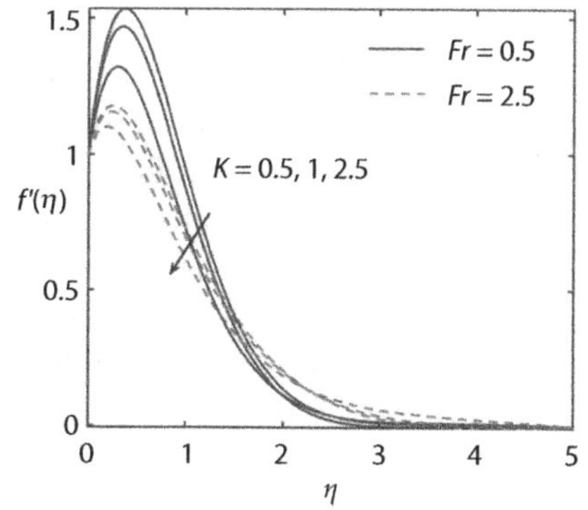

Figure 4.3 Impact of K and Fr over $f'(\eta)$.

the porous matrix. It is also evident that a rise in the Fr values declines the motion of the Casson fluid. This can be related to an additional flow resistance offered by a non-Darcian term; increasing the values of Fr implies dominance of inertial effects, and that is why fluid velocity decreased. It should be noted that the fluid velocity initially increases in the vicinity of the stretching sheet and attains its peak value near the surface of the stretching sheet $(\eta \approx 0.5)$, and then the velocity of the fluid declines toward the free stream.

The impression of the magnetic parameter (M) and the Prandtl number $(Pr = 5,13)$ over the velocity distribution is illustrated in Figure 4.4. The augmented values of M lead to an enhancement in the resistive force (Lorentz force) that resists the motion of Casson fluid due to which the fluid velocity degrades throughout the domain. Physically, the Prandtl number represents the ratio between momentum diffusivity and thermal

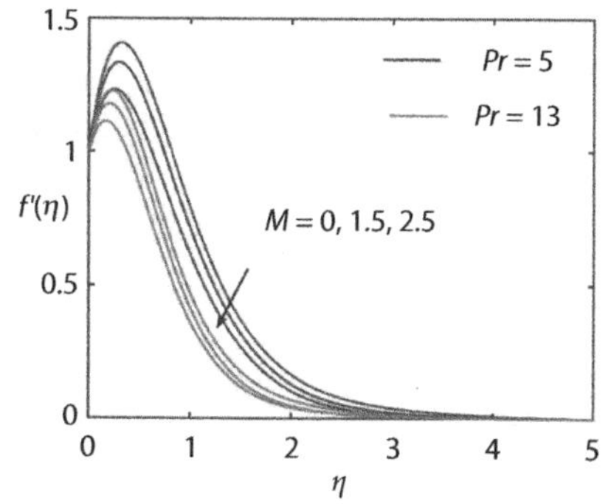

Figure 4.4 Velocity response to changes in magnetic parameter (M) and Prandtl number (Pr).

diffusivity, describing how quickly heat is transferred as compared to momentum. Amplified values of Pr weakens the thermal buoyancy force, which results in the degradation of fluid velocity. The augmented values of the Prandtl number also enhance the fluid's viscosity due to which the velocity of the fluid for $Pr = 13$ is less than the fluid's velocity for $Pr = 5$.

Figure 4.5 describes the influence of the Casson fluid parameter (β) and wall suction/injection parameter (F_w) on velocity distribution. The results reveal a dual effect of increasing β on velocity. Near the wall (small η), velocity increases as β increases, which is attributed to the weakening of non-Newtonian effects. However, toward the free stream, the velocity of the fluid degrades for the augmented magnitudes of β. The application of wall suction $(F_w > 0)$ or injection $(F_w < 0)$ plays a critical role in the momentum transfer. The velocity of the fluid is higher when fluid is injected $(F_w = 0.5)$ as compared to the velocity of the fluid when suction is applied at the surface of the stretching sheet.

Figure 4.6 illustrates how the magnetic parameter affects the temperature distribution of the Casson fluid for two different Prandtl numbers ($Pr = 5$ and 13). Temperature of the Casson fluid enhances for the augmented values of M. This is due to the resistive force experienced by the fluid molecules to travel, which in turn warms the fluid molecules due to less convection of heat energy; as a result, the Casson fluid's temperature rises. It is also evident that a higher Prandtl number corresponds to lower temperature profile, while a lower Prandtl number leads to a higher temperature of the Casson fluid. Physically, the Prandtl number is inversely related to the heat diffusivity, and higher values of Pr correspond to lower thermal diffusivity due to which the temperature of the Casson fluid is higher for $Pr = 5$ as compared to $Pr = 13$.

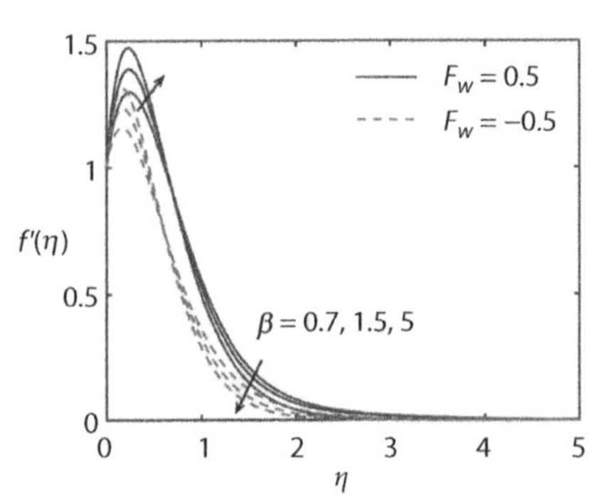

Figure 4.5 Velocity response to changes in Casson fluid parameter (β) and wall suction/injection (F_w).

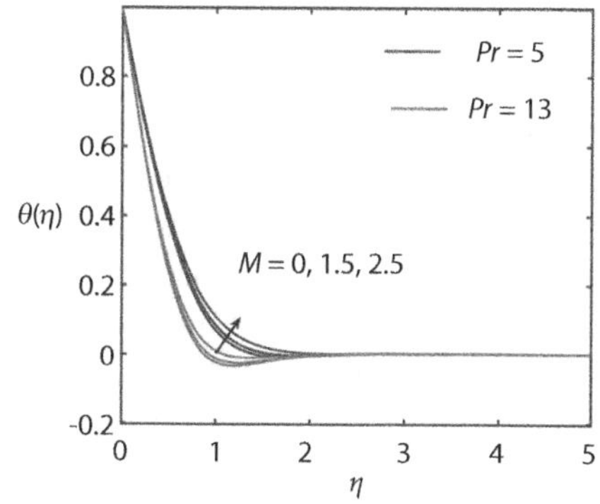

Figure 4.6 Impact of *M* and *Pr* over $\theta(\eta)$.

The impact of Fr and Ec over the temperature of Casson fluid is depicted in Figure 4.7. The disparity of the Forchheimer number and Eckert number on temperature distribution is depicted in Figure 4.5. It is detected that as Fr increases $(Fr = 2,3,4)$, the temperature rises, and the thermal boundary layer expands. This behaviour is attributed to the stronger inertial resistance associated with a higher Fr, which hinders convective heat loss and results in greater heat retention within the fluid. The Eckert number plays a similar role. Physically, the Eckert number quantifies the contribution of enthalpy and kinetic energy. Increased Ec results in greater temperature rise due to increased kinetic energy, while decreased Ec causes lower temperature levels in the boundary layer. Because Ec measures the dissipation of kinetic energy to heat *via* viscous dissipation, a greater Ec implies more intense internal heat generation, thus increasing the temperature in general. Conversely, for smaller Ec, fewer energies are dissipated, and the fluid cools faster.

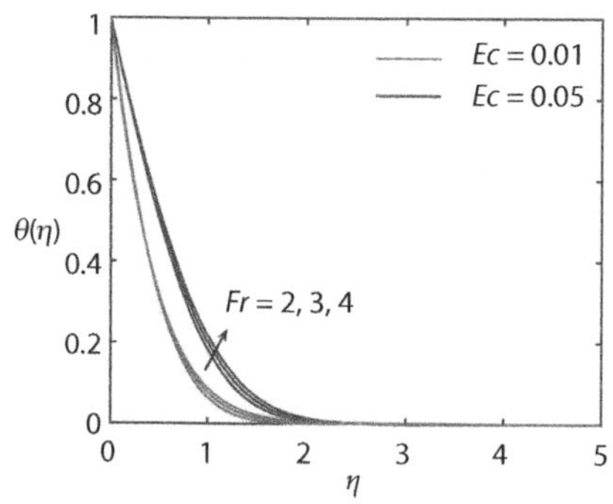

Figure 4.7 Effect of Forchheimer number (Fr) and Eckert number (Ec) on temperature distribution.

4.6 Conclusions

The two-dimensional laminar, steady-state flow of a Casson fluid across a permeable vertical stretching sheet saturated in a non-uniform porous matrix has been investigated under the stimulus of Lorentz force. The Darcy–Forchheimer relation has been incorporated to study momentum distribution in the porous matrix. In this analysis, heat transfer is affected by ohmic heating, viscous dissipation, and heating due to the presence of a porous matrix. The solution of the nonlinear ODEs is obtained by using the bvp4c MATLAB routine, and the solution for the velocity and temperature of the nanofluid has been plotted by varying the nondimensional parameters.

The following are the key outcomes of the current investigation:

- Porous parameter and inertial drag force both suppress the motion of the Casson fluid.
- Lower magnitudes of the Prandtl number boost the fluid's momentum, while Lorentz force degrades fluid velocity in the Casson model.
- Dual behaviour of β has been observed; near the surface of the stretching sheet velocity augments, while higher values of β declines the fluid's movement toward the free stream.
- Magnetic parameter and Forchheimer amplify the thermal profile of the Casson fluid, while augmented values of the Prandtl number decline the thermal profile of the Casson fluid.

Nomenclature

Symbols	Units		Symbols	Units	
(u,v)	ms^{-1}	Fluid velocity components along the x and y -axes	Pr		Prandtl number
B_0	$A.m^{-1}$	Magnetic field strength	C_b		Drag coefficient

(*Continued*)

(*Continued*)

Symbols	Units		Symbols	Units	
C_p	$(J.kg^{-1}.K^{-1})$	Surface concentration	β_T		Thermal expansion coefficient
Ec		Eckert number	β_C		Solutal expansion coefficient
f'		Dimensionless velocity	ρ	Kg / m^3	Fluid density
Fr		Forchheimer number	κ_{eff}		Effective thermal conductivity
g	m / s^2	Acceleration due to gravity	ϵ_0		Porosity
Gr_T		Thermal Grashof number	k_0		Permeability
K		Porous medium permeability	μ	$Kg^{-1}sec^{-1}$	Dynamic viscosity
M		Magnetic parameter	ν	m^2 / sec	Kinematic viscosity
Re		Reynolds number	θ		Dimensionless temperature
T	K	Temperature	η		Similarity variable
(d,d^*)		Constants	λ		Ratio of thermal conductivity
Fw		Suction/injection	σ		Electrical conductivity
f		Nondimensional stream function			
β		Casson fluid parameter			
T_∞	K	Ambient temperature			

AI disclosure statement: The author acknowledges the use of AI-based tools for minor language editing or grammatical correction. However, all intellectual contributions, analyses, and conclusions presented in this chapter are entirely the author's own.

References

1. Forchheimer, P.H., Wasserbewegungdurch boden. *Z. Ver. Dtsch. Ing.*, 45, 1782–1788, 1901.
2. Muskat, M., *The flow of homogeneous fluids through porous media*, vol. 763, p. 100, JW Edwards. Inc., Ann Arbor, Michigan, 1946.
3. Hussain, M.A. and Ahmed, S., Micropolar Hydromagnetic Fluid Over a Vertical Surface in Darcian Regime: An Analytical Approach. *J. Nanofluids.*, *12*, 6, 1614–1620, 2023. https://doi.org/10.1166/jon.2023.2044.
4. Ahmed, K.K., Ahmed, S., Chamkha, A.J., Impact of Inclined Magnetic Force on Bio-Fluid in Permeable Bifurcated Arteries: Analytical Approach. *J. Nanofluids.*, *12*, 2, 332–340, 2023. https://doi.org/10.1166/jon.2023.2000.
5. Hayat, T., Abbas, Z., Pop, I., Asghar, S., Effects of radiation and magnetic field on the mixed convection stagnation-point flow over a vertical stretching sheet in a porous medium. *Int. J. Heat Mass Transfer*, *53*, 1–3, 466–474, 2010. https://doi.org/10.1016/j.ijheatmasstransfer.2009.09.010.
6. Dessie, H. and Kishan, N., MHD effects on heat transfer over stretching sheet embedded in porous medium with variable viscosity, viscous dissipation and heat source/sink. *Ain Shams Eng. J.*, *5*, 3, 967–977, 2014. https://doi.org/10.1016/j.asej.2014.03.008.
7. Das, B. and Ahmed, S., Numerical modeling of bioconvection and heat transfer analysis of Prandtl nanofluid in an inclined stretching sheet: a finite difference scheme, in: *Numerical Heat Transfer, Part A: Applications*, pp. 1–21, 2023.
8. Ahmed, S. and Das, B., Darcy Forchheimer flow of Williamson fluid past a stretching sheet with variable thermal conductivity. *Lat. Am. Appl. Research-An Int. J.*, 54, 2, 253–261, 2024.
9. Vafai, K., Convective flow and heat transfer in variable-porosity media. *J. Fluid Mechanics*, 147, 233–259, 1984. https://doi.org/10.1017/S002211208400207X.
10. Chandrasekhara, B.C. and Vortmeyer, D., Flow model for velocity distribution in fixed porous beds under isothermal conditions. *Waerme und Stoffuebertragung*, *12*, 2, 105–111, 1979. https://doi.org/10.1007/BF01002325.
11. Ahmad, S., Ashraf, M., Ali, K., Simulation of thermal radiation in a micropolar fluid flow through a porous medium between channel walls. *J. Therm. Anal. Calorim.*, *144*, 941–953, 2021. https://doi.org/10.1007/s10973-020-09542-w.
12. Hazarika, S. and Ahmed, S., Physical Insights on Bio-Convection in Prandtl Nanofluid over an Inclined Stretching Sheet in Non-Darcy Medium:

Numerical Simulation, in: *Scientia Iranica*, 2023, https://doi.org/10.24200/sci.2023.62949.8125.

13. Ahmed, S. and Hazarika, N.J., Variation of Thermal Conductivity on Two-Dimensional Hydromagnetic Flow of Thermophoretic Forces: Impact of Suction/Blowing. *J. Nanofluids.*, *12*, 1, 261–270, 2023. https://doi.org/10.1166/jon.2023.1997.
14. Anuar, N.S., Bachok, N., Turkyilmazoglu, M., Arifin, N.M., Rosali, H., Analytical and stability analysis of MHD flow past a nonlinearly deforming vertical surface in Carbon Nanotubes. *Alexandria Eng. J.*, *59*, 1, 497–507, 2020. https://doi.org/10.1016/j.aej.2020.01.024.
15. Khan, I., Chinyoka, T., Ismail, E.A., Awwad, F.A., Ahmad, Z., MHD flow of third-grade fluid through a vertical micro-channel filled with porous media using semi-implicit finite difference method. *Alexandria Eng. J.*, *86*, 513–524, 2024. https://doi.org/10.1016/j.aej.2023.11.070.
16. Bég, O.A., Khan, M.S., Karim, I., Alam, M.M., Ferdows, M., Explicit numerical study of unsteady hydromagnetic mixed convective nanofluid flow from an exponentially stretching sheet in porous media. *Appl. Nanosci.*, 4, 8, 943–957, 2014. https://doi.org/10.1007/s13204-013-0275-0.
17. Khan, M., Karim, I., Rahman, M., Arifuzzaman, S.M., Biswas, P., Williamson fluid flow behaviour of MHD convective-radiative Cattaneo–Christov heat flux type over a linearly stretched-surface with heat generation and thermal-diffusion. *Front. Heat Mass Transf.*, 9, 1, 1–11, 2017. http://dx.doi.org/10.5098/hmt.9.15.
18. Rasheed, H.U., Islam, S., Zeeshan, Khan, W., Khan, J., Abbas, T., Numerical modeling of unsteady MHD flow of Casson fluid in a vertical surface with chemical reaction and Hall current. *Adv. Mech. Eng.*, 14, 3, 2022. doi:10.1177/16878132221085429.
19. Chandrasekhara, B.C., Namboodiri, P.M.S., Hanumanthappa, A.R., Similarity solutions for buoyancy induced flows in a saturated porous medium adjacent to impermeable horizontal surfaces. *Wärme-und Stoffübertragung*, 18, 17–23, 1984.

5

Exploring Non-Newtonian Fluid Dynamics in Porous Media: A CNT-Water Diven Analytical Approach in Vertical Channels

Sahin Ahmed[1*], Nava Jyoti Hazarika[2], Eny Tayang[1] and Dhananjay Yadav[3]

[1]*Department of Mathematics, Rajiv Gandhi University, Rono Hills, Itanagar, Arunachal Pradesh, India*
[2]*Department of Mathematics, Tyagbir Hem Baruah College, Jamugurihat, Sonitpur, Assam, India*
[3]*Department of Mathematical and Physical Sciences, University of Nizwa, Nizwa, Oman*

Abstract

This study presents an analytical investigation into the motion and thermal behavior of a non-Newtonian water-based nanofluid containing carbon nanotubes (CNTs) within a vertical channel embedded in a porous medium. The research incorporates the effects of porosity (K_r), suction (S), magnetohydrodynamics (M), nanoparticle volume fraction (ϕ), heat generation (Q_0), and the Schmidt number (Sc) to provide a comprehensive understanding of the transport phenomena. The originality of this work lies in the simultaneous analytical treatment of these effects on non-Newtonian CNT-based nanofluids, offering new insights into flow control and thermal regulation. Using convergent series solution method, the leading mathematical formulations have been analyzed analytically and are investigated over the prime fluid variables *via* MATLAB plots. Considering CNT in this investigation develops the thermal conductivity that raises the fluid temperature, though an increase in ϕ also raises viscosity, resulting in a reduction in velocity. The Schmidt number (Sc), representing mass diffusivity effects, plays a crucial role in modifying concentration gradients, influencing both heat and mass diffusion features. The analytical approach offers accurate solutions for engineering applications involving non-Newtonian fluids, CNT dispersion, and external forces.

**Corresponding author*: sahin.ahmed@rgu.ac.in

Dhananjay Yadav, Mukesh Kumar Awasthi and Harith Mohamed Al-Azri (eds.) Non-Newtonian Fluids for Industrial Applications: Modeling and Simulations, (103–124) © 2026 Scrivener Publishing LLC

This method helps in understanding fluid behavior, enhancing performance, and improving the efficiency of systems in various industrial and biomedical fields.

Keywords: Non-Newtonian fluids, CNT nanofluid, porosity, heat generation, thermal diffusion

Nomenclature

$\bar{u}$	x -component dimensional velocity, m/s	$\bar{Q}_o$ R	generative heat radiation parameter
$\bar{T}$	nanofluid dimensional temperature, K	Sc	Schmidt number
$\bar{C}$	nanofluid dimensional concentration, mol/m3	t	time, s
u	x -component nondimensional velocity	d	distance, m
$\bar{T}_0$	temperature at $y = 0,$ K	**Greek symbols**	
$\bar{C}_0$	concentration at $y = 0,$ mol/m3	θ	nondimensional temperature of the nanofluid
$\bar{T}_a$	temperature at $y = a,$ K	ψ	nondimensional concentration of the nanofluid
$\bar{C}_a$	concentration at $y = a,$ mol/m3	ε	reference constant due to perturbation
B_0	strength of magnetic field	σ	electric conductivity, Sm-1
g	acceleration due to gravity	β	Casson parameter
(ρC_p)	heat capacitance, J/K	ρ	density, kg/m3
Pe	Peclet number	μ	viscosity, kg·m-1·s-1

Re	Reynold's number	κ	thermal conductivity, W/(m•K)
Gr	Free convection parameter	ω	angular velocity
Gm	Grashof number due to mass diffusion	α	thermal diffusivity, m2/s
M	Hartmann number	ϕ	solid volume fraction of nanoparticles
S	suction $(S>0)$,	$(\rho\beta)$	thermal expansion coefficient
$\underline{D}$ $\underline{k_r}$ $\overline{C_r}$	mass diffusivity porosity chemical reaction	γ	Casson parameter
Subscripts		**Abbreviations**	
f	base fluid	**CNT**	carbon nanotube
nf	nanofluid	**SWCNT**	single-wall carbon nanotube
		MWC NT	multiwall carbon nanotube

5.1 Introduction

The introduction of nanoparticles into base fluids to create nanofluids has led to significant advancements in heat transfer applications. Hybrid nanofluids, which combine two or more types of nanoparticles, offer enhanced thermal and rheological properties compared to single-component nanofluid. With applications in cooling systems, lubrication, and biomedical devices, understanding the dynamics of hybrid nanofluids in channel flows has become essential. Hybrid nanofluids are formed by dispersing two or more types of nanoparticles in a basic fluid medium. Hybrid nanofluids have received much attention owing to their significant ability to improve heat transfer. They have been found to be potentially applicable in various areas such as generator cooling, lubrication, solar heating, welding, electronic cooling, cooling and heating in buildings, vehicle thermal management, space aircrafts, biomedical and drug reduction, and so on. A landmark work has been initiated by Choi [1] in connection with the improvement of thermal conductivity of nanofluid and better heat transport. Maxwell [2] presented a theoretical note on thermal conductivity in the year 1873. Flows with nanoparticles through porous media were initiated by Mahdi [3] in terms of porosity, permeability, and thermal conductivity, and they also incorporated the enhancement of heat transmission *via* thermoproperties. The study [4] on channel flow of nanofluids in porous materials under the influence of external forces is a dynamic field of research with significant applications in engineering, biomedicine, and energy.

In a physical point of view, the variation of pressure and without change of viscosity are corelated with Newtonian fluid of nanoparticles, which is added to a base fluid, whereas the shear thinning and thickening properties have been seen in non-Newtonian fluids where viscosity is changeable. The Ostwald–de Waele power-law model is a classical framework widely used to characterize the rheological behavior of non-Newtonian fluids [5]. In a comparative investigation, Kim *et al.* [6] examined the differences in convective behavior and heat transfer performance between Newtonian and non-Newtonian nanofluids, particularly under transient buoyant natural convection conditions. Their study highlighted enhanced convection intensity and thermal transport in non-Newtonian fluids. Lamsaadi *et al.* [7] reported that shear-thinning nanofluids promote stronger fluid motion and improved convective heat transfer relative to Newtonian fluids. Additionally, Hojjat *et al.* [8] analyzed forced convection within a uniformly heated circular cylinder and found that nanofluids exhibited significantly higher local and average heat transfer coefficients than the base fluid. Turan *et al.* [9], through simulation of laminar natural convection in a square enclosure with differentially heated walls, demonstrated that the mean Nusselt number increases with the Rayleigh number in non-Newtonian nanofluids. Chandi and Raj [10] concluded that shear-thinning fluids exhibit superior heat transfer capabilities compared to Newtonian fluids (where the power-law index n=1n = 1n=1). In the context of Rayleigh–Bénard convection, Kaddiri *et al.* [11] numerically identified a critical Rayleigh number that grows with the power-law index. Furthermore, Shojaeian *et al.* [12] revealed that the thermal behavior and entropy generation in non-Newtonian fluids are significantly influenced by the variability of their thermophysical properties.

Given their exceptional thermal properties and growing importance in nanotechnology, carbon nanotubes (CNTs) have become a leading approach for enhancing heat transfer in numerous 21st-century applications. CNTs are a distinct allotrope of carbon, belonging to the fullerene family, and structurally positioned between closed-shell buckyballs and planar graphene sheets. The two primary forms employed in nanofluids are single-walled carbon nanotubes (SWCNTs) and multiwalled carbon nanotubes (MWCNTs), each offering unique thermal and mechanical characteristics. Several researchers have explored CNT-based flows under various physical conditions. For instance, Hayat *et al.* [13] and Hussain *et al.* [14] investigated the Darcy–Forchheimer flow of CNTs influenced by rotating discs and channels. In the biomedical context, Khalid *et al.* [15] presented a study on magnetohydrodynamic (MHD) blood flow through porous media infused with CNTs, incorporating temperature effects. Additionally, Khan *et al.* [16] analyzed heat transfer and fluid motion of CNTs along a plate with Navier slip conditions. The works of Kandasamy *et al.* [17], Mayer *et al.* [18], and Pramanik [19] delve into broader aspects of fluid dynamics, examining nanofluid behavior—particularly SWCNTs and MWCNTs—under the influence of Casson fluid dynamics, magnetic fields, heat radiation, porous structures, and chemical reactions. These studies offer deeper insights into the complex interplay of forces governing heat and fluid flow in nanofluid systems. Furthermore, Aman *et al.* [20] conducted an analysis using four different molecular liquids to investigate free convection heat transfer enhancement in Maxwell nanofluids containing CNTs.

The influence of chemical reactions is widely recognized as a critical factor in the analysis of heat and mass transfer across various domains of science and engineering. Such effects are particularly significant in applications including nuclear reactor safety, combustion processes, solar thermal systems, metallurgical operations, and chemical manufacturing. The incorporation of chemical reaction parameters into flow models enhances the understanding and control of concentration profiles, rendering hybrid nanofluids especially suitable for catalytic and reactive systems. In this study, a chemical reaction parameter is integrated into the governing equations to provide deeper insights into concentration variations within the channel, following the approach adopted by Aman *et al.* [20]. Numerous researchers have contributed to this field [21–23]. For example, Nadeem and Akbar [24] investigated the role of heat and chemical reactions in Walter's B fluid model for blood flow through a tapered artery. Hayat and Abbas [25] explored the channel flow of Maxwell fluids under the influence of chemical reactions. Abdul-Kahar *et al.* [26] applied scaling group transformations to study the effects of chemical reaction and thermal radiation on boundary-layer flow of a nanofluid over a porous vertical stretching surface. Similarly, Kameswaran *et al.* [27] examined homogeneous–heterogeneous reactions in nanofluid flow over a porous stretching sheet. Uddin *et al.* [28] conducted a numerical analysis on magnetoconvective slip flow over a non-isothermal, permeable, radiating plate in Darcian porous media with chemical reactions. Rashidi *et al.* [29] utilized the optimal homotopy analysis method (OHAM) to model free convective nanofluid flow past a chemically reacting horizontal plate in porous media. More recently, Srinivas *et al.* [30] assessed mass transfer in a viscous flow through expanding or contracting porous pipes under the influence of chemical reactions, while Uddin *et al.* [31] analyzed MHD free convective boundary layer flow of a chemically reacting nanofluid from a convectively heated, permeable vertical surface.

A *porous medium* refers to a solid matrix containing interconnected or partially connected voids, allowing fluids to pass through its internal structure. Fluid flow through porous media is integral to a wide range of applications, including crude oil extraction, geophysical and environmental processes, mineral recovery, and large-scale chemical operations involving filters, catalysts, and adsorbents. Alazmi and Vafai [32] provided a comprehensive review of the various practical applications of fluid flow through porous materials, highlighting its relevance in geophysics, metal processing, die casting, agricultural irrigation, industrial fluid dispersion, and oil recovery techniques. Choudhury *et al.* [33] examined the influence of the Dufour effect on magnetohydrodynamic (MHD) dissipative flow over a vertical plate embedded in a porous medium. In a related study, Thakur and Sood [34] analyzed the impact of internal heat sources on the unsteady MHD convective flow of a Williamson nanofluid through a porous structure. Kodi *et al.* [35] investigated the role of mixed convection and MHD forces in Maxwell nanofluid flow past a porous vertical cone, incorporating variable thermal conductivity and diffusion effects. Additionally, Krishna and Chamkha [36] explored the combined effects of Hall and ion slip currents on MHD rotating boundary layer flow of nanofluids over an infinite vertical plate within a porous medium. These investigations collectively contribute to a deeper understanding of the complex interactions between fluid dynamics, thermal transport, and magnetic effects in porous domains.

An additional aspect addressed in this study is the influence of generative (or internal) heat within channel flow, which originates from energy sources intrinsic to the fluid or suspended nanoparticles. This internal heat generation plays a pivotal role in enhancing the thermal energy distribution across the flow domain, significantly influencing temperature profiles and diminishing reliance on external heat sources. By adjusting the generative heat parameter, thermal engineers can effectively optimize energy consumption in hybrid nanofluid systems, thereby improving the efficiency of heating and cooling mechanisms in various industrial processes. The present work incorporates this parameter into the governing equations to explore its effect on thermal behavior, building upon recent advancements in thermal fluid dynamics, such as those reported by Hazarika *et al.* [40]. The role of heat generation in heat transfer continues to be a critical focus due to its implications in diverse physical and engineering contexts. For instance, Ziabakhsh *et al.* [41] studied micropolar fluid flow under the influence of internal heat generation, providing further insights into its impact on thermal transport characteristics.

The originality of this investigation is to introduce the influence of porosity, chemical reaction, and generative heat on the hybrid nanofluid flow with CNTs and to make a comparative analysis of Newtonian and non-Newtonian fluid. Using a similarity variable, the basic governing equations of the proposed model are reframed into a system of dimension-one on-line ordinary differential equations and then numerically computed using the MATLAB software. With the following research inquiries, physical parameters are discussed:

1. The three-dimensional modelling of such flows, which can provide more precise predictions of flow behavior and heat transfer characteristics, has a research deficit.
2. A unique contribution to the literature would be to comprehend how the porosity, chemical reaction, and generative heat affect the flow of nanofluids in the channel.
3. What effects do the magnetic field, suction parameter, Schmidt number, and various other parameters have on the hybrid nanofluid's velocity, temperature, and concentration?
4. How does the volume fraction of nanoparticles affect the temperature and velocity profiles?

5.2 Mathematical Formulation

Here, we take into account an electrically conducting Casson fluid in a vertical channel (Figure 5.1) with width 'a', static walls at $y = 0$ and $y = a$, and a uniform magnetic field with intensity B_0 that is applied transversely to the channel.

The fundamental modelled equations are ([20, 40])

$$\rho_{nf}\left(\frac{\partial \bar{u}}{\partial \bar{t}}+\bar{v}\frac{\partial \bar{u}}{\partial \bar{y}}\right)=\left\{\begin{array}{c} -\dfrac{\partial \bar{p}}{\partial \bar{x}}+\mu_{nf}\left(1+\dfrac{1}{\gamma}\right)\dfrac{\partial^2 \bar{u}}{\partial \bar{y}^2}-\sigma_{nf}B_0^2\bar{u} \\ +(\rho\beta)_{nf}\,g(\bar{T}-\bar{T}_a)+(\rho\beta)_{nf}\,g(\bar{C}-\bar{C}_a)-\dfrac{\nu}{\bar{k}_r}\bar{u} \end{array}\right\}, \tag{5.1}$$

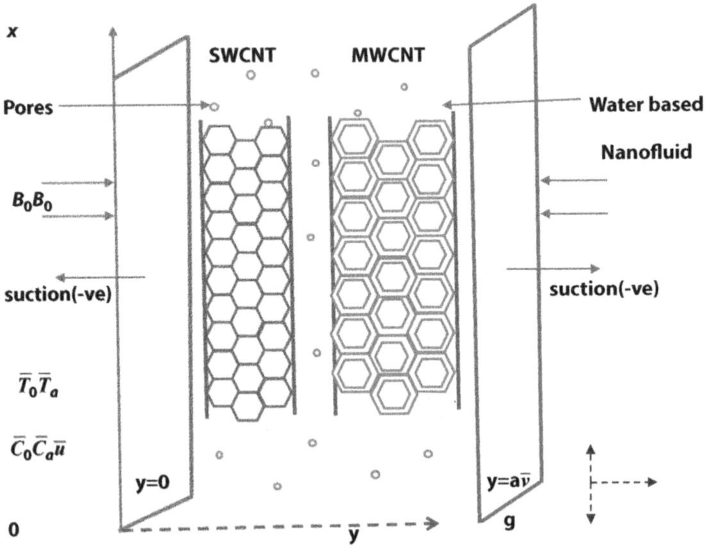

Figure 5.1 Physical configuration and coordinate system.

$$\left(\rho C_p\right)_{nf}\left(\frac{\partial \bar{T}}{\partial \bar{t}}+\bar{v}\frac{\partial \bar{T}}{\partial \bar{y}}\right)=-\frac{\partial \bar{q}_r}{\partial \bar{y}}+\kappa_{nf}\frac{\partial^2 \bar{T}}{\partial \bar{y}^2}+\bar{Q}_o\left(\bar{T}-\bar{T}_a\right), \tag{5.2}$$

$$\left(\frac{\partial \bar{C}}{\partial \bar{t}}+\bar{v}\frac{\partial \bar{C}}{\partial \bar{y}}\right)=D\frac{\partial^2 \bar{C}}{\partial \bar{y}^2}-\bar{C}_r\left(\bar{C}-\bar{C}_a\right) \tag{5.3}$$

In view of the authors of [37–39], κ_{nf} and μ_{nf} may be defined as follows:

$$\left\{\begin{array}{c} \bar{v}=-v_0,\ \ \mu_{nf}=\dfrac{\mu_f}{(1-\phi)^{2.5}},\ \ (\rho)_{nf}=(1-\phi)(\rho)_f+\phi(\rho)_{CNT}, \\ (\rho\beta)_{nf}=(1-\phi)(\rho\beta)_f+\phi(\rho\beta)_{CNT},\ \ (\rho C_p)_{nf}=(1-\phi)(\rho C_p)_f+\phi(\rho C_p)_{CNT}, \\ \dfrac{\kappa_{nf}}{\kappa_f}=\dfrac{(1-\phi)+2\phi\left(\dfrac{\kappa_{CNT}}{\kappa_{CNT}-\kappa_f}\right)ln\left(\dfrac{\kappa_{CNT}+\kappa_f}{2\kappa_f}\right)}{(1-\phi)+2\phi\left(\dfrac{\kappa_f}{\kappa_{CNT}-\kappa_f}\right)ln\left(\dfrac{\kappa_{CNT}+\kappa_f}{2\kappa_f}\right)},\sigma_{nf}=\sigma_f\left[1+\dfrac{3(\sigma-1)\phi}{(\sigma+2)-(\sigma-1)\phi}\right] \end{array}\right\} \tag{5.4}$$

The radiative heat flux is given by ([20, 40])

$$\frac{\partial \bar{q}}{\partial \bar{y}} = -4\alpha^2\left(\bar{T} - \bar{T}_\infty\right) \tag{5.5}$$

Here are several nondimensional parameters ([20, 40]):

$$\left\{\begin{array}{c} x = \frac{\bar{x}}{d},\quad y = \frac{\bar{y}}{d},\quad u = \frac{\bar{u}}{U_0},\quad t = \frac{\bar{t}U_0}{d},\quad \theta = \frac{\bar{T} - \bar{T}_a}{\bar{T}_0 - \bar{T}_a},\quad \psi = \frac{\bar{C} - \bar{C}_a}{\bar{C}_0 - \bar{C}_a}, \\ p = \frac{\bar{p}d}{\mu U_0},\quad \omega = \frac{\bar{\omega}d}{U_0},\quad \frac{\partial p}{\partial x} = \lambda\exp(i\omega t),\quad Re = \frac{U_0 d}{\nu_f}, \\ M^2 = \frac{\sigma_f d^2 B_0^2}{\mu_f},\quad Gr = \frac{d^2 g\beta_f\left(\bar{T}_0 - \bar{T}_a\right)}{U_0\mu_f},\quad Gm = \frac{d^2 g\beta_f\left(\bar{C}_0 - \bar{C}_a\right)}{U_0\mu_f}, \\ R^2 = \frac{4d^2\alpha^2}{\kappa_f},\quad Pe = \frac{U_0 d(\rho C_p)_f}{k_f},\quad \lambda_n = \frac{\kappa_{nf}}{\kappa_f},\quad Sc = \frac{U_0 d}{D},\quad S = \frac{v_0}{U_0}, \\ k_r = \frac{\bar{k}_r\rho_{nf}U_O}{vd},\quad Q_o = \frac{\bar{Q}_O d}{\left(\rho C_p\right)_{nf}U_O},\quad C_r = \frac{\bar{C}_r d}{U_O} \end{array}\right\} \tag{5.6}$$

The boundary conditions are ([20, 40])

$$\left\{\begin{array}{ll} \bar{u}(0,t) = 0, & \bar{u}(a,t) = 0 \\ \bar{T}(0,t) = \bar{T}_0, & \bar{T}(a,t) = \bar{T}_a \\ \bar{C}(0,t) = \bar{C}_0, & \bar{C}(a,t) = \bar{C}_a \end{array}\right\} \tag{5.7}$$

Equations 5.1–5.3 reduce to the dimensionless form with the aid of Equations 5.4–5.6:

$$\phi_1 Re\left(\frac{\partial u}{\partial t} - S\frac{\partial u}{\partial y}\right) = \left\{\begin{array}{c} \varepsilon\lambda exp(i\omega t) + \phi_2\left(1 + \frac{1}{\gamma}\right)\frac{\partial^2 u}{\partial y^2} \\ -M^2\phi_3 u + \phi_4 Gr\theta + \phi_4 Gm\psi - u/kr \end{array}\right\}, \tag{5.8}$$

$$\phi_5 Pe \frac{1}{\lambda_n}\left(\frac{\partial\theta}{\partial t} - S\frac{\partial\theta}{\partial y}\right) = \frac{\partial^2\theta}{\partial y^2} + \frac{R^2}{\lambda_n}\theta + Q_0\theta, \tag{5.9}$$

$$\left(\frac{\partial\psi}{\partial t} - S\frac{\partial\psi}{\partial t}\right) = \frac{1}{Sc}\frac{\partial^2\psi}{\partial y^2} - C_r\varphi \tag{5.10}$$

The forms of Equation 5.7 in nondimensions are

$$\left\{\begin{matrix} u(0,t)=0, & u(1,t)=0 \\ \theta(0,t)=1, & \theta(1,t)=0 \\ \psi(0,t)=1, & \psi(1,t)=0 \end{matrix}\right\} \tag{5.11}$$

Rewriting Equations 5.8–5.10 as

$$a_0\left(\frac{\partial u}{\partial t} - S\frac{\partial u}{\partial y}\right) = \varepsilon\lambda exp(i\omega t) + a_1\frac{\partial^2 u}{\partial y^2} - a_2 u + a_3\theta + a_4\psi - a_5 u, \tag{5.12}$$

$$b_0\left(\frac{\partial\theta}{\partial t} - S\frac{\partial\theta}{\partial y}\right) = \frac{\partial^2\theta}{\partial y^2} + b_1\theta + Q_0\theta, \tag{5.13}$$

$$\left(\frac{\partial\psi}{\partial t} - S\frac{\partial\psi}{\partial y}\right) = \frac{1}{Sc}\frac{\partial^2\psi}{\partial y^2} - c_2\varphi \tag{5.14}$$

The perturbed solutions of Equations (5.12–5.14) with boundary conditions 5.11 are supposed to have the following forms:

$$F(y,t) = F_0(y) + \sum_{j=1}^{\infty}\varepsilon^j e^{i\omega t}F_j(y), \tag{5.15}$$

where F stands for $u, \theta, or\, \psi$.

Equation 5.15 is incorporated into Equations (5.12–5.14) to form the following system of ODEs:

$$\frac{d^2u_0(y)}{dy^2} + a_6 \frac{du_0(y)}{dy} - a_7 u_0(y) = -a_8\theta_0(y) - a_9\psi_0(y), \quad (5.16)$$

$$\frac{d^2u_1(y)}{dy^2} + a_6 \frac{du_1(y)}{dy} - a_{10} u_1(y) = -\lambda_1 - a_8\theta_1(y) - a_9\psi_1(y), \quad (5.17)$$

$$\frac{d^2\theta_0(y)}{dy^2} + b_2 \frac{d\theta_0(y)}{dy} + b_1\theta_0(y) + Q_0\theta_0(y) = 0, \quad (5.18)$$

$$\frac{d^2\theta_1(y)}{dy^2} + b_2 \frac{d\theta_1(y)}{dy} - b_3\theta_1(y) = 0, \quad (5.19)$$

$$\frac{d^2\psi_0(y)}{dy^2} + d_0 \frac{d\psi_0(y)}{dy} - d_1\varphi_0(y) = 0, \quad (5.20)$$

$$\frac{d^2\psi_1(y)}{dy^2} + d_0 \frac{d\psi_1(y)}{dy} - d_2\psi_1(y) - d_1\psi_1(y) = 0 \quad (5.21)$$

The converted boundary conditions (5.11) are

$$\left\{\begin{array}{llll} u_0(0)=0, & u_0(1)=0, & u_1(0)=0, & u_1(1)=0 \\ \theta_0(0)=1, & \theta_0(1)=0, & \theta_1(0)=0, & \theta_1(1)=0 \\ \psi_0(0)=1, & \psi_0(1)=0, & \psi_1(0)=0, & \psi_1(1)=0 \end{array}\right\} \quad (5.22)$$

The solutions to Equations 5.16 to 5.21 under boundary conditions 5.22 lead to

$$u(y,t)=\left\{\begin{matrix} E_1e^{r_1y}+E_2e^{-r_2y}+P_1e^{n_1y}+P_2e^{-n_2y}+P_3e^{m_1y}+P_4e^{-m_2y} \\ +\varepsilon e^{i\omega t}\left(E_3e^{r_3y}+E_4e^{-r_4y}+\dfrac{\lambda_1}{a_{10}}\right) \end{matrix}\right\} \tag{5.23}$$

$$\theta(y,t)=\frac{e^{n_1}e^{-n_2y}-e^{-n_2}e^{n_1y}}{e^{n_1}-e^{-n_2}} \tag{5.24}$$

$$\psi(y,t)=\frac{e^{m_1}e^{-m_2y}-e^{-m_2}e^{m_1y}}{e^{m_1}-e^{-m_2}} \tag{5.25}$$

5.3 Validity and Accuracy

The accuracy of the present non-Newtonian nanofluid model is verified by comparing the results of fluid velocity with published results initiated by Hazarika and Ahmed [40]. Very good agreement between the present and the other results confirms the accuracy of the method used. An outstanding agreement is obtained between the current and previous results by comparing the present outcomes with the foregoing ones, which clearly

Table 5.1 Velocity distribution of non-Newtonian nanofluid for base water at $\lambda=1,\ t=2,\ \omega=2,\ R=2,\ Gr=0.5,\ Re=2,\ Pe=3.5,\ Sc=2,\ S=2,\ M=2$.

Hazarika and Ahmed [40]				**Present work**		
y	**S = 0.5**	**S = 1**	**S = 3**	**S = 0.5**	**S = 1**	**S = 3**
0.0	1	1	1	1	1	1
0.2	0.088181	0.06369	0.010219	0.71618	0.37142	0.030199
0.4	0.025201	0.010683	0.0001566	0.564975	0.161791	0.0011512
9	0.00712	0.001661	0	0.36344	0.057866	0.0000379
0.8	0.002002	0.000253	0	0.21713	0.019297	0
1.0	0.000561	0.000038	0	0.12563	0.00624825	0

reflects the convergence of the perturbation method. Table 5.1 demonstrates that the velocity reduces as suction (S) rises.

5.4 Results and Discussion

To gain a deeper physical understanding of the problem, we have conducted numerical simulations to analyze the nondimensional velocity, temperature, and species concentration for non-Newtonian fluids. By assigning specific parameter values relevant to the study, we have explored their influence on fluid behavior. The results, illustrated in Figures 5.2–5.9, highlight the existence of dual solutions, demonstrating the complex and nonlinear nature of non-Newtonian fluid dynamics under the given conditions. The default parameters are $\lambda = 1$, $R = 0.5$, $Gr = 2$, $Gm = 2$, $Re = 2$, $Pe = 3.5$, $Sc = 0.78$, $S = 2$, $M = 2$, $\gamma = 1.0$, $t = 5$, $\omega = 2$, $K_r = 0.1$, $C_r = 0.1$.

The nanofluid temperature $\theta(y,t)$ and nanofluid velocity $u(y,t)$ in a non-Newtonian fluid flowing through a vertical channel in porous media under the influence of heat generation (Q_0) are presented in Figures 5.2 and 5.3.

Figure. 5.2 shows that as the heat generation parameter (Q_0) increases, the nanofluid temperature (θ) increases throughout the domain. This is expected, as additional heat is being supplied to the fluid, leading to higher energy accumulation. The temperature profiles are plotted for two different values of the suction parameter (S). Higher suction $(S = 5.0)$ leads to a higher temperature than lower suction $(S = 0.5)$. This could be because suction enhances the removal of cooler fluid layers near the walls, allowing hotter fluid to dominate. In Figure 5.3, an increase in heat generation enhances the velocity profile. This is due to the thermal buoyancy effect,

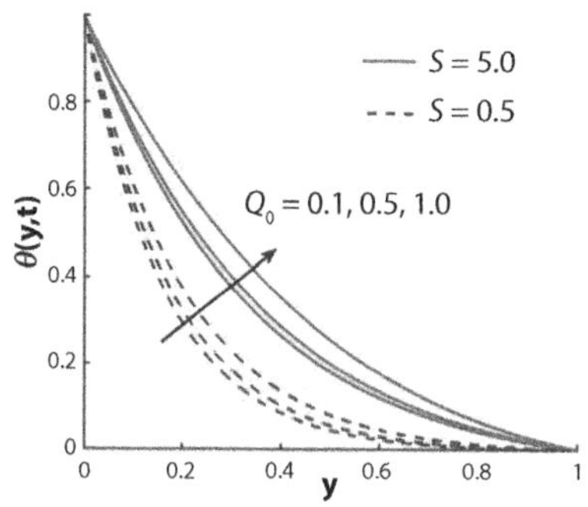

Figure 5.2 Variation of S and Q_0 on the fluid temperature.

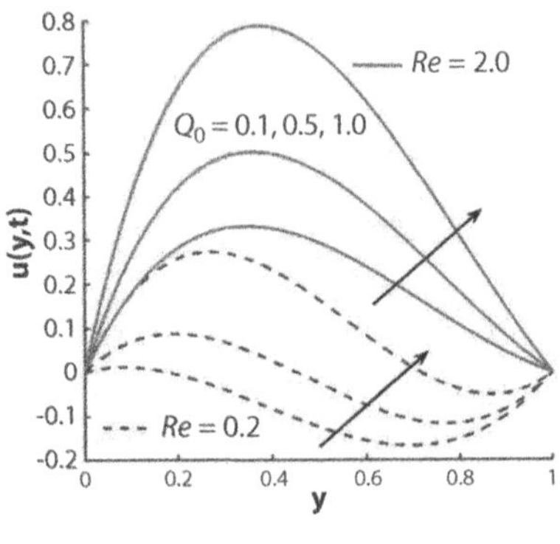

Figure 5.3 Variation of *Re* and Q_0 on the fluid velocity.

which increases the driving force of the fluid motion. The velocity is plotted for two different values of *Re* ($Re = 2.0$ and $Re = 0.2$). A higher Reynolds number ($Re = 2.0$) results in a more pronounced velocity peak, while a lower Reynolds number ($Re = 0.2$) leads to a smaller, oscillatory velocity profile. This suggests that at low Reynolds numbers, the porous medium's resistance is more dominant.

The nanofluid concentration $\psi(y,t)$ and velocity $u(y,t)$ in a non-Newtonian fluid flowing through a vertical channel in porous media under the influence of a chemical reaction (C_r) and different Schmidt numbers (*Sc*) are presented in Figures 5.4 and 5.5. The concentration decreases as C_r increases. This is because higher reaction rates lead to more rapid consumption of the solute, thereby reducing its concentration in the fluid. Higher Schmidt numbers ($Sc = 1.2$, dashed blue lines) correspond to lower diffusivity of the solute, meaning that mass transfer is slower. As a result, the concentration decreases more rapidly. Lower Schmidt numbers ($Sc = 0.2$,

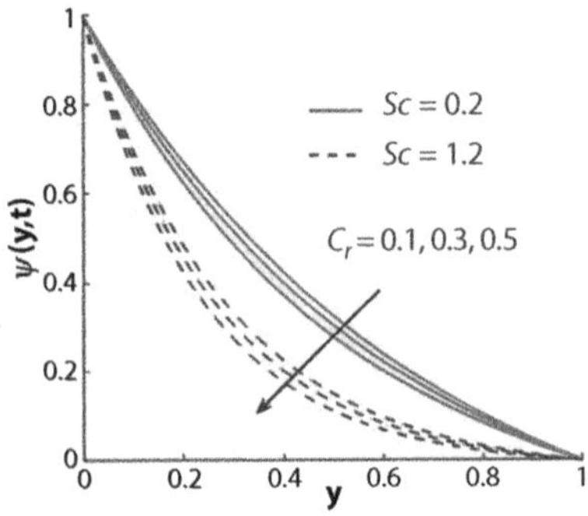

Figure 5.4 Variation of C_r and *Sc* on the concentration (ψ).

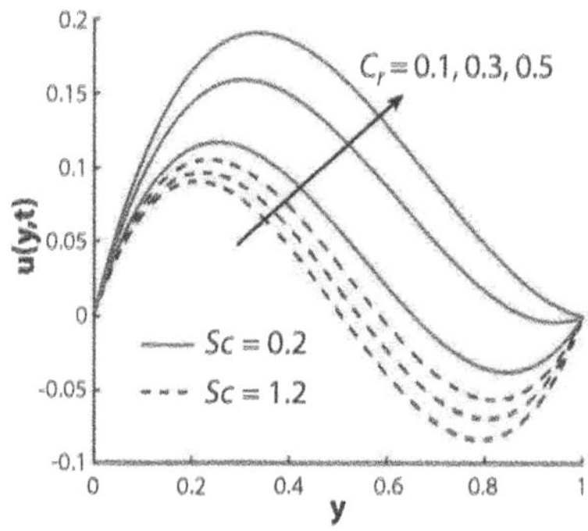

Figure 5.5 Variation of C_r and Sc on the fluid velocity (u).

solid red lines) allow for greater diffusion, leading to a more spread-out concentration profile. Increasing C_r enhances the velocity. This is likely due to the change in buoyancy forces caused by concentration variations, leading to increased flow momentum. For lower $Sc = 0.2$ (solid red lines), velocity is higher due to enhanced mass diffusion, which reduces the resistance from solute gradients. For higher $Sc = 1.2$ (dashed blue lines), velocity decreases, possibly due to the reduced mass diffusion, which dampens buoyancy-driven flow effects.

In Figures 5.6 and 5.7, the analysis focuses on how the nanoparticle volume fraction (ϕ) and Reynolds number (Re) affect the surface velocity and surface temperature of a non-Newtonian nanofluid in a vertical channel filled with porous media.

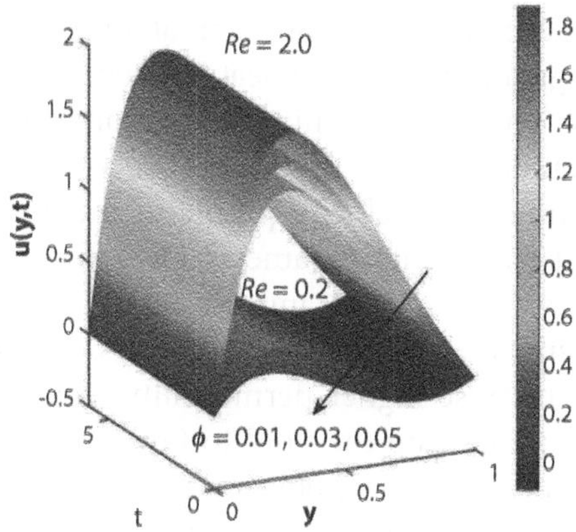

Figure 5.6 Variation of ϕ and Re on the fluid velocity (u).

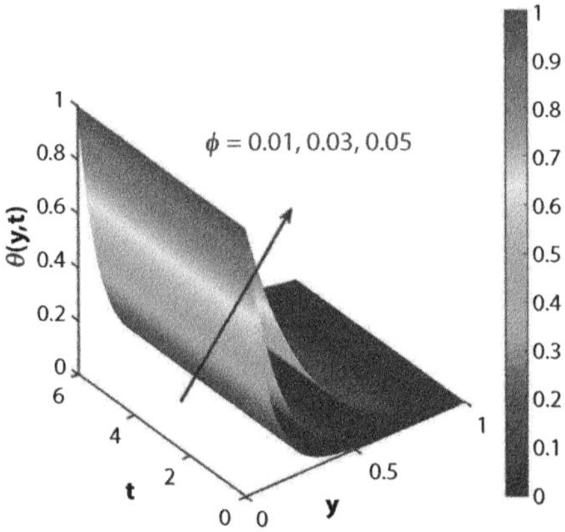

Figure 5.7 Variation of ϕ at $Re=0.2$ on the fluid temperature (θ).

Increasing ϕ (from 0.01 to 0.05) generally reduces the surface velocity due to the increase in effective viscosity of the fluid. The presence of nanoparticles increases resistance to motion, making it harder for the fluid to move along the surface. At a low Reynolds number **($Re=0.2$),** the velocity reduction due to ϕ is more significant because viscous forces dominate. At a higher Reynolds number **($Re=2.0$),** the effect of ϕ on velocity reduction is less pronounced since inertia becomes the governing force. Increasing ϕ leads to higher surface temperatures due to the enhanced thermal conductivity of the nanofluid. Since nanoparticles facilitate heat transfer, the surface temperature rises as ϕ increases, leading to more uniform temperature distribution along the channel walls. The increase in temperature is more prominent at a higher Reynolds number **($Re=2.0$)** due to greater convective heat transfer. The thermal boundary layer thickens as ϕ increases, meaning heat is conducted more effectively through the nanofluid.

Figure 5.8 illustrates the velocity profile of the non-Newtonian fluid influenced by the permeability parameter (K_r) for two Reynolds numbers: $Re=0.2$ (low inertia flow) and $Re=2.0$ (high inertia flow). An increase in (K_r) enhances the fluid velocity for both Reynolds numbers, which is expected because higher permeability reduces the resistance of the porous medium, allowing the fluid to flow more freely. The effect is more pronounced for higher Reynolds numbers **($Re=2.0$),** where inertia plays a more dominant role in driving the flow. In contrast, for **$Re=0.2$,**

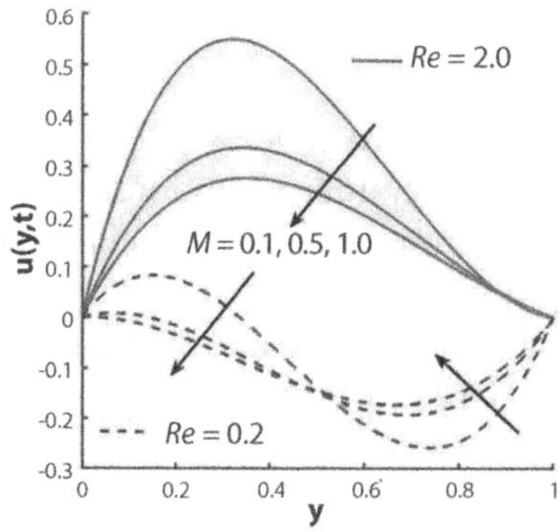

Figure 5.8 Variation of Re and M on the fluid velocity (u).

the velocity remains relatively lower due to the weaker inertial effects in a slow-moving fluid.

Figure 5.9 examines the effect of the magnetic field parameter $\boldsymbol{M}$ on the velocity of the non-Newtonian fluid. As $\boldsymbol{M}$ increases, the velocity decreases significantly for both $\boldsymbol{Re = 0.2}$ and $\boldsymbol{Re = 2.0}$. This happens because the Lorentz force acts as a resistive force, opposing the motion of the fluid. The reduction in velocity is more prominent in the low Reynolds number $\boldsymbol{(Re = 0.2)}$ case, as the magnetic damping effect is stronger when inertial forces are weaker. For higher Reynolds numbers $\boldsymbol{(Re = 2.0)}$, the fluid still experiences resistance due to the Lorentz force, but the overall velocity remains higher compared to **Re=0.2**, indicating that inertia partially counteracts the magnetic drag.

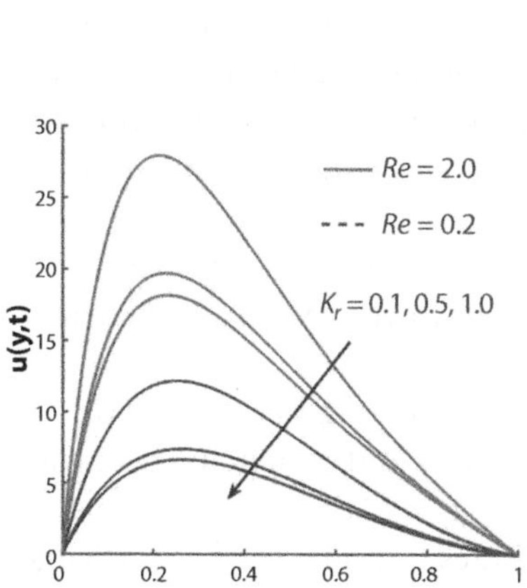

Figure 5.9 Variation of Re and K_r on the fluid velocity (u).

5.5 Conclusion

The study of non-Newtonian nanofluids in porous media shows a delicate balance between viscosity, inertia, and heat transfer mechanisms. By adjusting nanoparticle volume fraction and considering non-Newtonian behavior, one can optimize heat dissipation and fluid flow in engineering applications such as cooling systems, biomedical flows, and industrial thermal management. Here is a concise summary of the main findings of this investigation:

- A low Reynolds number results in a more stable flow, reducing turbulence and enhancing heat retention in the system.
- The inclusion of carbon nanotubes (CNTs) significantly enhances thermal conductivity and flow stability, improving heat transfer efficiency.
- Magnetic and buoyancy forces play a crucial role in controlling fluid motion, demonstrating their potential for flow regulation in engineering applications.
- Non-Newtonian fluids exhibit shear-thinning or shear-thickening effects, affecting how the velocity responds to ϕ and Re.
- Non-Newtonian behavior alters the heat transfer characteristics, leading to either enhanced or suppressed thermal convection.
- In shear-thinning fluids, temperature gradients near the surface are reduced, enhancing heat transfer.
- In shear-thickening fluids, temperature gradients are steeper, leading to higher localized heating near the walls.
- Increasing nanoparticle volume fraction (ϕ) reduces surface velocity but increases surface temperature due to improved thermal conductivity.
- At a low Reynolds number ($Re = 0.2$), viscous effects dominate, leading to lower velocity and weaker convective heat transfer.
- At a high Reynolds number ($Re = 2.0$), inertia takes over, increasing velocity and enhancing heat transfer efficiency.

The study contributes foundational knowledge to fluid mechanics, showing how hybrid nanofluids can be engineered to achieve specific performance outcomes based on their thermal and rheological characteristics. Future research can extend this study by incorporating experimental validation, time-dependent effects, and more complex flow geometries. Exploring smart nanomaterials, such as functionalized CNTs, may further enhance thermal regulation and flow control.

AI disclosure statement: The author acknowledges the use of AI-based tools for minor language editing or grammatical correction. However, all intellectual contributions, analyses, and conclusions presented in this chapter are entirely the author's own.

Bibliography

1. Choi, S.U., Enhancing thermal conductivity of fluids with Nanoparticles. *Develop. Appl. non-Newtonian Flows*, 231, 99–105, 1995.
2. Maxwell, J.C., *Treatise of Electricity and Magnetism*, 3rd Edition, vol. 1, Oxford University Press, 435–449, 1892.
3. Mahdi, R.A., Mohammed, H.A., Munisamy, K.M., Saeid, N.H., Review of convection heat transfer and fluid flow in porous media with Nanofluid. *Renew. Sustain. Energ. Rev.*, 41, 715–34, 2015.
4. Nasrin, R., Alim, M.A., Chamkha, A.J., Effect of the heating wall position on forced convection along two-sided open enclosure with porous medium utilizing Nanofluid. *Int. J. Energ. Technol.*, 5, 1–13, 2013.
5. Chhabra, R.P., *Bubbles, drops, and particles in non-Newtonian fluids*, 2nd, Taylor and Francis Ltd, Boca Raton, 2006.
6. Kim, G., Hyun, J., Kwak, H.S., Transient buoyant convection of a power-law non-Newtonian fluid in an enclosure. *Int. J. Heat Mass Transf.*, 46, 3605–3617, 2003.
7. Lamsaadi, M., Naïmi, M., Hasnaoui, M., Natural convection of non-Newtonian power law fluids in a shallow horizontal rectangular cavity uniformly heated from below. *Heat Mass Transf./ Waerme- und Stoffuebertragung*, 41, 239–249, 2005.
8. Hojjat, M., Etemad, S., Bagheri, R., Thibault, J., Rheological characteristics of non-Newtonian nanofluids: experimental investigation. *Int. Commun. Heat Mass Transf.*, 38, 2, 144–148, 2011.
9. Turan, O., Sachdeva, A., Chakraborty, N., Poole, R., Laminar natural convection of power-law fluids in a square enclosure with differentially heated sidewalls subjected TP constant wall heat flux. *J. Non-Newtonian Fluid Mech.*, 166, 1049–1063, 2011.
10. Chandi, S. and Raj, C.P., Laminar natural convection from heated square cylinder immersed in power-law liquids. *J. Non-Newtonian Fluid Mech.*, 166, 811–830, 2011.
11. Kaddiri, M., Naïmi, M., Raji, A., Hasnaoui, M., Rayleigh-Bènard convection of non-Newtonian power-law fluids with temperature-dependent viscosity. *ISRN Thermodyn.*, 2012.

12. Shojaeian, M., Yildiz, M., Kosar, A., Convective heat transfer and second law analysis of non-Newtonian fluid flows with variable thermophysical properties in circular channels. *Int. Commun. Heat Mass Transf.*, 60, 21–31, 2015.
13. Hayat, T., Haider, F., Muhammad, T., Alsaedi, A., On Darcy-Forchheimer flow of carbon nanotubes due to a rotating disk. *Int. J. Heat Mass Transfer*, 112, 248–254, 2017. https://dx.doi.org:10.1016/j.ijheatmasstransfer.2017.04.123.
14. Hussain, S.T., Haq, R.U., Khan, Z.H., Nadeem, S., Water driven flow of carbon nanotubes in a rotating channel. *J. Mol. Liq.*, 214, 136–144, 2016. https://dx.doi.org: org/10.1016/j.molliq.2015.11.042 0167-7322.
15. Khalid, A., Khan, I., Khan, A., Shaed, S., Tlilie, I., Case study of MHD blood flow in a porous medium with CNTS and thermal analysis. *Case Stud. Therm. Eng.*, 12, 374–380, 2018. https://dx.doi.org:10.1016/j.csite.2018.04.004.
16. Khan, W.A., Khan, Z.H., Rahi, M., Fluid flow and heat transfer of carbon nanotubes along a plate with Navier slip boundary. *Appl. Nanosci.*, 4, 633–641, 2014. https://dx.doi.org:10.1007/s13204-013-0242-9.
17. Kandasamy, R., Muhaimin, I., Mohammad, R., Single walled carbon nanotubes on MHD unsteady flow over a porous wedge with thermal radiation with variable stream conditions. *Alex. Eng. J.*, 55, 275–285, 2016. https://dx.doi.org/10.1016/j.aej.2015.10.006 1110-0168.
18. Mayer, J., Mckrell, T., Grote, K., The influence of multi-walled carbon nanotubes on single-phase heat transfer and pressure drop characteristics in the transitional flow regime of smooth tubes. *Int. J. Heat Mass Transfer*, 58, 1–2, 597–609, 2013. https://dx.doi:10.1016/j.ijheatmasstransfer.2012.11.074.
19. Pramanik, S., Casson fluid flow and heat transfer past an exponentially porous stretching surface in presence of thermal radiation. *Ain Shams Eng. J.*, 5, 205–212, 2014. https://dx.doi.org:/10.1016/j.asej.2013.05.003.
20. Aman, S., Khan, I., Ismail, Z., Salleh, M.Z., Al-Mdalla, Q.M., Heat transfer enhancement in free convection flow of CNTs Maxwell nanofluids with four different types of molecular liquid. *Sci. Rep. Vol.*, 7, 2445, 2017a. https://dx.doi: 10.1038/s41598-017-01358-3.
21. Vajravelu, K., Prasad, K.V., Rao, N.S.P., Diffusion of a chemically reactive species of a power-law fluid past a stretching surface. *Comput. Math. Appl.*, 62, 93–108, 2011.
22. Cortell, R., MHD flow and mass transfer of an electrically conducting fluid of a second grade in a porous medium over a stretching sheet with chemically reactive species. *Chem. Eng. Process. Process Intensif.*, 46, 721–728, 2007.
23. Bridges, C. and Rajagopal, K.R., Pulsatile flow of a chemically reacting nonlinear fluid. *Comput. Math. Appl.*, 52, 1131–1144, 2006.
24. Nadeem, S. and Akbar, N.S., Influence of heat and chemical reactions on Walter's B fluid model for blood flow through a tapered artery. *J. Taiwan Inst. Chem. Eng.*, 42, 67–75, 2011.
25. Hayat, T. and Abbas, Z., Channel flow of a Maxwell fluid with chemical reaction. *Z. Angew. Math. Phys.*, 59, 124–144, 2008.

26. Abdul-Kahar, R., Kandasamy, R., Muhaimin, Scaling group transformation for boundary-layer flow of a nanofluid past a porous vertical stretching surface in the presence of chemical reaction with heat radiation. *Comput. Fluids*, 52, 15–21, 2011.
27. Kameswaran, P.K., Shaw, S., Sibanda, P., Murthy, P.V.S.N., Homogeneous–heterogeneous reactions in a nanofluid due to a porous stretching sheet. *Int. J. Heat Mass Transf.*, 57, 465–472, 2013.
28. Uddin, M.J., Khan, W.A., Ismail, A.I., Lie group analysis and numerical solutions for magneto convective slip flow along a moving chemically reacting radiating plate in porous media with variable mass diffusivity. *Heat Transf.-Asian Res.*, 2014. http://dx.doi.org/10.1002/htj.21161.
29. Rashidi, M.M., Momoniat, E., Ferdows, M., Basiriparsa, A., Lie group solution for free convective flow of a nanofluid past a chemically reacting horizontal plate in porous media. *Math. Probl. Eng.*, 2014, 1–21, 2014.
30. Srinivas, S., Subramanyam Reddy, A., Ramamohan, T.R., Mass transfer effects on viscous flow in an expanding or contracting porous pipe with chemical reaction. *Heat Transf. Asian Res.*, 44, 552–567, 2015.
31. Uddin, M.J., Beg, O.A., Aziz, A., Ismail, A.I.M., Group analysis of free convection flow of a magnetic nanofluid with chemical reaction. *Math. Probl. Eng.*, 2015, 1–11, 2015.
32. Alazmi, B. and Vafai, K., Analysis of variable porosity, thermal dispersion, and local thermal nonequilibrium on free surface flows through porous media. *J. Heat Transfer*, 126, 3, 389–399, 2004.
33. Choudhury, K., Agarwalla, S., Ahmed, N., Diffusion-thermo effect on MHD dissipative flow past a porous vertical plate through porous media. *Heat Transfer*, 51, 7, 6836–6855, 2022.
34. Thakur, A. and Sood, S., Effect of prescribed heat sources on convective unsteady MHD flow of Williamson nanofluid through porous media: Darcy-Forchheimer model. *Int. J. Appl. Comput. Math.*, 8, 2, 74, 2022.
35. Kodi, R., Ganteda, C., Dasore, A., Kumar, M.L., Laxmaiah, G., Hasan, M.A., Islam, S., Razak, A., Influence of MHD mixed convection flow for Maxwell nanofluid through a vertical cone with porous material in the existence of variable heat conductivity and diffusion. *Case Stud. Therm. Eng.*, 44, 102875, 2023.
36. Krishna, M.V. and Chamkha, A.J., Hall and ion slip effects on MHD rotating boundary layer flow of nanofluid past an infinite vertical plate embedded in a porous medium. *Results Phys.*, 15, 102652, 2019.
37. Maxwell, J.C. and Tompson, J.J., *A Treatise on Electricity and Magnetism*, Oxford University Press, Oxford, UK, 1904.
38. Brinkman, H.C. and Luis, M., The viscosity of concentrated suspensions and solutions. *J. Chem. Phys.*, 20, 571, 1952.
39. Xuan, Y., Li, Q., Hu, W., Aggregation structure and thermal conductivity of nanofluids. *AIChE J.*, 49, 1038–1043, 2003, https://dx.doi.org/10.1002/aic.690490420.

40. Hazarika, S., Ahmed, S., Chamkha, A.J., Heat Transfer of Casson Fluid in Poiseuille Flow of Carbon Nanotubes: A Power Series Approach. *J. Nanofluids*, 12, 1–9, 2023. https://dx.doi:10.1166/jon.2023.1995.
41. Ziabakhsh, Z., Domairry, G., Bararnia, H., Analytical solution of non-Newtonian micropolar fluid flow with uniform suction/blowing and heat generation. *J. Taiwan Inst. Chem. Eng.*, 40, 4, 443–451, 2009.

6

Non-Newtonian Fluid Flow in Porous Media

Yeddula Rameswara Reddy[1], Damodara Reddy Annapureddy[1], K. Jyothi[2], Raghunath Kodi[3]*, Dhananjay Yadav[4] and Ramachandra Reddy Vaddemani[5]

[1]*Mechanical Engineering Department, JNTUACE Pulivendula, Constitute College of Jawaharlal Nehru Technological University Ananthapuramu, Andhra Pradesh, India*
[2]*Department of Humanities and Basic Sciences, G. Pulla Reddy Engineering College, Kurnool, Andhra Pradesh, India*
[3]*Department of Humanities and Sciences, St. Johns College of Engineering and Technology, Yemmiganur, Andhra Pradesh, India*
[4]*Department of Mathematical & Physical Sciences, University of Nizwa, Nizwa, Oman*
[5]*Department of Humanities and Sciences, K.S.R.M. College of Engineering, Kadapa, Andhra Pradesh, India*

Abstract
The objective of this article is to declare the determinations of an exhaustive examination into the effects of diffusive heating and the rate of chemical reaction of unsteady mixed convection Maxwell fluid that has been created by a permeable exponentially stretchy sheet. Using a similitude modification, the governing equations of continuity, momentum, and energy are reduced to the forms of ordinary differential equations. Utilizing the Runge–Kutta fourth-order method combined with the shooting strategy allows for the numerical solutions of these equations to be obtained. The findings that

**Corresponding author*: kraghunath25@gmail.com; ORCID: 0000-0002-1622-3634
Yeddula Rameswara Reddy: ORCID: 0000-0003-2801-0668
Damodara Reddy Annapureddy: ORCID: 0000-0001-8434-8108
K. Jyothi: ORCID: 0000-0003-0010-4566
Dhananjay Yadav: ORCID: 0000-0001-8404-2053
Ramachandra Reddy Vaddemani: ORCID: 0000-0002-5157-3525

Dhananjay Yadav, Mukesh Kumar Awasthi and Harith Mohamed Al-Azri (eds.) Non-Newtonian Fluids for Industrial Applications: Modeling and Simulations, (125–146) © 2026 Scrivener Publishing LLC

were obtained for the different flow parameters that influence the flow are graphically depicted, and there is additional investigation into the effects that these factors have on the velocity, temperature, and concentration profiles. The subsequent table demonstrates the values of the skin contention factor, the Nusselt numeral, and the Sherwood numeral. All of these coefficients are displayed in the table. It is glimpsed that raising the magnetic field parameter results in a decrease in the velocity values, while increasing the porous medium has a minor influence on the velocity increase. This is an important finding that has been observed.

Keywords: Rate of chemical reaction, viscoelastic fluid, porous media, stretching sheet

6.1 Introduction

Non-Newtonian fluids are characterized by a viscosity that varies with the applied shear stress or shear rate, unlike Newtonian fluids that maintain constant viscosity. These fluids display complex flow behavior such as shear-thinning, shear-thickening, or viscoelastic properties, making them suitable for various industrial and medical applications. When non-Newtonian fluids flow through porous media, their interactions with the solid matrix are influenced by the changing viscosity, which affects the permeability, pressure drop, and overall flow dynamics. This is particularly important in fields like enhanced oil recovery, where the flow of non-Newtonian fluids through porous rock formations can optimize displacement efficiency. Additionally, in biomedical applications, understanding the flow of non-Newtonian fluids through porous tissues is crucial for modeling blood circulation and developing drug delivery systems. Lavanya *et al.* [1] present a comprehensive analysis of Carreau nanofluid flow over an inclined vertical plate with Cattaneo–Christov heat flux. The authors conducted numerical simulations and statistical analyses to understand the thermal and flow behavior of Carreau nanofluids under different conditions. The results contribute to the better understanding of heat transfer in nanofluids and could be applied in heat exchangers and other thermal systems.

Several studies have contributed to the understanding of these dynamics in porous media. For instance, Pasha *et al.* [2] analyzed the effects of rotational forces and thermal diffusion on unsteady magnetohydrodynamic (MHD) flow of viscoelastic fluids through porous media. Reddy *et al.* [3] explored the effects of aligned magnetic fields on MHD convection in Casson fluids over inclined plates. Hussain *et al.* [4] focused on the thermophysical aspects of non-Newtonian hybrid nanofluids over stretched plates, providing valuable insights into flow and heat transfer characteristics. Further studies by Valiveti *et al.* [5] have examined heat and mass transfer in MHD flow through porous media, emphasizing the influence of thermal diffusion and chemical reactions.

Additionally, the influence of Hall currents and thermal radiation on MHD rotating flows of water-based nanofluids in porous media has been studied by Prasad *et al.* [6]. Jyothi *et al.* [7] have also conducted a neural network–assisted analysis of MHD boundary layer flow and thermal radiation effects on SWCNT nanofluids. Bhargava *et al.* [8] examined the mixed convection of hybrid nanofluid flow through porous media over a stretched surface. These studies highlight the growing significance of non-Newtonian fluid flow through porous media in both theoretical and applied contexts, opening new avenues for research and practical applications.

Maxwell fluids are a class of viscoelastic fluids that exhibit both elastic and viscous characteristics, characterized by a time-dependent viscosity. These fluids follow a linear stress–strain relationship, incorporating a relaxation time that describes the fluid's stress response to deformation. Maxwell fluids are particularly useful in modeling materials that exhibit both viscous and elastic behaviors, such as polymers, lubricants, and biological fluids. Their applications span across various industries, including food processing, biomedical engineering, and polymer production, where understanding shear thinning and viscoelastic properties is crucial. Additionally, Maxwell fluids are relevant in advanced materials design, microfluidics, and energy systems. Recent studies, such as those by Al Khayer and Chakraborty [9], Sudarmozhi *et al.* [10], and Oyelami *et al.* [11], have explored the behavior of Maxwell fluids under various conditions, such as stratification, Joule heating, and chemical reactions. These studies contribute to a deeper understanding of the flow dynamics and heat transfer in complex systems involving Maxwell fluids. Further applications of Maxwell fluids include modeling of MHD flows in porous media, as discussed by Kamran *et al.* [12], and hybrid nanofluids, as seen in the work by Prajapati and Meher [13], which highlight the interplay between magnetic fields and viscoelastic properties in enhancing thermal management and fluid flow in industrial processes.

Activation energy is the minimum energy required to initiate a chemical reaction or physical process, representing the energy barrier that reactants must overcome to be transformed into products. It plays a fundamental role in reaction kinetics, thermodynamics, and material science, affecting the rate at which reactions occur. The concept is essential in various fields, including chemistry, biology, and engineering, where it influences reaction rates, energy efficiency, and material properties. For instance, in catalytic reactions, activation energy governs the efficiency of the catalyst, as demonstrated by studies such as those by Zhang *et al.* [14] and

Ramachandra Reddy *et al.* [15], which examine the impact of activation energy on MHD flow dynamics and nanofluid systems. In biological systems, enzymes lower activation energy to enable biochemical processes, as seen in metabolic pathways. Additionally, activation energy plays a crucial role in industrial applications like combustion, material synthesis, and energy storage systems. Recent research by Li *et al.* [16] and Kumar *et al.* [17] highlights its influence on fluid flow in MHD systems and nanofluids, underlining the importance of understanding activation energy in optimizing processes such as energy conversion and fluid transport.

Magnetohydrodynamics (MHD) is the study of the behavior of electrically conductive fluids in the presence of magnetic fields. It integrates the principles of fluid mechanics, electromagnetism, and thermodynamics to explore the dynamics of conducting fluids such as plasmas, molten metals, and certain types of nanofluids. The MHD effects are crucial in a wide array of applications where the fluid is exposed to a magnetic field, and its conductivity influences flow dynamics and heat transfer characteristics. In energy generation, MHD is applied in MHD generators and advanced cooling systems, leveraging the interaction between magnetic fields and conductive fluids to enhance power efficiency. In industrial processes, such as metal processing, the MHD theory helps in controlling fluid flow within reactors, pumps, and pipelines. Additionally, MHD is used in space science to understand solar winds, magnetic fields in the interstellar medium, and the behavior of plasmas. Recent studies, such as those by Yadav *et al.* [18] and Raghunath *et al.* [19], have explored MHD flow in porous media, emphasizing the effects of temperature variations, Hall current, and thermal radiation. Furthermore, the interaction of MHD with nanofluids, as explored by Raghunath *et al.* [20], presents potential in optimizing heat transfer and fluid transport, especially in energy and manufacturing industries. Studies on the impact of thermodynamic forces and buoyancy on MHD flow [21] and the role of thermal radiation in fluid dynamics [22] are crucial for advancing technology in heat exchangers, energy storage, and propulsion systems. Recent work by Yedhiri *et al.* [23] and HariBabu *et al.* [24] further demonstrates the importance of MHD in modeling complex systems such as chemically reacting and rotating fluid flows. In addition, the significance of MHD in the development of energy-efficient devices, as seen in studies by Raghunath *et al.* [25] and Khan *et al.* [26], highlights its role in optimizing fluid dynamics in systems with thermal and radiation effects. The study of MHD flow under varying magnetic field influences, as seen in the work of Raghunath *et al.* [27, 28] and Raju *et al.* [29], continues to provide insights for applications in industrial processes and environmental control systems.

The primary objective of this work is to investigate the consequences of stimulation energy and diffusive heating on an unstable viscoelastic fluid flow that is moving through a porous sheet that is being stretched vertically. By using similarity transformations, which may be acquired from the governing equations, it is conceivable to facilitate a framework of controlling pdes and the associated perimeter requirements for an approach to odes. This reduction can be accomplished while maintaining the integrity of the framework. The fourth order and shooting approach developed by Runge-Kutta are used in order to solve these problems. An investigation and analysis of the impacts of various flow-related variables on velocity, temperature, and concentration outlines are carried out with the use of a graphical depiction.

6.2 Problem Formulation

The bi-dimensional viscoelastic indestructible laminar flow as well as the heat transmission of an unconventional electrically conductive viscoelastic liquid across a porous media are something that we take into consideration. Furthermore, we take into consideration the combined impacts of convective boundary conditions and concurrent convection inside the space. A capacitive field that is livery is applied in the opposite overseeing of the direction in which the fluid is flowing. Where c and χ are positive constants, the stretched sheet of velocity is represented by the variables u and v. Furthermore, all of the many physical characteristics that are associated with the formulas are evaluated to ensure that they are consistent. One may see this modeling's geometry, which can be seen in Figure 6.1,

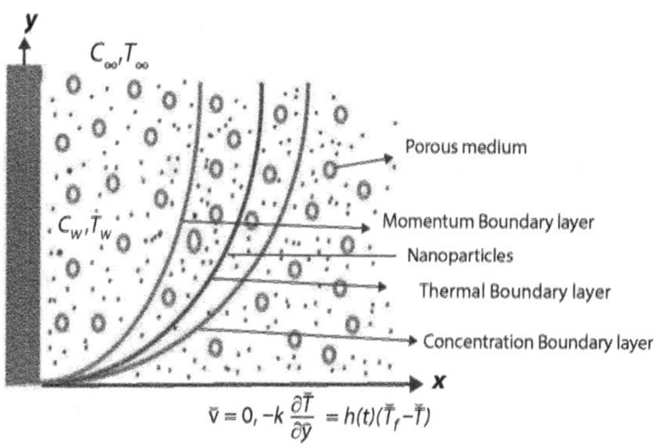

Figure 6.1 The problem's physical structure in its entirety.

which is available for viewing. For both energy and temperature outlines as well as concentration profiles, the Boussinesq approximation is often used. Following is a written representation of the continuity, momentum, energy, and concentration equations that control this sort of flow when there is a chemical reaction present:

$$\frac{\partial u}{\partial x}+\frac{\partial u}{\partial y}=0 \tag{6.1}$$

$$\frac{\partial u}{\partial t}+u\frac{\partial u}{\partial x}+v\frac{\partial u}{\partial y}=\nu\frac{\partial^2 u}{\partial y^2}-\lambda_1\left(u^2\frac{\partial^2 u}{\partial x^2}+v^2\frac{\partial^2 u}{\partial y^2}+2uv\frac{\partial^2 u}{\partial x\,\partial y}\right)$$

$$+g\beta_T\left(T-T_\infty\right)+g\beta_C\left(C-C_\infty\right)-\sigma\frac{B_0(t)}{\alpha}-\frac{\vartheta}{k}u \tag{6.2}$$

$$\frac{\partial T}{\partial t}+u\frac{\partial T}{\partial x}+v\frac{\partial T}{\partial y}=\alpha_m\left(\frac{\partial^2 T}{\partial y^2}\right)+\frac{D_m k_T}{c_s c_p}\frac{\partial^2 C}{\partial y^2} \tag{6.3}$$

$$\frac{\partial C}{\partial t}+u\frac{\partial C}{\partial x}+v\frac{\partial C}{\partial y}=D_B\left(\frac{\partial^2 C}{\partial y^2}\right)--k_r^2\left(C-C_\infty\right)\left(\frac{T}{T_\infty}\right)^m\exp\left(\frac{-E_a}{K^*T}\right) \tag{6.4}$$

The current issue's related boundary conditions are

$$u=U_w(x,t),\quad v=0,\quad -k\frac{\partial T}{\partial y}=h(t)\left(T_f-T_\infty\right),\ C=C_w \quad \text{at } y=0$$

$$u\to 0,\ v\to 0,\quad T\to T_\infty\quad C\to C_\infty \quad \text{as } y\to\infty \tag{6.5}$$

The following similarity transformations are implemented:

$$u=\frac{ax}{1-\chi t}f'(\eta),\ \ \eta=\sqrt{\frac{\nu a}{1-\chi t}}y,\ \phi(\eta)=\frac{C-C_\infty}{C_w-C_\infty},\ \theta(\eta)=\frac{T-T_\infty}{T_w-T_\infty},$$

$$h(t)=\frac{d}{\sqrt{1-\chi t}}.\quad v=-\sqrt{\frac{a}{(1-\chi t)}}f(\eta) \tag{6.6}$$

Using similarity transformation Equations 6.6, Equations 6.2–6.5 take the form

$$f''' - f'^2 + ff'' - (M+\beta)f' + \lambda(\theta + N\phi) - \delta\left(f' + \frac{1}{2}\eta f''\right) + \Lambda\left(2ff'f'' - f^2 f'''\right) = 0 \quad (6.7)$$

$$\theta'' - \delta \Pr \frac{1}{2}\eta\theta' - \Pr f\theta' + \Pr D_u \phi' = 0 \quad (6.8)$$

$$\phi'' - \delta \Pr \frac{1}{2}\eta\phi' - Scf\phi' - K_E(1+\theta)^m \exp\left(\frac{-E}{1+\theta}\right)\phi = 0 \quad (6.9)$$

These boundary conditions (Equation 6.5) are the ones that relate to them.

$$\begin{aligned} &f(\eta) = S,\ \theta'(\eta) = -\gamma(1-\theta(\eta)), \quad at\ \eta = 0 \quad \phi(\eta) = 1, \quad at\ \eta = 0 \\ &f(\eta) \to 0,\ \phi(\eta) \to 0\ \ \theta(\eta) \to 0, \quad as\ \eta \to \infty \end{aligned} \quad (6.10)$$

To overcome the complexity of elaborated problem, we consider the below nondimensional similarity transformations:

$$\begin{aligned} &\delta = \frac{c}{\chi},\ Q = \frac{Q_0}{\alpha\rho C_p},\ \Lambda = c\lambda_1,\ \lambda = \left(\frac{g\beta_T(T_w - T_\infty)x}{U_w^2}\right), \\ &N = \frac{\beta_c(C_w - C_\infty)}{\beta_T(T_w - T_\infty)},\ \beta = c\lambda_1,\ M = \frac{\sigma B_0^2}{\rho Cp}, \\ &\Pr = \frac{v}{\alpha} = \frac{v\rho C_p}{k},\ Sc = \frac{v}{D_B},\ \gamma = \left(\frac{h_f}{k}\sqrt{\frac{v}{\alpha}}\right),\ Du = \frac{D_M k_T(C_w - C_\infty)}{C_S C_p v a^2 (T_w - T_\infty)}, \\ &\delta_1 = \frac{T - T_\infty}{T_\infty},\ K_E = \frac{k_r^2}{c},\ E = \frac{E_a}{K^* T_\infty}. \end{aligned}$$

6.3 Physical Quantities

The skin friction coefficients, the reduced Nusselt numeral, and the Sherwood numbers are characterized, respectively, as follows:

$$Cf = \frac{\tau_w}{\rho U_w^2}, \quad Nu_x = \frac{x q_w}{k(T_w - T_\infty)}, \quad Sh_x = \frac{x q_m}{D_B(C_w - C_\infty)} \tag{6.11}$$

6.4 Code Validation

Within the context of the confined example, we evaluated the correctness of the present results with the preceding literature and found that there was a wonderful agreement. The next table presents a comparison of the numerical results for the Sherwood number values with the publicized result of Indira and Srinivas Raju [30]. All of the results are in remarkable agreement with one another as shown in Table 6.1.

6.5 Result and Discussion

Examination of the MHD flow of viscoelastic liquid with convective perimeter layer is the focus of the study that is currently being done. Additionally taken into consideration are heat transport and viscous dissipation. Calculating the outcome that was produced allows one to demonstrate the impact of a number of relevant physical elements. As a result of these physical parameters, the conclusion is expanded by Figures 6.2–6.16, and the value of entrenched flow variables is calculated and shown in Table 6.2.

Figures 6.2 and 6.3 both show the speed and temperature outlines for a variety of the Lorentz force M. These profiles are presented in a

Table 6.1 A contrast of the current Sherwood number with the Sherwood number findings that were released over the time period 0 equals E=Q=δ1=0.

Sh	Indira and Srinivasa Raju [30]	Present study
0.20	0.47563	0.47873
0.40	0.44584	0.44884
0.60	0.48473	0.48873

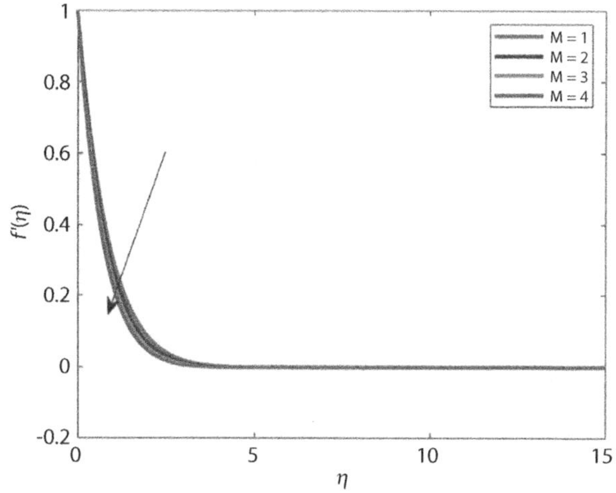

Figure 6.2 Velocity profiles and the influence of the magnetic parameter M.

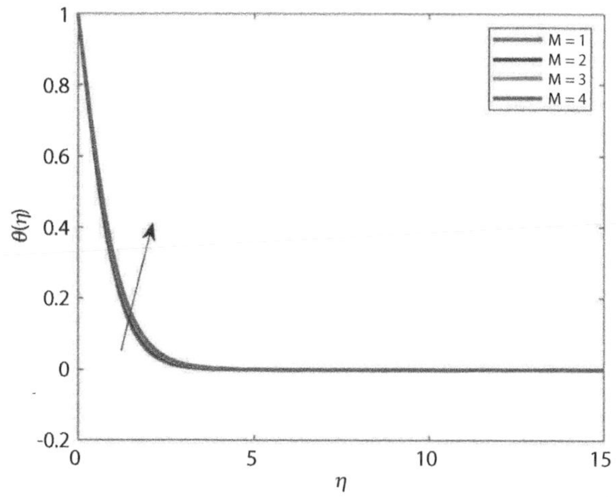

Figure 6.3 Temperature profiles and the influence of the magnetic parameter M.

relative order. Figure 6.2 makes it abundantly evident that a decrease in velocity is the result of a rise in the Lorentz force M. Figure 6.3 illustrates the impacts of the Lorentz force, which are observed to be responsible for the elevation of the temperature profiles. Because of the existence of the Lorentz force, the force that is exerted on the velocity field is slowed down, and as a result, the velocity profiles, which are affected by the magnetic

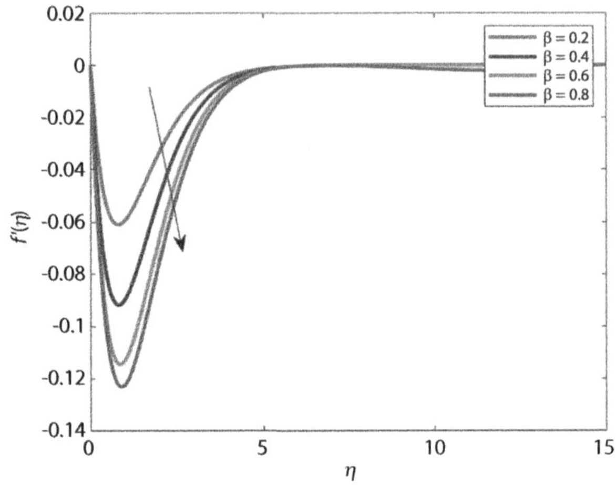

Figure 6.4 Velocity profiles and the influence of the porous parameter β.

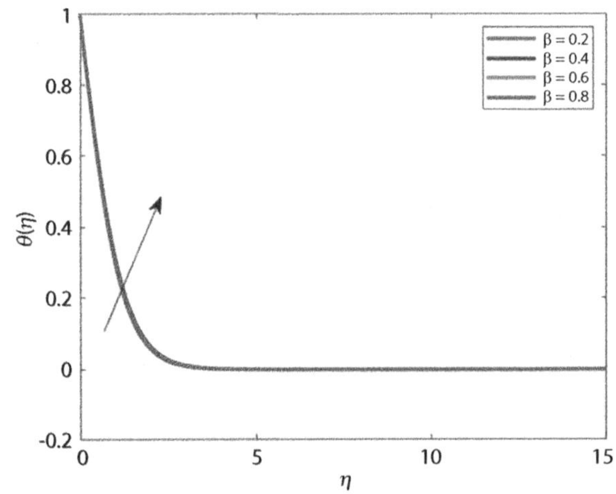

Figure 6.5 Temperature profiles and the influence of the porous parameter β.

parameter, are reduced. As a result, it is likely that the temperature will continue to rise.

For the purpose of analyzing the $f'(\eta)$ and $\theta(\eta)$ outlines while considering the numerical value of the Darcy pours factor β, Figures 6.4 and 6.5 are shown. An observation has been made that the velocity field is a function that decreases as the porosity increases. In reality, when the value of

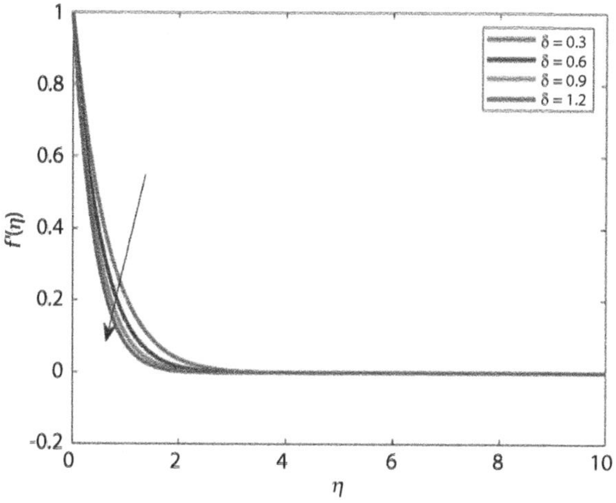

Figure 6.6 Velocity profiles and the influence of the unsteady parameter δ.

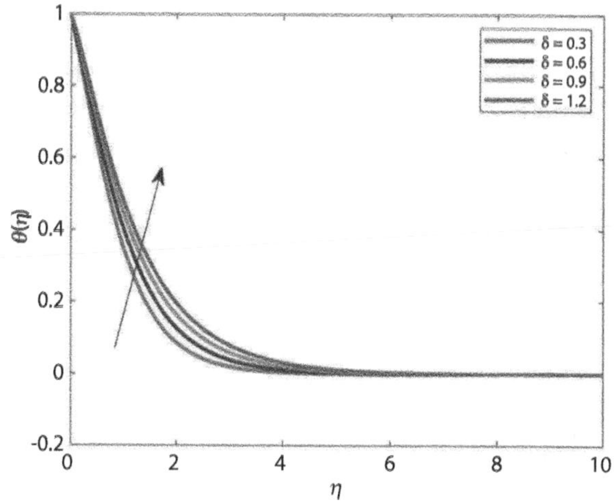

Figure 6.7 Temperature profiles and the influence of the unsteady parameter δ.

β is increased, it results in a greater restriction and an enhancement in the thickness of the porous medium, which ultimately leads to a decrease in the fluid flow motion. Therefore, as the value of β is raised, the velocity of the fluid diminishes at a lower rate. The behavior seen in θ (η) is contrary to what was expected.

As shown in Figures 6.6–6.8, the development of f′(η), θ (η), and ϕ (η) is assessed across a wide range of values of an unsteadiness factor δ. As can

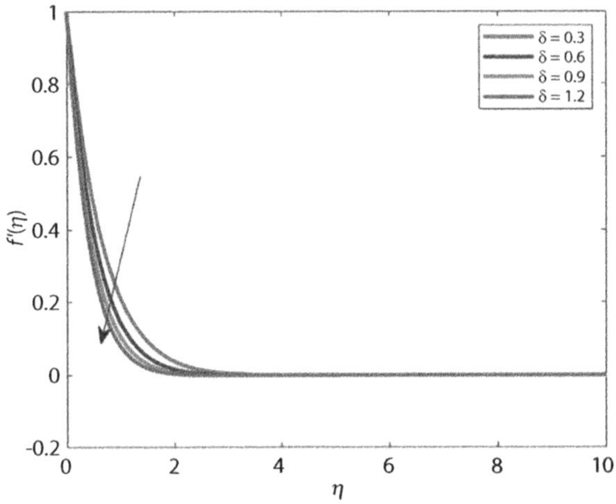

Figure 6.8 Concentration profiles and the influence of the unsteady parameter δ.

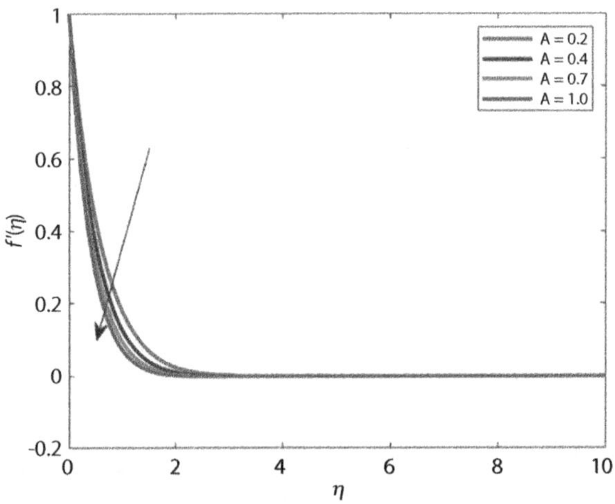

Figure 6.9 Velocity profiles and the influence of the viscoelastic fluid parameter Λ.

be seen, an increasing in the unsteady parameter results in a decrease in the velocity field and concentration, which serves as the boundary layer for the momentum and the solute. Increasing the value of δ, which serves as the thermal boundary layer, results in an increase in the temperature of nanoparticles inside the system.

In Figures 6.9–6.11, the influence of the viscoelastic liquid factor Λ on $f'(\eta)$, $\theta(\eta)$, and $\phi(\eta)$ is brought into concern. These statistics make it abundantly transparent that the declining function of the liquid factor is being

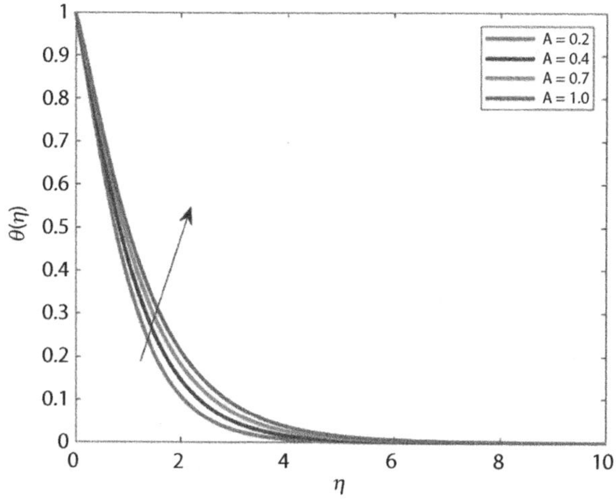

Figure 6.10 Temperature profiles and the influence of the viscoelastic fluid parameter Λ.

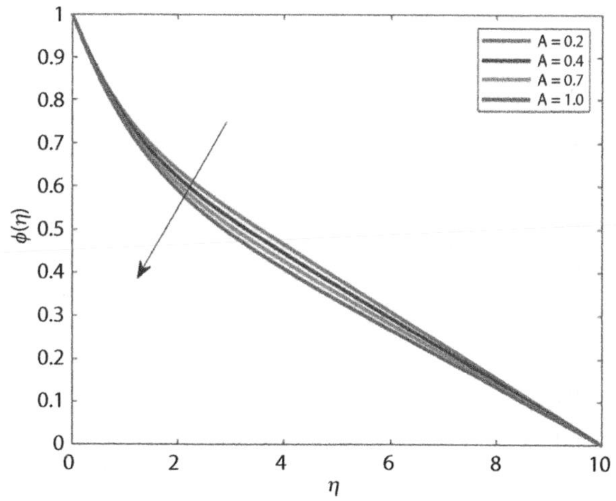

Figure 6.11 Concentration profiles and the influence of the viscoelastic fluid parameter Λ.

seen clearly. It is possible to see that the impact of improving the estimates of liquid specifications is to decrease the value of $f'(\eta)$, which is the field, and is connected with the decrease in the consistency of the perimeter layer by reducing. The $\theta(\eta)$ and $\varphi(\eta)$ both exhibit a behavior that is contrary to one another as the value of the fluid factors increases.

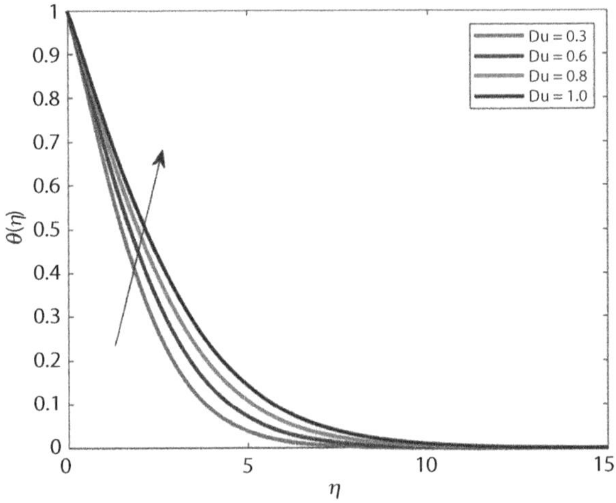

Figure 6.12 Temperature profiles and the influence of the diffusion thermo parameter Du.

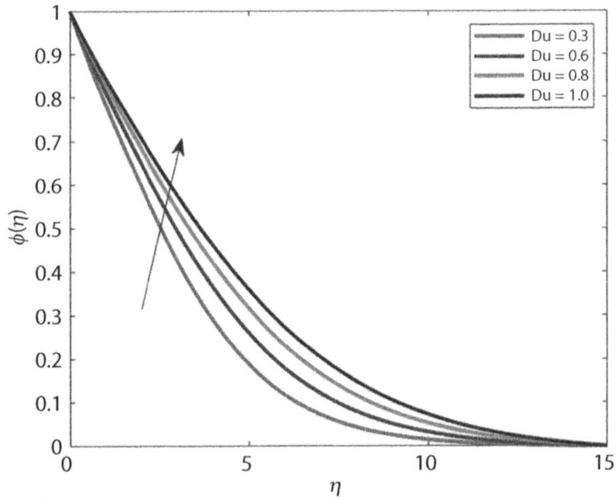

Figure 6.13 Concentration profiles and the influence of the diffusion thermo parameter Du.

The influence of diffusion heating on $\theta(\eta)$ and $\varphi(\eta)$ curves, respectively, is shown in Figures 6.12 and 6.13. Increases in diffusion heating are accompanied by increases in the energy or temperature profiles as well as the concentration profiles. There is a clear correlation between an increase in diffusion heating and an increase in temperature.

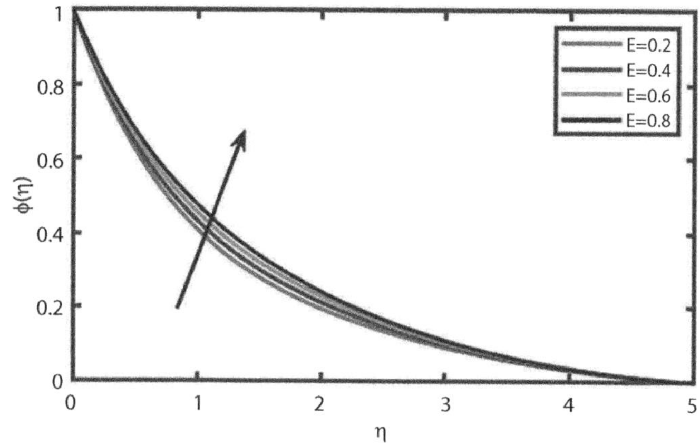

Figure 6.14 Concentration profiles and the influence of the activation energy E.

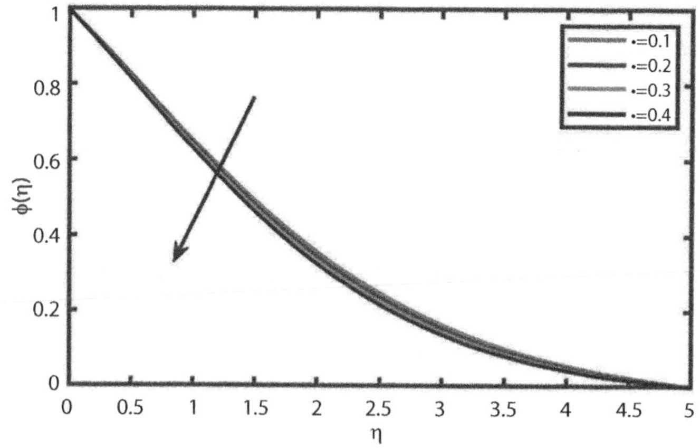

Figure 6.15 Concentration profiles and the influence of the chemical reaction rate δ.

As can be seen in Figure 6.14, the activation energy (E) has a significant impact on the concentration field. The concentration profile is shown to grow as the value of E increases, as seen in the graph. It is because of the activation energy snowballing value that the Arrhenius function is degraded. This results in the chemical reaction proceeding at a slower pace, which impacts the overall rate of the reaction. This method will directly result in increased attention on the task at hand. The data shown in Figure 6.15 illustrate that as the rate of chemical reaction (δ1) is increased, there is a corresponding drop in the concentration profile. There is a correlation between a fast chemical reaction rate and a fallout solute border layer that is denser.

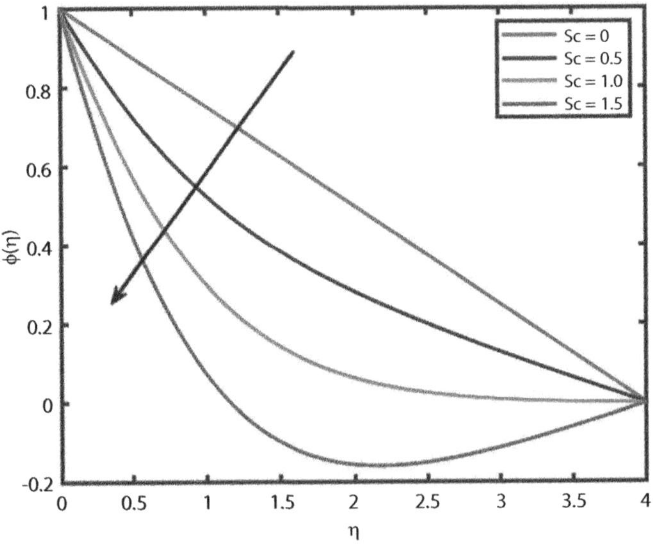

Figure 6.16 Concentration profiles and the influence of the Schmidt number Sc.

Table 6.2 Computation of $Re_x^{1/2} Cf_x$, $Re_x^{-1/2} Nu_x$ *and* $Re_x^{-1/2} Sh_x$ for different parameters.

δ	β	Sc	E	Du	M	Λ	$Re_x^{1/2} Cf_x$	$Re_x^{-1/2} Nu_x$	$Re_x^{-1/2} Sh_x$
					1		1.1195	0.8972	0.2531
					2		1.4445	0.7636	0.6544
					3		1.8705	0.5434	0.9837
0.5							1.6505	0.8272	0.7384
1.0							1.1893	0.7268	0.9878
1.5							0.9905	0.5634	1.3534
	0.2						1.2343	0.9847	0.3647
	0.4						1.4282	0.7633	0.1095
	0.7						1.6298	0.5396	0.0944
		0.5					1.1085	0.1133	0.1248
		1.0					1.3595	0.1097	0.1184

(Continued)

Table 6.2 Computation of $\mathrm{Re}_x^{1/2} Cf_x$, $\mathrm{Re}_x^{-1/2} Nu_x$ *and* $\mathrm{Re}_x^{-1/2} Sh_x$ for different parameters. (*Continued*)

δ	β	Sc	E	Du	M	Λ	$\mathrm{Re}_x^{1/2} Cf_x$	$\mathrm{Re}_x^{-1/2} Nu_x$	$\mathrm{Re}_x^{-1/2} Sh_x$
		1.5					1.7154	0.0893	0.2128
			0.2				1.2503	0.1126	0.2174
			0.4				1.4929	0.1085	0.2178
			0.6				1.82387	0.0774	0.3081
				0.3			1.0086	0.7106	0.3756
				0.6			1.0475	0.6823	0.2985
				0.8			1.0985	0.6555	0.2763
						0.2	1.5134	0.1093	0.3782
						0.4	1.5993	0.0992	0.1094
						0.7	1.9183	0.0463	0.2146

Figure 6.16 provides a visual representation of the Schmidt numeral, which has influence on the concentration profile. A drop in the Schmidt number causes the concentration profile to lessen, which is the effect of the decline.

6.6 Conclusion

The following is a condensed version of the conclusions that were derived from the numerical results:

- In the first place, an increase in the magnetic parameter M leads to a rise in the temperature, concentration, and transvers velocity.
- The concentration that was produced has increased as the values of the activation energy parameters have increased correspondingly.
- The behavior has been shown to have a consequent concentration field with increases in the chemical reaction rate.
- When the diffusive heating parameter is increased, both the temperature of the fluid and its concentration have been shown to improve.

AI disclosure statement: The author confirms that no artificial intelligence (AI)-generated content or AI-assisted writing tools were used in the preparation of this chapter (including summaries and the abstract). All text, analysis, and conclusions are the result of the author's original work.

References

1. Lavanya, B., Girish Kumar, J., Raghunath, K., Rameswara Reddy, Y., Jayachandra Babu, M., A comprehensive study of Carreau nanofluid flow over an inclined vertical plate with Cattaneo-Christov heat flux: numerical simulation and statistical analysis, in: *Radiation Effects and Defects in Solids*, pp. 1–18, 2025, https://doi.org/10.1080/10420150.2025.2484722.
2. Ali Pasha, A., K. Al Mesfer, M., Kareem, M.W., Raghunath, K., Danish, M., Patil, S., Ramachandra Reddy, V., Effects of rotational forces, and thermal diffusion on unsteady magnetohydrodynamic (MHD) flow of a viscoelastic fluid through porous media with isothermal inclined plates. *Case Stud. Therm. Eng.*, 69, 105977, 2025. https://doi.org/10.1016/j.csite.2025.105977.
3. Ramachandra Reddy, V., Raghunath, K., Imran Khan, P., Sudhakaru, K., Thermo-physical and aligned magnetic field effects on unsteady MHD convection in Casson fluids over isothermal inclined plates in the presence of joule heating and viscous dissipation. *Therm. Adv.*, 3, 100030, 2025. https://doi.org/10.1016/j.thradv.2025.100030.
4. Hussain, S., Deepthi, V.V.L., Omeshwar Reddy, V., Raghunath, K., Giulio, L., Thermo physical aspects of three dimensional non-Newtonian Fe3O4/Al2O3 water based hybrid nanofluid with rotational flow over a stretched plate. *Int. J. Heat Technol.*, 43, 1, 67–75, 2025. https://doi.org/10.18280/ijht.430108.
5. Valiveti, S., Jetti, M., Yallalla, M.R., Kodi, R., Lorenzini, G., Analysis of heat and mass transfer in unsteady magnetohydrodynamic MHD Casson fluid flow over isothermal inclined plates with thermal diffusion and heat source effects. *Power Eng. Eng. Thermophys.*, 3, 3, 195–208, 2024. https://doi.org/10.56578/peet030305.
6. Prasad, V.R., Varma, N.U.B., Jamuna, B., Suresh, M.S., Sudhakaru, K., Raghunath, K., Effects of hall current and thermal diffusion on unsteady MHD rotating flow of water based Cu, and TiO2 nanofluid in the presence of thermal radiation and chemical reaction. *Multiscale Multidiscip. Model. Exp. Des.*, 8, 161, 2025. https://doi.org/10.1007/s41939-025-00736-z.
7. Jyothi, K., Sailakumari, A., Ramachandrareddy, V., Raghunath, K., Neural network-assisted analysis of MHD boundary layer flow and thermal radiation effects on SWCNT nanofluids with Maxwellian and non-Maxwellian models. *Multiscale Multidiscip. Model. Exp. Des.*, 8, 155, 2025. https://doi.org/10.1007/s41939-025-00752-z.

8. Bhargava, K.N.V.C., Ibrahim, S.M., Raghunath, K., Magneto-hydrodynamic mixed convection chemically rotating and radiating 3D hybrid nanofluid flow through porous media over a stretched surface. *Math. Modell. Numer. Simul. Appl.*, 4, 4, 495–513, 2024. https://doi.org/10.53391/mmnsa.1438636.
9. Al Khayer, S.J. and Chakraborty, S., Effect of Stratification and Joule Heating on MHD Dusty Viscoelastic Fluid Flow through Inclined Channels in Porous Medium in Presence of Molecular Diffusivity. *East Eur. J. Phys.*, 1, 221–231, 2024.
10. Sudarmozhi, K., Iranian, D., Alqarni, M.M., Khan, M.S., Mahmoud, E.E., Pradhan, R., Haque, M.M., Exploring the Steady Flow of a Viscoelastic Fluid Passing over a Porous Perpendicular Plate Subjected to Heat Generation and Chemical Reactions. *J. Math.*, 2024, 2024.
11. Oyelami, F.H., Tharapatia, G., Ayegbusi, F.D., Onwubuoya, C., Dynamics of Tangent Hyperbolic Fluid Past a Semi-infinite Plate with the Significance of Joule Heating, Thermal Radiation and Soret-Dufour Mechanisms. *CFD Lett.*, 16, 2, 162–183, 2024.
12. Kamran, A., Azhar, E., Atif, S.M., Magnetic dipole interactions with tangent hyperbolic ferromagnetic fluid over a moving plate: A numerical study. *Numer. Heat Transfer, Part B*, 1–17, 2024.
13. Prajapati, V.J. and Meher, R., Analysis of MHD tangent hyperbolic hybrid nanofluid flow with different base fluids over a porous stretched sheet. *J. Taibah Univ. Sci.*, 18, 1, 2300851, 2024.
14. Zhang, L., Ramachandra Reddy, V., Aruna, G., Suneetha, B., Raghunath, K., 3D-MHD mixed convection in a Darcy–Forchheimer–Maxwell fluid: Thermo diffusion, diffusion-thermo effects, and activation energy influence. *Case Stud. Therm. Eng.*, 61, 104916, 2024. https://doi.org/10.1016/j.csite.2024.104916.
15. Ramachandra Reddy, V., Sreedhar, G., Raghunath, K., Effects of Hall Current, Activation Energy and Diffusion Thermo of MHD Darcy–Forchheimer Casson Nanofluid Flow in the Presence of Brownian Motion and Thermophoresis. *J. Adv. Res. Fluid Mech. Therm. Sci.*, 105, 2, 129–45, 2023. https://doi.org/10.37934/arfmts.105.2.129145.
16. Li, S., Raghunath, K., Alfaleh., A., *et al.*, Effects of activation energy and chemical reaction on unsteady MHD dissipative Darcy–Forchheimer squeezed flow of Casson fluid over horizontal channel. *Sci. Rep.*, 13, 2666, 2023. https://doi.org/10.1038/s41598-023-29702-w.
17. Suresh Kumar, Y., Hussain, S., Raghunath, K., Ali, F., Guedri, K., Eldin, S.M., Ijaz Khan, M., Numerical analysis of magnetohydrodynamics Casson nanofluid flow with activation energy, Hall current and thermal radiation. *Sci. Rep.*, 13, 4021, 2023. https://doi.org/10.1038/s41598-023-28379-5.
18. YadavKumar Awasthi, D.M., Ragoju, R., Bhattacharyya, K., Raghunath Hassan, K.M., Wang, J., Impact of temperature-reliant thermal conductivity and viscosity variations on the convection of Jeffrey fluid in a rotating

cellular porous layer. *Proc. R. Soc.*, 480, 2301, A.48020240206, 2024. http://doi.org/10.1098/rspa.2024.0206.

19. Raghunath, K., Damodhar Reddy, A., Ramachandra Reddy, V., Rotating Mixed Convective Casson Fluid Flow Past Inclined Porous Plates with the Effects of Hall and Ion Slip, Radiation Absorption, and Diffusion Thermo, in: *Proceedings of the 2nd International Conference on Nonlinear Dynamics and Applications (ICNDA 2024), Volume 2. ICNDA 2024. Springer Proceedings in Physics*, vol. 315, A. Saha and S. Banerjee (Eds.), Springer, Cham, 2024, https://doi.org/10.1007/978-3-031-69134-8_34.
20. Raghunath, K., Ramachandra Reddy, V., Haribabu, K., Samad, N., Fernandez-Gamiz, U., Thermodynamic and buoyancy force effects of Cu and TiO2 nanoparticles in engine oil flow over an inclined permeable surface. *J. King Saud Univ. Sci.*, 36, 10, 103434, 2024. https://doi.org/10.1016/j.jksus.2024.103434.
21. Yadav, D., Kumar, A.M., Ravi, R., Bhattacharyya, K., Raghunath Wang, K.J., The impact of rotation on the onset of cellular convective movement in a Casson fluid saturated permeable layer with temperature dependent thermal conductivity and viscosity deviations. *Chin. J. Phys.*, 91, 262–277, 2024. https://doi.org/10.1016/j.cjph.2024.07.020.
22. Sridevi, D., Ramana Murthy, C.V., Raghunath, K.,., An investigation into the impact of thermal radiation and chemical reactions on the flow through porous media of a Casson hybrid nanofluid including unstable mixed convection with stretched sheet in the presence of thermophoresis and Brownian motion. *Open Phys.*, 22, 1, 20240043, 2024. https://doi.org/10.1515/phys-2024-0043.
23. Rani, Y.S., KalyanKumar, P., Raghunath, K., Asmat, F., Unsteady MHD rotating mixed convective flow through an infinite vertical plate subject to Joule heating, thermal radiation, Hall current, radiation absorption. *J. Therm. Anal. Calorim.*, 149, 8813–8826, 2024. https://doi.org/10.1007/s10973-024-12954-7.
24. HariBabu, K., Raghunath, K., Charankumar, G., Lorenzini, G., Heat and Mass Transfer on Unsteady MHD Chemically Reacting Rotating Flow of Jeffrey Fluid Past an Inclined Plates under the Impact of Hall Current, Diffusion Thermo and Radiation Absorption. *J. Adv. Res. Fluid Mech. Therm. Sci.*, 111, 2, 225–241, 2024. https://doi.org/10.37934/arfmts.111.2.225241.
25. Raghunath, K., Ali, F., Khalid, M., Abdullaeva, B.S., Altuijri, R., Ijaz Khan, M., Heat and mass transfer on MHD flow of Jeffrey nanofluid based on Cu and TiO2 over an inclined plate and diffusion-thermo and radiation absorption effects. *Pramana – J. Phys.*, 97, 202, 2023. Indian Academy of Sciences. https://doi.org/10.1007/s12043-023-02673-3.
26. Ijaz Khan, M., Ravikumar, S., Raghunath, K., HalaHejazi, A., Chen, G., Tao, L., Understanding Prandtl fluid flow in conduits with slip boundary conditions: Implications for engineering and physiology. *Phys. Fluids*, 35, 113106, 2023. doi:10.1063/5.0174196.

27. Raghunath, K., Mohana Ramana, R., Veeranna, V., Ijaz Khan, M., Abdullaev, S., Tamam, N., Hall current and thermal radiation effects of 3D rotating hybrid nanofluid reactive flow *via* stretched plate with internal heat absorption. *Results Phys.*, 53, 106915, 2023. https://doi.org/10.1016/j.rinp.2023.106915.
28. Raghunath, K., Ramachandra Reddy, V., Ijaz Khan, M., Abdullaev, S.S., Habibullah, Boudjemline, A., Boujelbene, M., Bouazzi, Y., Unsteady magneto-hydro-dynamics flow of Jeffrey fluid through porous media with thermal radiation, Hall current and Soret effects. *J. Magn. Magn. Mater.*, 582, 171033, 2023. https://doi.org/10.1016/j.jmmm.2023.171033.
29. Raju, K.V., Mohanaramana, R., Reddy, S.S., Raghunath, K., Chemical Radiation and Soret Effects on Unsteady MHD Convective Flow of Jeffrey Nanofluid Past an Inclined Semi-Infinite Vertical Permeable Moving Plate. *Commun. Math. Appl.*, 14, 1, 237–255, May 2023.
30. Indira, S. and Srinivasa Raju, R., Analytical Study of MHD Mixed Convection Flow for Maxwell Nanofluid Through a Vertical Cone with Porous Material in the Presence of Variable Thermal Conductivity and Soret, Dufour Effects. *J. Adv. Res. Fluid Mech. Therm. Sci.*, 106, 2, 129–142, 2023.

7

Effect of Couple Stresses on Thermal Convection of Navier–Stokes–Voigt Fluid in Porous Media

Sunil[1] *, Sweta Sharma[1], Deepak Kumar[1] and Poonam Sharma[2]

[1]Department of Mathematics and Scientific Computing, National Institute of Technology Hamirpur, Hamirpur, Himachal Pradesh, India
[2]Department of Mathematics, NSCBM Govt. College, Hamirpur, Himachal Pradesh, India

Abstract

This chapter examines the influence of couple stresses on thermal convection in a Navier–Stokes–Voigt (NSV) fluid layer within porous media, with a focus on industrial applications where non-Newtonian fluid characteristics are essential. By incorporating couple stresses into the NSV model, the analysis highlights viscoelastic and microstructural effects critical to applications such as polymer processing, bioengineering, and petroleum extraction, where fluid behavior significantly deviates from Newtonian assumptions. The study employs linear analysis using normal mode techniques and nonlinear analysis through the energy method to examine stability and convection onset. Additionally, energy decay analysis provides insight into stability predictions, offering parameters that influence convection dynamics. Findings from this investigation assist in optimizing the thermal management and performance of non-Newtonian fluid systems in heat exchangers, geotechnical processes, and enhanced oil recovery. Understanding how couple stresses alter flow and heat transfer in porous environments enhances modeling accuracy, ultimately supporting efficient engineering applications and improved operational outcomes.

**Corresponding author*: sunilnithmr@gmail.com; ORCID: 0000-0001-6738-6815
Sweta Sharma: ORCID: 0009-0001-9784-8643
Deepak Kumar: ORCID: 0009-0009-8205-2704

Dhananjay Yadav, Mukesh Kumar Awasthi and Harith Mohamed Al-Azri (eds.) Non-Newtonian Fluids for Industrial Applications: Modeling and Simulations, (147–176) © 2026 Scrivener Publishing LLC

Keywords: Couple stresses, thermal convection, Navier–stokes–Voigt fluid, porous media, Non-Newtonian fluids, stability analysis, viscoelastic effects, energy method

7.1 Introduction

Non-Newtonian fluids, characterized by complex flow properties that significantly diverge from Newtonian assumptions, play a crucial role across industries, including polymer processing, bioengineering, and petroleum extraction. These fluids' behaviors—such as viscoelasticity, shear dependency, and unique stress responses—require specialized modeling approaches that transcend traditional methods [12, 19]. This chapter explores how couple stresses, a microstructural effect intrinsic to certain non-Newtonian materials, influence thermal convection in NSV fluid within porous media. By investigating stability and heat transfer behaviors under these conditions, the analysis provides insights applicable to industrial contexts, supporting the development of optimized, high-performance systems in sectors where precise thermal and flow control are essential.

Building on foundational principles of continuum mechanics and rheology, this chapter applies the NSV model to capture the viscoelastic behavior of non-Newtonian fluids under conditions of thermal convection. The NSV model, widely recognized for its ability to simulate viscoelasticity through a stress–relaxation mechanism, was initially developed by Oskolkov [31, 32] and has since been expanded to address boundary value problems, as explored by Zvyagin and Turbin [61]. In recent studies, Straughan [45, 46, 48] has extended the study to diverse stability and convective phenomena, including competitive double-diffusive convection, thermosolutal convection, and the investigation of instability thresholds in Kelvin–Voigt fluids of variable order. Di Plinio *et al.* [14] further contributed to the model's refinement by examining memory effects, while Layton and Rebholz [24] explored variations in relaxation times, and Berselli and Bisconti [10] analyzed its structural stability. Recent studies have also highlighted the model's adaptability to specific physical contexts, such as those involving Soret and slip boundary conditions, as analyzed by Badday and Harfash [5]; stability problems with generalized Navier–Stokes–Voigt theories, as investigated by Straughan [50]; and magnetohydrodynamic stability, as explored by Kavitha *et al.* [21]. Notably, Baranovskii [6] advanced the understanding of the NSV model by examining strong solutions for the incompressible form, while Shankar and Shivakumara [37] analyzed natural convection stability in a vertical layer of NSV fluids, adding practical insights into applications requiring detailed stability assessment. Sharma *et al.* [40] further explored thermosolutal convection stability within a rotating NSV fluid, and Basavarajappa and Bhatta [9] expanded the model's relevance by performing a nonlinear stability analysis on the Rayleigh–Bénard problem for NSV fluids. Additionally, Straughan [49] investigated convection under the generalized Maxwell–Cattaneo law, as well as nonlinear stability for scenarios with temperature-dependent viscosity, which broadened the NSV model's utility in complex industrial processes. This chapter builds on these contributions by adopting an NSV framework incorporating couple stresses to examine thermal convection stability within porous media, advancing the field's understanding of non-Newtonian fluid dynamics with practical applications in areas such as polymer processing and bioengineering.

Couple stress is a measure of internal rotational forces within a material, represented by a tensor that accounts for moments per unit area. It extends traditional stress analysis by considering the influence of microstructural rotations and torques on the material's mechanical behavior. In recent times, there has been a growing interest in exploring the importance of couple stresses in various domains, including material sciences, microfluidics, and biological systems. The concept of couple stresses in fluids was first introduced by Stokes [42], who highlighted their relevance to fluid dynamics. Goel *et al.* [18] further investigated the hydromagnetic stability of an unbounded couple-stress binary fluid mixture under vertical temperature and concentration gradients, revealing the impact of couple stresses on the stability of rotating systems subject to external forces. Lin [26] explored the linear stability of rotor-bearing systems modeled by couple-stress fluids, illustrating the effects of these fluids on mechanical stability in such systems. Later studies, such as those by Weng [59], delved into the influence of couple-stress effects on the effective viscosity of magnetic fluids, demonstrating how microstructural properties alter the fluid's resistance to shear in the presence of a magnetic field. Sunil *et al.* [54] examined the global stability of thermal convection in couple-stress fluids, highlighting the distinct stability behaviors exhibited by these fluids in comparison to classical Newtonian fluids. This was extended by Sunil *et al.* [55], who considered temperature- and pressure-dependent viscosity in their study of the global stability of thermal convection in couple-stress fluid, showing how such factors influence the fluid's stability. Sunil and Devi [51] further investigated the global stability of thermal convection in couple-stress fluid with rigid boundaries, uncovering the complexities of heat transfer and stability in confined systems. Shivakumara and Kumar [41] studied linear and weakly nonlinear triple-diffusive convection in couple-stress fluid layer, revealing how multiple diffusion processes impact convective stability. The stability of couple-stress fluid flow through horizontal porous media was explored by Shankar *et al.* [38], who demonstrated the role of porous media in influencing fluid stability and heat transfer. Mahajan and Nandal [27] contributed further by studying the stability of penetrative convection in couple-stress fluid, offering insights into how variations in fluid properties affect convective heat transport in different environments. Pan *et al.* [33] examined the transitions and bifurcations in couple-stress fluid-saturated porous media using a thermal non-equilibrium model, contributing to the understanding of complex flow regimes in porous media. Choudhary *et al.* [13] studied stability in a couple-stress fluid layer with variable viscosity, focusing on the effects of different conducting boundaries on fluid stability. Recently, Afluk and Harfash [2] analyzed the stability and instability of thermosolutal convection in Brinkman–Darcy–Kelvin–Voigt fluids with couple-stress effects, emphasizing the importance of these effects in both porous media and complex fluid systems. Together, these studies underscore the growing significance of couple-stress effects in fluid dynamics over the past several decades. Their implications extend to industrial applications, material processing, and environmental engineering, where understanding the intricate behavior of these fluids can lead to improved designs and more efficient processes.

A porous medium is a material composed of a solid matrix with interconnected voids, which allows fluids to pass through it, playing an essential role in many natural and industrial processes. The study of fluid dynamics in porous media is critical for a variety of applications, including groundwater flow, petroleum extraction, heat transfer in geothermal systems, and filtration. The foundational work of Muskat [29] on the flow of fluids through porous media was pivotal in shaping the understanding of these systems. Whitaker [60] further expanded the theory of fluid motion in porous media, introducing important advances in the description of flow behavior, particularly in industrial and engineering contexts. Later, Qin and Kaloni [34] examined the nonlinear stability problem in rotating porous layers, shedding light on the role of rotation in convective heat transfer within porous media. Sharma *et al.* [39] further explored thermosolutal instability in rotating fluids within porous media, examining how both thermal and solutal gradients influence convection—an important consideration for environmental and industrial systems. The 2000s saw more focused studies on the behavior of complex fluids in porous media, such as the work of Straughan [43], who investigated the sharp nonlinear stability thresholds in rotating porous convection, crucial for understanding convection patterns in geophysical systems.

Kim *et al.* [22] contributed to expanding heat transfer research by studying thermal instability in viscoelastic fluids within porous media. Meanwhile, Sunil *et al.* [52] analyzed the effects of suspended particles on the convection of couple-stress fluids in porous media, enhancing the understanding of fluid dynamics in systems containing particulate matter. Sunil *et al.* [53] examined thermosolutal convection in ferromagnetic fluids saturating porous media, with significant implications for material processing and energy systems. A comprehensive reference on convection in porous media was provided by Nield and Bejan [30], which offered an in-depth theoretical framework for heat and mass transfer within porous structures. The *Handbook of Porous Media* by Vafai [58] further consolidated the growing body of knowledge, serving as a valuable resource for both theoretical research and practical applications. More recently, the research focus has shifted toward understanding the stability of complex fluid systems in porous media. Roy *et al.* [36] explored the effects of double diffusion and viscous dissipation on oscillatory convection in a viscoelastic fluid-saturated porous layer, advancing the study of nonlinear convection phenomena. Thakur *et al.* [56] analyzed the effects of couple stresses on the stability of magnetized ferrofluids in porous media, a topic of increasing importance in industrial fluid control and energy systems. Thakur *et al.* [57] also examined the effect of rotation on ferroconvection in porous media, focusing on the role of couple stress forces and offering new insights into the dynamics of rotating systems. The work of Kumar and Tyagi [23] on hydromagnetic thermosolutal instability of nanofluids in porous media further underscores the increasing complexity of fluid dynamics research in porous media, with applications in nanotechnology and energy systems. Afluk and Harfash [3] studied the instability of thermosolutal convection in a Brinkman–Darcy–Kelvin–Voigt fluid, a work that advances the understanding of heat and mass transfer in porous structures and is particularly relevant for energy systems and industrial applications. These diverse studies illustrate the growing complexity of porous media research and its importance in optimizing a range of systems, including oil recovery, geothermal energy, water filtration, and advanced heat exchangers.

At the core of thermal convection lies the physics of buoyancy-driven fluid motion, where temperature gradients induce density variations, causing warmer, less dense fluid to rise while cooler, denser fluid descends [1]. This mechanism plays a vital role in natural and engineering systems, such as atmospheric circulation, heat exchangers, and polymer processing. Understanding thermal convection is essential for designing efficient thermal management systems in fields like bioengineering and geophysics. The classical works of Chandrasekhar [11] and Drazin and Reid [15] laid the groundwork for understanding the behavior of thermal convection in simple fluid systems. They explored the onset of convection and the role of Rayleigh number as a key parameter for stability analysis in fluid flows. As fluid dynamics research advanced, the focus shifted toward more complex fluids. Studies by Rosenblat [35] and Li and Khayat [25]extended these ideas by examining convection in viscoelastic fluids, where additional rheological properties influence the development of convection patterns and the selection of amplitude in oscillatory convection. The study of complex fluids, such as viscoelastic and couple-stress fluids, has become increasingly important, especially in porous media applications. For example, Amendola and Fabrizio [4] investigated the effects of fading memory in viscoelastic convection, showing how these materials can alter the stability thresholds for convective flow. These insights have important implications for understanding convection in more structured and heterogeneous environments, such as biological tissues and polymeric systems. Further work by Joseph [20] enhanced the understanding of stability mechanisms in non-Newtonian systems, offering a deeper analysis of the behavior of fluids with complex rheological characteristics. Energy-based methods, introduced by Galdi and Padula [17], have provided new tools for stability analysis, especially in non-Newtonian and viscoelastic fluids. These methods allow for a more comprehensive understanding of convective stability, offering criteria for assessing global stability in fluid flows. More recently, Straughan [44] expanded upon these energy-based techniques to further refine the understanding of stability in thermally driven flows. The study of Rayleigh–Bénard convection has also seen significant developments, with researchers such as Malashetty and Swamy [28] investigating double-diffusive convection and extending classical principles to more complex scenarios, such as those involving multiple species or different types of diffusion. This chapter aligns with these classical principles while leveraging contemporary analytical models, such as weighted residuals, comprehensively discussed by Finlayson [16], to deepen the analysis of stability mechanisms in thermally driven flows.

In this chapter, the effect of couple stresses on thermal convection in a Navier–Stokes–Voigt fluid within porous media is examined, considering various boundary conditions: free–free, rigid–rigid, and rigid–free. Section 7.2 begins with geometrical configuration and formulation of the fundamental set of governing equations, which are presented in their basic form. These equations are subsequently perturbed and nondimensionalized to facilitate further analysis. In Section 7.3, a nonlinear analysis is conducted using the energy method to demonstrate energy decay, and the variational principle is applied to derive the eigenvalue problem.

The boundary conditions corresponding to this analysis are also outlined. Section 7.4 focuses on the linear analysis, employing normal mode analysis to formulate the eigenvalue problem, and investigates the validity of the principle of exchange of stabilities. Section 7.5 discusses the solution methodology utilized for solving the eigenvalue problem. Section 7.6 presents the results, which are illustrated graphically to highlight key findings. Section 7.7 provides a summary of the primary conclusions and insights drawn from the study. The final section explores potential real-world applications of this work across various fields in the future.

7.2 Geometrical Configuration and Mathematical Formulation

This section provides a structured framework to analyze the effect of couple stresses on thermal convection in Navier–Stokes–Voigt fluid within porous media. We begin by establishing the governing equations that describe the system's behavior, followed by defining the basic state to represent equilibrium conditions. Perturbation analysis is then applied to explore deviations and identify stability thresholds. The equations are transformed into dimensionless form, simplifying the problem and enhancing the applicability of the results. Finally, boundary conditions are incorporated to ensure the model reflects realistic physical situations, offering a comprehensive understanding of fluid flow behavior in porous media.

7.2.1 Governing Equations

The governing equations describe the behavior of Navier–Stokes–Voigt fluids in porous media, incorporating couple stresses and thermal convection. These equations are derived from the conservation of mass, momentum, and energy, which account for fluid dynamics and heat transfer. The Navier–Stokes–Voigt model extends the classical equations by adding higher-order viscous effects, as discussed by Straughan [47], with the regularization term $-\hat{\lambda}\nabla^2 \frac{\partial \mathbf{q}}{\partial t}$. The modified constitutive equation for couple stress effects follows Stokes [42], while temperature effects are modeled using the Boussinesq approximation [7, 8], providing a unified framework for fluid and thermal behavior. In the Boussinesq approximation, density is treated as constant in all terms of the equations of motion, except for the one in the external force. To model fluid flow in porous media, we use the Darcy–Brinkman model, which accounts for both porous medium and boundary layer flow, including shear stress near solid boundaries. Unlike Darcy's law, which neglects viscous effects in the porous matrix, the Darcy–Brinkman model, as explained by Nield and Bejan [30], is more accurate in high permeability or thin channel situations. This model is important in various fields such as soil mechanics and bioengineering.

Consider an infinite layer of an incompressible Navier–Stokes–Voigt fluid saturating a porous medium, confined between two surfaces separated by a thickness *d*, and subjected to heating from below. Additionally, the fluid is influenced by couple stress forces. The Darcy–Brinkman model describes fluid flow in porous media, characterized by a porosity e and permeability $K_{1.}$ The force of gravity **g** acts in the negative direction of *z*-axis. The temperature *T* varies from T_o at $z = 0$ to T_d at $z = d$, maintaining a uniform temperature gradient $\left(\beta = \left|\frac{dT}{dz}\right|\right)$. The geometric setup is depicted in Figure 7.1.

The governing equations for the system under consideration (applying the Boussinesq approximation) are given as follows [2, 46, 50]:

Equation of continuity

$$\nabla \cdot \mathbf{q} = 0. \tag{7.1}$$

Equations of motion

$$\frac{\rho_0}{\epsilon}\left(\frac{\partial \mathbf{q}}{\partial t} - \left(\hat{\lambda}\nabla^2\right)\frac{\partial \mathbf{q}}{\partial t}\right) = -\nabla p - \frac{1}{K_1}\left(\mu - \mu'\nabla^2\right)\mathbf{q} + \tilde{\mu}\nabla^2\mathbf{q} + \rho_o g\alpha T\hat{\mathbf{k}}. \tag{7.2}$$

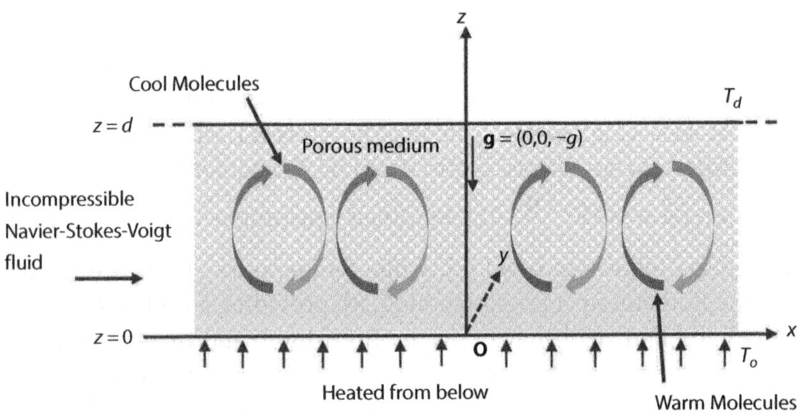

Figure 7.1 Geometrical configuration of the problem.

Equation of energy

$$\left(\rho_o C_o\right)_m \frac{\partial T}{\partial t} + \left(\rho_o C_o\right)_f (\mathbf{q}\cdot\nabla)T = \nabla\cdot(k\nabla T). \tag{7.3}$$

Here, we denote the quantities as follows: $\mathbf{q} = (u, v, w)$ for velocity, ρ for density, K_1 for permeability of porous medium, ϵ for porosity, ρ_0 for reference density, p for pressure, $\hat{\lambda}$ for the Kelvin–Voigt coefficient, μ' for the coefficient of couple stress viscosity, μ for the coefficient of viscosity, t for time, $\tilde{\mu}$ for effective viscosity, α for the coefficient of thermal expansion, k for thermal conductivity, and C_0 for specific heat for constant pressure. Also, $(\rho_0 C_0)_m = (1-\epsilon)(\rho_0 C_0)_s + \epsilon(\rho_0 C_0)_f$ and $\kappa = (1-\epsilon)\kappa_s + \kappa_f$. The subscript '$m$', '$s$', and '$f$' refers to fluid–solid mixture, solid, and fluid, respectively.

7.2.2 Basic State and Perturbation Equations

The basic state represents the equilibrium conditions of the system, where the fluid is assumed to be at rest and thermally uniform. In this state, there are no velocity or temperature fluctuations, and the system is in a stable configuration. The basic state serves as the reference point from which perturbations, such as small deviations in velocity or temperature, are introduced to examine the onset of instability and thermal convection. By defining the basic state, we can isolate the effects of perturbations and better understand how deviations from equilibrium lead to dynamic behavior in the system.

For this analysis, the basic state is assumed to be steady and is represented as follows:

$$\begin{aligned}\mathbf{q} &= (0,0,0) = \mathbf{q}_b,\ T = -\beta z + T_0 = T_b(z),\ \rho = \rho_o(1+\alpha\beta z)\\ &= \rho_b(z),\ p = p_b(z).\end{aligned} \tag{7.4}$$

Here, subscript 'b' denotes the basic state.

To investigate the stability or instability of the Navier–Stokes–Voigt fluid with couple stress effects in porous medium, we perturb the system with disturbances $\mathbf{q}'$, p', and θ in velocity, pressure, and temperature, respectively. The system of equations, with the apostrophes removed from the perturbed quantities, becomes

$$\nabla\cdot\mathbf{q} = 0, \tag{7.5}$$

$$\frac{\rho_o}{\epsilon}\left(\frac{\partial \mathbf{q}}{\partial t}-(\hat{\lambda}\nabla^2)\frac{\partial \mathbf{q}}{\partial t}\right)=-\nabla p-\frac{1}{K_1}\left(\mu-\mu'\nabla^2\right)\mathbf{q}+\mu\nabla^2\mathbf{q}+\rho_o g\alpha\theta\hat{\mathbf{k}}, \tag{7.6}$$

$$A\frac{\partial \theta}{\partial t}+(\mathbf{q}\cdot\nabla)\theta=\beta w+\kappa\nabla^2\theta, \tag{7.7}$$

where $A=\dfrac{(\rho_o C_o)_m}{(\rho_o C_o)_f}$ denotes the ratio of volumetric heat capacity, and $\kappa=\dfrac{k}{(\rho_o C_o)_f}$ denotes the thermal diffusivity.

7.2.3 Dimensionless Perturbation Equations

To simplify the governing equations and make them more broadly applicable, we introduce dimensionless variables. Nondimensionalization eliminates arbitrary units and scales the equations to reveal the key parameters that govern the behavior of the system. This process involves selecting characteristic length, velocity, temperature, and various other scales, allowing for the reduction of the equations to dimensionless form. By doing so, we can identify the key dimensionless numbers, such as the Rayleigh number, which play a crucial role in determining the onset of convection and the stability of the system. The dimensionless equations provide a clearer understanding of the underlying physical mechanisms and enhance the comparison between different fluid flow situations. To nondimensionalize the perturbed equations, we use the following characteristic scales:

$$z=dz*,\ t=\frac{d^2}{\kappa}t*,\ \mathbf{q}=\frac{\kappa}{d}\mathbf{q}*,\ p=\frac{\mu\kappa}{K_1}p*,\ \theta=\frac{\beta d}{Ra^{1/2}}\theta*,\ \hat{\lambda}=t\kappa\lambda*, \tag{7.8}$$

and the following nondimensional parameters are introduced:

$$Ra=\frac{\alpha g\beta d^4}{\nu\kappa},\ \Pr=\frac{\nu}{\kappa},\ F=\frac{1}{\nu}\frac{\mu'}{\rho_o d^2},\ V_r=\frac{\tilde{\mu}}{\mu},\ V_a=\frac{\epsilon\nu d^2}{\kappa K_1},$$
$$D_a=\frac{K_1}{d^2},\ B_r=\frac{\tilde{\mu}}{\mu}\frac{K_1}{d^2}=V_r D_a,\ Ra_D=\frac{\alpha g\beta\rho_o K_1 d^2}{\mu\kappa}=D_a Ra, \tag{7.9}$$

namely, Rayleigh number, Prandtl number, couple stress parameter, the ratio of effective viscosity to dynamic viscosity, Vadasz number, Darcy number, Darcy–Brinkman number, and Darcy–Rayleigh number, respectively.

The nondimensional parameters introduced each highlight a distinct physical aspect of fluid flow in porous media. The Rayleigh number (Ra) represents the influence of thermal buoyancy, driving convection within the fluid. The Prandtl number (Pr) compares the rates of momentum and thermal diffusion, indicating the relative thickness of velocity and temperature boundary layers. The couple stress parameter (F) reflects the effects of fluid microstructure, crucial for capturing the influence of couple stresses in non-Newtonian flows. The viscosity ratio (V_r) expresses the relationship between effective viscosity and dynamic viscosity, impacting flow resistance. The Vadasz number (V_a) captures the balance between inertial and viscous forces, while the Darcy number (D_a) represents the permeability of the porous medium, affecting how easily fluid flows through it. The Darcy–Brinkman number (Br) quantifies the influence of shear stress within the porous flow, particularly significant near boundaries. Lastly, the Darcy–Rayleigh number (Ra_D) combines thermal effects with porous media properties, governing the onset and behavior of convection within the porous structure.

Using scales mentioned in Equation 7.8 and nondimensional parameters in Equation 7.9 in the perturbed Equations 7.5–7.7, we derive the dimensionless perturbation equations (dropping *s) as follows:

$$\nabla \cdot \mathbf{q} = 0, \tag{7.10}$$

$$\frac{1}{V_a}\left(\frac{\partial \mathbf{q}}{\partial t} - (\lambda \nabla^2)\frac{\partial \mathbf{q}}{\partial t}\right) = -\nabla p = (F + Br)\nabla^2 \mathbf{q} - \mathbf{q} + Ra_D^{1/2}\theta \hat{\mathbf{k}}, \tag{7.11}$$

$$A\left(\frac{\partial \theta}{\partial t} + (\mathbf{q} \cdot \nabla)\theta\right) = Ra_D^{1/2} w + \nabla^2 \theta. \tag{7.12}$$

These dimensionless equations provide a simplified yet comprehensive representation of the physical system, where each nondimensional parameter reveals specific influences on the stability and convection behavior in the fluid. This formulation enables a clearer understanding of how various factors—such as thermal buoyancy, viscous effects, and couple stresses—affect the flow dynamics in porous media. By eliminating arbitrary units, the nondimensional equations also enhance the comparability of results across different physical scenarios, making this framework versatile for analyzing thermal convection and stability thresholds in similar systems.

7.2.4 Boundary Conditions

The boundary conditions play a crucial role in accurately capturing the dynamics of thermal convection within the porous medium, as they confine and influence the fluid's behavior at its limits. In this analysis, we consider three types of boundary surfaces: free–free, rigid–free, and rigid–rigid, each imposing distinct constraints on fluid motion and heat transfer. The free–free boundary condition allows fluid movement without resistance at both surfaces, suitable for situations where minimal external forces interact with the fluid. The rigid–free condition introduces one fixed boundary, representing surfaces where the fluid is partially confined, while the rigid–rigid condition confines the fluid completely, restricting movement at both surfaces. These boundary conditions are essential to simulate realistic scenarios and assess how different constraints impact the onset and stability of thermal convection.

The boundary conditions (BCs) are

$$w = 0, \theta = 0, \text{ at } z = 0, 1. \tag{7.13}$$

together with the fact that w and θ satisfy the plane tiling periodicity in the xy-plane.

7.3 Nonlinear Analysis

This section focuses on capturing the nonlinear interactions that arise in the thermal convection of Navier–Stokes–Voigt fluids with couple stresses in porous media. Nonlinear analysis is critical for understanding the system's behavior under realistic conditions, where initial disturbances grow and interact, creating complex flow patterns and potentially leading to intricate dynamics. By incorporating nonlinear terms, this analysis accounts for higher-order interactions among temperature, velocity, and couple stresses, offering a detailed view of stability thresholds and the onset of convection. This approach allows us to examine the amplitude and structure of emerging convection patterns, providing a thorough understanding of fluid flow and heat transfer within porous media under different physical and boundary conditions.

7.3.1 Conditional Energy Stability

Conditional energy stability in nonlinear analysis ensures that a system's energy remains bounded and behaves physically over time. It prevents unphysical energy growth or instability, especially in complex systems with nonlinear dynamics. This concept is essential for both theoretical stability and numerical simulations, ensuring realistic, stable solutions. The effect of couple stresses on the stability of Navier–Stokes–Voigt fluid is effectively examined using the energy method. This analysis pertains to an infinite spatial region, whereas many practical convection studies typically involve a finite region of disturbances. Hence, without loss of generality, $\mathbf{q} = 0$ is assumed with the boundary conditions given in Equation 7.13, as thoroughly explained by Straughan [44]. On multiplying Equation 7.11 by $\mathbf{q}$ and Equation 7.12 by θ and then integrating over V, we get

$$\frac{1}{2V_a}\frac{d}{dt}\|\mathbf{q}\|^2 + \frac{\lambda}{2V_a}\frac{d}{dt}\|\nabla\mathbf{q}\|^2 = -\|\mathbf{q}\|^2 - (F - Br)\|\nabla\mathbf{q}\|^2 + Ra_D^{1/2}\langle w\theta\rangle, \tag{7.14}$$

$$\frac{A}{2}\frac{d}{dt}\|\theta\|^2 = -\|\nabla\theta\|^2 + Ra_D^{1/2}\langle w\theta\rangle. \tag{7.15}$$

Here, $\langle\cdot\rangle$ and $\|\cdot\|$ denote the inner product and norm of vectors defined in $L^2(V)$, respectively.

By using Equations 7.14 and 7.15 along with the boundary conditions, we construct the energy E(t), and the change in energy over time is

$$\frac{dE}{dt} = I_o - D_o. \tag{7.16}$$

Here,

$$E = \frac{A}{2}\|\theta\|^2 + \frac{\lambda_1}{2V_a}\|\mathbf{q}\|^2 + \frac{\lambda\lambda_1}{2V_a}\|\nabla\mathbf{q}\|^2, \tag{7.17}$$

$$I_o = (1+\lambda_1)Ra_D^{1/2}\langle w\theta\rangle, \tag{7.18}$$

$$D_o = \|\nabla\theta\|^2 + \lambda_1\|\mathbf{q}\|^2 + \lambda_1(F+Br)\|\nabla\mathbf{q}\|^2. \tag{7.19}$$

Here, λ_1, I_o, and D_o represent the positive coupling parameter, the production term, and the dissipation term, respectively.

Define

$$m = \max_H \frac{I_o}{D_o},$$

where H is the space of admissible functions.

Then, $m < 1$ is required, for

$$\frac{dE}{dt} \le -D_o a_o. \tag{7.20}$$

Here, $a_o = 1 - m$, $(a_o > 0)$.

Using Poincare inequality in Equation 7.19, we get

$$D_o \ge \mathbf{K}^* E(t), \tag{7.21}$$

where $\mathbf{K}^* = 2\pi^2 \min\left(\frac{V_a}{\pi^2}, \frac{V_a(F+Br)}{\lambda\pi^2}, \frac{1}{A}\right)$

Using Inequality 7.21 in Equation 7.20 and then integrating, we get

$$E(t) \le e^{-a_o\mathbf{K}^* t} E(0). \tag{7.22}$$

Therefore, energy E decays exponentially fast, ensuring nonlinear stability for all values of $E(0)$. Also, we have seen that couple stress parameter, medium permeability, and Kelvin–Voigt parameter plays significant role for energy decay if $\frac{V_a(F+Br)}{\lambda\pi^2} < \left(\frac{V_a}{\pi^2}, \frac{1}{A}\right)$. If the left side of this inequality is smaller, it indicates that the system maintains stability and slows down energy dissipation, leading to more stable fluid behavior.

7.3.2 Variational Principle

The variational principle is a method for deriving governing equations by extremizing a functional, such as kinetic energy or dissipation, with respect to the velocity and pressure fields. In nonlinear analysis, it is crucial for stability analysis, where small perturbations to a base flow are examined. By applying the variational principle, one can derive eigenvalue problems to assess the stability of flow patterns, determining the growth rates of disturbances and identifying stable or unstable regimes, especially in complex, nonlinear systems like turbulence.

We utilize the calculus of variations to assess the maximum problem at the critical argument $m = 1$. Taking transformations $w = \frac{\hat{w}}{\sqrt{\lambda_1}}$, $\mathbf{q} = \frac{\hat{\mathbf{q}}}{\sqrt{\lambda_1}}$ for Equations 7.18 and 7.19, the associated Euler–Lagrange equation $\delta I_o - \delta D_o = 0$ gives the results

$$\left(\frac{1+\lambda_1}{\sqrt{\lambda_1}}\right) Ra_D^{1/2}\theta - 2\mathbf{q} + 2(F + Br)\nabla^2\mathbf{q} = 0, \tag{7.23}$$

$$\left(\frac{1+\lambda_1}{\sqrt{\lambda_1}}\right) Ra_d^{1/2} w + 2\nabla^2\theta = 0. \tag{7.24}$$

By applying curl curl on Equation 7.23 and subsequently extracting the third component, we obtain the following result:

$$\left(\frac{1+\lambda_1}{\sqrt{\lambda_1}}\right) Ra_d^{1/2}\nabla_1^2\theta - 2\nabla^2 w + 2(F + Br)\nabla^4 w = 0, \tag{7.25}$$

where $\nabla_1^2 = \frac{\partial^2}{\partial x^2} + \frac{\partial^2}{\partial y^2}$ is the horizontal Laplacian.

The plane tiling form is now assumed as

$$\{w, \theta\} = \{W(z), \Theta(z)\}\, \Psi(x, y) \tag{7.26}$$

where 'Ψ' denotes the shape of the stability cell [11], and $\Psi(x, y)$ satisfies

$$\nabla_1^2\Psi = -a^2\Psi. \tag{7.27}$$

Here, 'a' represents the wave number.

Using plane tiling form Equation 7.26 in Equations 7.25 and 7.24, the eigenvalue problems we obtain are

$$-2(D^2-a^2)W+2(F+Br)(D^2-a^2)^2W-\left(\frac{1+\lambda_1}{\sqrt{\lambda_1}}\right)Ra_D^{1/2}a^2\Theta=0, \tag{7.28}$$

$$2(D^2-a^2)\Theta+\left(\frac{1+\lambda_1}{\sqrt{\lambda_1}}\right)Ra_D^{1/2}W=0, \tag{7.29}$$

with the boundary conditions

$$W=0,\ D^2W=0,\ \Theta=0,\ \text{at}\ z=0,1\ \text{(for free–free)}, \tag{7.30}$$

$$W=0,\ DW=0,\ \Theta=0,\ \text{at}\ z=0,1\ \text{(for rigid–rigid)}, \tag{7.31}$$

$$\left.\begin{aligned}&W=0,\ DW=0,\ \Theta=0,\ at\ z=0,\\&W=0,\ D^2W=0,\ \Theta=0,\ at\ z=1\end{aligned}\right\}\text{(for rigid-free)}. \tag{7.32}$$

7.4 Linear Analysis

Linear analysis is used to identify the onset of instability by determining critical conditions where disturbances start to grow, simplifying equations by ignoring nonlinear terms, allowing analytical solutions, and highlighting the dominant modes that drive initial instability. We linearize the nondimensional perturbed Equations 7.11 and 7.12 for the linear analysis by neglecting nonlinear terms. Normal mode analysis method simplifies linear analysis by reducing complex PDEs to ODEs, allowing examination of each disturbance mode independently. This method identifies the growth or decay of disturbances, applies boundary conditions effectively, isolates the most unstable modes, and offers insight into physical mechanisms driving stability and instability. Utilizing this method, we assume solutions of the form $\mathbf{q}(\mathbf{x})=\mathbf{q}(\mathbf{x})e^{\sigma t}$, $p(\mathbf{x},t)=p(\mathbf{x})e^{\sigma t}$, $\theta(\mathbf{x},t)=\theta(\mathbf{x})e^{\sigma t}$. Subsequently, the linearized nondimensional perturbed equations are derived as follows:

$$\frac{\sigma}{V_a}(1-\lambda\nabla^2)\mathbf{q} = -\nabla p + (F+Br)\nabla^2\mathbf{q} - \mathbf{q} + Ra_D^{1/2}\theta\hat{\mathbf{k}}, \tag{7.33}$$

$$A\sigma\theta = \nabla^2\theta + Ra_D^{1/2}w. \tag{7.34}$$

Here, σ represents the growth rate, which characterizes the rate at which disturbances or instabilities within a system evolve over time. When the growth rate is positive, disturbances grow exponentially, indicating an unstable system. Conversely, when the growth rate is negative, disturbances decay, suggesting a stable system. This parameter is essential for determining the conditions that give rise to phenomena such as convection or oscillatory behavior.

The curl–curl operator is used to simplify equations by eliminating the pressure term and easing the application of boundary conditions. It decouples velocity components, making it more straightforward to examine stability and instability behaviors in both nonlinear and linear analyses. By applying curl curl on Equation 7.33 and subsequently extracting the third component, we obtain the following result:

$$\frac{\sigma}{V_a}(1-\lambda\nabla^2)\nabla^2 w = (F+Br)\nabla^4 w - \nabla^2 w + Ra_D^{1/2}\nabla_1^2\theta. \tag{7.35}$$

Using plane tiling form Equation 7.26 in Equations 7.35 and 7.34, the eigenvalue problems we obtain are

$$\begin{aligned}\frac{\sigma}{V_a}(D^2-a^2)W - \frac{\sigma\lambda}{V_a}(D^2-a^2)^2W &= -(D^2-a^2)W \\ &\quad +(F+Br)(D^2-a^2)^2W - a^2Ra_D^{1/2}\Theta,\end{aligned} \tag{7.36}$$

$$A\sigma\Theta = (D^2-a^2)\Theta + Ra_D^{1/2}W, \tag{7.37}$$

with the same boundary conditions as those provided in Equations 7.30–7.32.

7.4.1 Principle of Exchange of Stabilities

The principle of exchange of stability (PES) states that when a system undergoes perturbations, its stable equilibrium may transition to a new stable state, with stability being exchanged between the original and new states. In linear analysis, this principle helps predict when instability will occur and whether the system will shift to a more stable configuration. It simplifies the analysis by focusing on critical conditions for instability and stability transitions.

The PES is valid if a stationary pattern of motions prevails at the onset of instability [11]. To verify this, we have to show that σ is real.

Multiply Equation 7.36 by W^* and Equation 7.37 by Θ^* (* represents complex conjugate), and subsequently integrating the resulting expression over the z-domain while applying boundary conditions 7.30–7.32, we obtain the following results:

$$-\frac{\sigma}{V_a}\mathcal{I}_1-\frac{\sigma\lambda}{V_a}\mathcal{I}_1=\mathcal{I}_2+(F+Br)\mathcal{I}_2-a^2Ra_D^{1/2}\int_0^1 W^*\Theta dz, \tag{7.38}$$

$$A\sigma\mathcal{I}_3=-\mathcal{I}_4+Ra_D^{1/2}\int_0^1 W\Theta^* dz, \tag{7.39}$$

where

$$\mathcal{I}_1=\int_0^1(a^2|W|^2+|DW|^2)\,dz,\ \mathcal{I}_2=\int_0^1\left(a^4|W|^2+2a^2|DW|^2+|D^2W|^2\right)dz,$$
$$\mathcal{I}_3=\int_0^1\left(|\Theta|^2\right)dz,\quad \mathcal{I}_4=\int_0^1\left(a^2|\Theta|^2+|D\Theta|^2\right)dz. \tag{7.40}$$

Here, the integrals I_1, I_2, I_3, and I_4 are positive.

Adding Equation 7.38 with the product of a^2 and the complex conjugate of Equation 7.39, we get

$$-\left(\frac{\sigma}{V_a}+1\right)\mathcal{I}_1-\left(\frac{\sigma\lambda}{V_a}+(F+Br)\right)\mathcal{I}_2+a^2A\sigma^*\mathcal{I}_3+a^2\mathcal{I}_4=0. \tag{7.41}$$

Substituting $\sigma=\sigma_r+\iota\sigma_i$ into Equation 7.41 yields the imaginary part of the resulting equation as follows:

$$\sigma_i\left(\frac{\mathcal{I}_1}{V_a}+\frac{\lambda\mathcal{I}_2}{V_a}+a^2A\mathcal{I}_3\right)=0, \tag{7.42}$$

which implies that $\sigma_i = 0$, since $\left(\frac{\mathcal{J}_1}{V_a} + \frac{\lambda \mathcal{J}_2}{V_a} + a^2 A \mathcal{J}_3\right) > 0$, because $\mathcal{J}_1$, $\mathcal{J}_2$, $\mathcal{J}_3$ are positive, and both V_a and A are nonnegative physical quantities. This validates the PES, indicating the lack of oscillatory mode of convection. Thus, the shift from stability to instability must proceed through a stationary convection state. Consequently, the equations for the marginal state can be derived by setting $\sigma = 0$ in Equations 7.36 and 7.37. Therefore, the eigenvalue problem can be formulated as

$$-(D^2 - a^2)W + (F + Br)(D^2 - a^2)^2 W - a^2 Ra_d^{1/2}\Theta = 0, \quad (7.43)$$

$$(D^2 - a^2)\Theta + Ra_D^{1/2} W = 0, \quad (7.44)$$

with the same boundary conditions as those provided in Equations 7.30–7.32.

7.5 Solution Methodology

The Galerkin technique is a method for approximating solutions to differential equations by projecting them onto a finite basis. It ensures accuracy and efficiency, particularly in solving complex eigenvalue problems in engineering. The eigenvalue problems have been solved using a single-term Galerkin approach. Chandrasekhar [11] provided the exact solution for free–free bounding surfaces. For the rigid–rigid and rigid–free bounding surfaces, solutions that satisfy their respective boundary conditions are assumed. Here, the set of solutions that adhere to the boundary conditions 7.30–7.32 is defined as

$$W = \sum_{i=1}^{n} G_i W_i, \quad (7.45)$$

$$\Theta = \sum_{i=1}^{n} H_i \Theta_i, \quad (7.46)$$

where W_i, Θ_i are the basis functions and G_i, H_i are the constants.

For free–free bounding surfaces:
As per boundary conditions 7.30, the set of exact solutions is as follows:

$$W_i = \sin i(\pi z),\ \Theta_i = \sin i(\pi z). \tag{7.47}$$

For rigid–rigid bounding surfaces:
As per boundary conditions 7.31, the basis functions are as follows:

$$W_i = -0.5z^{2i} + 2.5z^{2i+2} - 3z^{2i+3} + z^{2i+4},\ \Theta_i = z^i - 2z^{i+2} + z^{i+3}. \tag{7.48}$$

For rigid–free bounding surfaces:
As per boundary conditions 7.32, the basis functions are as follows:

$$W_i = -z^{2i} + 3.25z^{2i+2} - 3.05z^{2i+3} + 0.8z^{2i+4},\ \Theta_i = z^i - 2z^{i+2} + z^{i+3}. \tag{7.49}$$

Using the solutions described in forms 7.45 and 7.46 into the eigenvalue problems of nonlinear stability and linear instability, and employing the Galerkin technique, yields a system of linear homogeneous equations:

$$E_{ji}G_i + F_{ji}H_i = 0, \tag{7.50}$$

$$P_{ji}G_i + Q_{ji}H_i = 0. \tag{7.51}$$

The set of the above two equations possesses a nontrivial solution when the determinant of the coefficient matrix is zero, which can be expressed as

$$\begin{vmatrix} \mathrm{E}_{ji} & \mathrm{F}_{ji} \\ \mathrm{P}_{ji} & \mathrm{Q}_{ji} \end{vmatrix} = 0. \tag{7.52}$$

Using Equation 7.52, we get the value of Darcy–Rayleigh numbers for both nonlinear and linear analyses. Subsequently, using relation $Ra = \dfrac{Ra_D}{D_a}$, we get Rayleigh numbers Ra_N (for nonlinear) and Ra_L (for linear).

7.6 Results and Discussion

In this work, we calculated the Rayleigh numbers for both nonlinear and linear eigenvalue problems using a method called the single-term

Galerkin method. For the nonlinear case, the Rayleigh numbers (Ra_N) were influenced by several factors, including F, V_r, D_a, and a, depending on the boundary conditions used. To find the optimal value for the parameter λ_1, we looked for the point where the Rayleigh number did not change as λ_1 was adjusted. This was done by setting the derivative of Ra_N with respect to λ_1 equal to zero. We found that $\lambda_1 = 1$ provided the optimal result for all boundary conditions tested. When we substituted $\lambda_1 = 1$ into the Rayleigh number expressions for each boundary condition, the results were the same as those obtained from the linear analysis (Ra_L). This shows that the nonlinear analysis can be simplified to the linear case under the optimal condition. The detailed results are presented below. We obtain the Rayleigh number for free–free boundary conditions as follows:

$$Ra_N = \frac{(a^2+\pi^2)^2\left(\frac{1}{D_a}+(V_r+\frac{F}{D_a})a^2+V_r\pi^2+\frac{\pi^2 F}{D_a}\right)}{a^2} = Ra_L. \quad (7.53)$$

We obtain the Rayleigh number for rigid–rigid boundary conditions as

$$Ra_N = \frac{1}{a^2}\left(\left(V_r+\frac{F}{D_a}\right)\left(1.0381a^6+36.9703a^4+792.905a^2+5222.93\right) + \frac{1}{D_a}\left(1.0381a^4+23.6086a^2+131.892\right)\right) = Ra_L. \quad (7.54)$$

We obtain the Rayleigh number for rigid–free boundary conditions as

$$Ra_N = \frac{1}{a^2}\left(\left(V_r+\frac{F}{D_a}\right)\left(1.07057a^6+36.3051a^4+514.057a^2+2566.47\right) + \frac{1}{D_a}\left(1.07075a^4+23.436a^2+127.027\right)\right) = Ra_L. \quad (7.55)$$

Here, the identical Rayleigh numbers ($Ra_N = Ra_L$) indicate that there are no subcritical instabilities, which strongly suggests that the system is globally stable. We also found that the Rayleigh number depends on factors such as the wave number, the Darcy number, and the couple stress parameter. However, the Kelvin–Voigt parameter does not affect the Rayleigh number, although it does play an important role in the rate at which energy decays in the system.

To find the critical Rayleigh number, which marks the point where convection starts, we first determined the critical wave numbers by solving the condition $\frac{dRa_N}{da}=0$. After finding these critical wave numbers, we substituted them into Equations 7.53–7.55 to calculate the corresponding critical Rayleigh numbers. The results for the critical Rayleigh numbers are then presented in graphical form.

Figure 7.2 shows the variations of the critical Rayleigh number (Ra_c) and the couple stress parameter (F) across different combinations of bounding surfaces (i.e., free–free, rigid–rigid, and rigid–free) at $D_a = 0.1$ and $V_r = 1$. The range of couple stress parameter F is taken from 0 to 1,with 0 indicating the absence of couples tresses [42]. The curves in the figure represent the onset of convection. The region below each curve indicates conditions under which the NSV fluid remains stable, while the region above each curve signifies instability. The figure also demonstrates that as the couple stress parameter F increases, the critical Rayleigh number (Ra_c) also increases. This means that the presence of couple stresses delays the onset of convection, acting as a stabilizing factor.

Figure 7.3 shows a variation of the critical Rayleigh number (Ra_c) and the Darcy number (D_a) across different combinations of bounding surfaces at F=0.005 and V_r=1. The range of Darcy number D_a is taken from 0 to 1 [30]. The region below each bounding surface curve represents stability for that particular bounding surface, while the region above each bounding surface curve signifies instability for that corresponding bounding surface. The figure clearly shows that as the Darcy number (D_a) increases, the critical Rayleigh number (Ra_c) decreases. This means that higher permeability of the medium makes the system more prone to instability or destabilizes it.

From Figures 7.2 and 7.3, we can conclude that the NSV fluid is most thermally stable when confined by rigid boundaries and least stable when both surfaces are free. This means that fluid motion is more restricted by rigid boundaries compared to when the surfaces are free to move. Additionally, these figures show that the curves for the Rayleigh number (Ra_N) match up with the Rayleigh number (Ra_L) for each type of boundary. This alignment suggests that there are no subcritical instability regions, which confirms that the system is globally stable, supporting the validity of our numerical results.

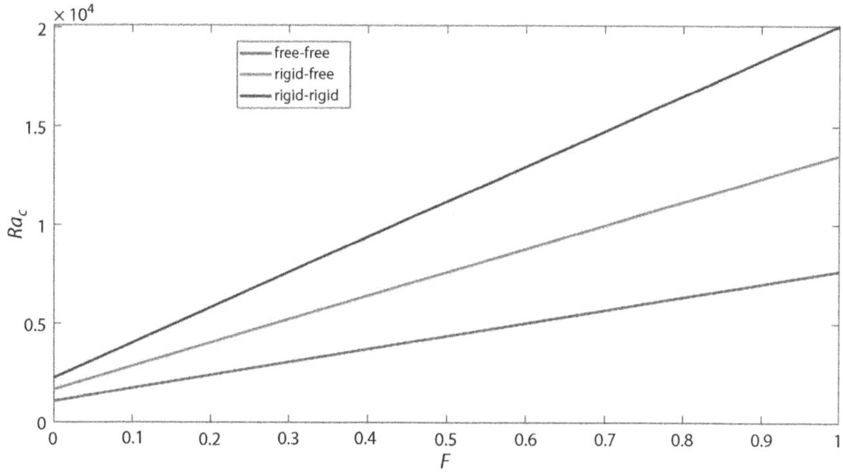

Figure 7.2 Critical Rayleigh number (Ra_c) vs couple stress parameter (F) at D_a =0.1 and V_r =1 for different combinations of bounding surfaces.

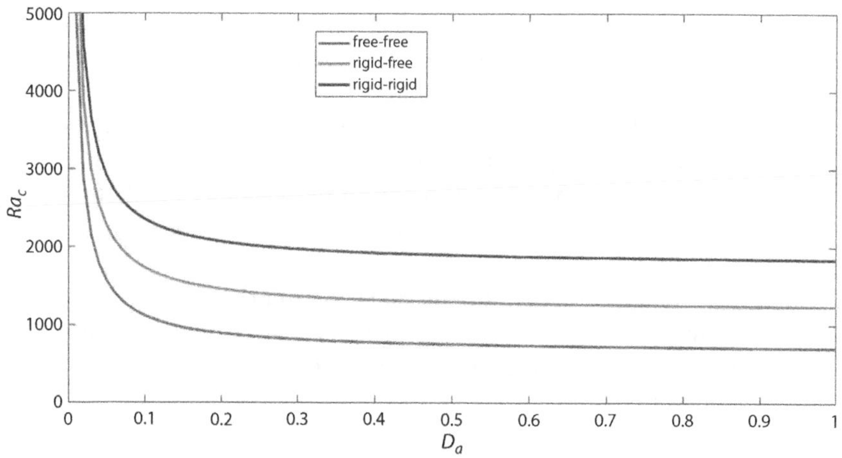

Figure 7.3 Critical Rayleigh number (Ra_c) vs Darcy number (D_a) at F=0.005 and V_r=1 for different combinations of bounding surfaces.

7.7 Conclusions

In this chapter, both linear and nonlinear analyses are employed to investigate the influence of couple stresses on thermal convection in an NSV fluid layer within porous media. The results of these analyses yield important insights into the thermal dynamics of NSV fluids, with applications in fields such as heat exchangers, enhanced oil recovery, and other processes involving complex fluid flows in porous media. The following conclusions summarize the key findings from this chapter:

- The Kelvin–Voigt parameter, medium permeability, and couple stress parameter are found to significantly influence the energy dissipation in the system. Notably, stationary convection is solely affected by the couple stress parameter and medium permeability.
- The Rayleigh numbers obtained from both nonlinear and linear analyses are identical, which implies the absence of subcritical instabilities in the system.
- The validation of PES confirms the absence of an oscillatory mode of convection, indicating a direct transition from conduction to steady convection without oscillatory or transient behavior.
- The thermal stability of the fluid is notably enhanced when the system is confined within rigid–rigid bounding surfaces. This suggests that such boundary conditions are particularly conducive to convection in Navier–Stokes–Voigt fluids, promoting stability under thermal perturbations.
- Couple stresses are found to delay the onset of thermal convection, implying their stabilizing role in resisting convective disturbances. In contrast, medium permeability accelerates the onset of thermal convection, highlighting its destabilizing effect on the fluid system.
- The Kelvin–Voigt parameter does not influence the Rayleigh numbers, indicating that its primary role is in modulating the energy dissipation rather than affecting the initiation of convection.

These findings provide insights into the complex interactions between material parameters and thermal convection, with implications for the design and control of systems involving non-Newtonian fluid dynamics.

7.8 Applications

The study of non-Newtonian fluid dynamics, particularly in porous media, has broad applications across several industries, including polymer processing, bioengineering, and petroleum extraction. This chapter's analysis of the effects of couple stresses on thermal convection in NSV fluid within porous media presents valuable insights for improving efficiency in these fields. Non-Newtonian fluids, which exhibit shear-dependent viscosity, viscoelasticity, and microstructural effects, require more sophisticated modeling than their Newtonian counterparts. The inclusion of couple stresses—internal rotational forces that account for moments per unit area—extends traditional models of fluid dynamics to account for microstructural torque effects, which are particularly relevant in the study of complex materials. The findings of this work have practical implications for industries where thermal and fluid flow behaviors are critical.

- **Polymer Processing: Enhancing Stability and Heat Transfer Control**

In polymer processing, where precise control over the flow and temperature of polymer melts is necessary, understanding how couple stresses affect convection stability can help optimize cooling rates, energy consumption, and material properties. The study's results indicate that couple stresses delay the onset of thermal convection, which can stabilize processes by preventing premature or erratic heat transfer, leading to more uniform processing conditions and reducing the likelihood of thermal stresses that could degrade product quality.

- **Bioengineering: Improving Thermal Stability in Tissue Engineering**

In the context of bioengineering, particularly in applications involving tissue engineering and the modeling of biological fluids, this research provides insights into how the viscoelastic nature of fluids can influence heat transport in porous structures. The enhanced thermal stability observed in rigid–rigid boundary conditions suggests that controlling the boundaries of bioreactors or synthetic tissue scaffolds could help regulate temperature gradients more effectively, which is crucial for maintaining optimal cell growth and metabolic activity.

- **Petroleum Extraction: Optimizing Thermal Convection in Enhanced Oil Recovery**

In petroleum extraction, the study's focus on porous media flow is directly applicable to oil and gas recovery processes. The effects of medium permeability and couple stresses on the thermal stability and convection behavior within reservoir rocks can inform better strategies for enhanced oil recovery (EOR). By modulating the permeability of the medium or adjusting the flow characteristics through the introduction of couple stress fluids, operators could potentially improve the efficiency of thermal EOR methods, such as steam injection, by controlling the onset and progression of thermal convection within the porous medium.

• **Geothermal Energy Systems: Enhancing Heat Transfer Efficiency**
The insights from this chapter are valuable for geothermal energy systems, where thermal convection within porous rocks plays a key role in heat transfer efficiency. The stabilization effects of couple stresses could enhance heat exchange within geothermal reservoirs, making energy extraction processes more efficient. Similarly, the study can inform the design of advanced heat exchangers by suggesting optimal fluid properties and boundary conditions that maximize thermal stability and reduce the potential for convective instabilities.

• **Microfluidic Systems and Material Synthesis: Controlling Heat Transfer Transitions**
The study's finding that couple stresses delay the onset of convection in NSV fluids suggests their potential use in systems where controlling the transition from conductive to convective heat transfer is critical. In applications such as microfluidic systems and material synthesis, where precise temperature control and stable flow regimes are required, the ability to tailor the fluid's rheological properties by introducing couple stresses could offer a way to finely tune system behavior for optimal performance.

Ultimately, this chapter highlights the importance of incorporating microstructural effects such as couple stresses into the analysis of non-Newtonian fluids in porous media. It paves the way for more accurate simulations and designs of industrial systems that require the precise control of both flow and thermal behavior. The ability to predict and manage convective instabilities in complex fluid systems will undoubtedly enhance the performance of technologies across a wide array of engineering and industrial applications.

AI disclosure statement: The author acknowledges the use of AI-based tools for minor language editing or grammatical correction. However, all intellectual contributions, analyses, and conclusions presented in this chapter are entirely the author's own.

References

1. Acheson, D.J., *Elementary Fluid Dynamics*, Oxford University Press, New York, 1990.
2. Afluk, Z.A. and Harfash, A.J., Stability and instability of thermosolutal convection in a Brinkman–Darcy– Kelvin–Voigt fluid with couple stress effect. *Phys. Fluids*, 36, 034108, 2024. doi: 10.1063/5.0196321.

3. Afluk, Z.A. and Harfash, A.J., Instability of thermosolutal convection of Kelvin-Voigt fluid in a porous medium using Darcy-Brinkman model. *J. Porous Media*, 28, 2025. doi: 10.1615/JPorMedia.2024050970.
4. Amendola, G. and Fabrizio, M., Thermal convection in a simple fluid with fading memory. *J. Math. Anal. Appl.*, 366, 2, 444–459, 2010.
5. Badday, A.J. and Harfash, A.J., The effects of the soret and slip boundary conditions on thermosolutal convection with a Navier–Stokes–Voigt fluid. *Phys. Fluids*, 35, 1, 2023.
6. Baranovskii, E.S., Strong solutions of the incompressible Navier–Stokes–Voigt model. *Mathematics*, 8, 181, 2020. doi: 10.3390/math8020181.
7. Barletta, A., Local energy balance, specific heats and the Oberbeck-Boussinesq approximation. *Int. J. Heat Mass Transf.*, 52, 21-22, 5266–5270, 2009.
8. Barletta, A., The Boussinesq approximation for buoyant flows. *Mech. Res. Commun.*, 124, 103939, 2022.
9. Basavarajappa, M. and Bhatta, D., Nonlinear stability analysis of Rayleigh-Bénard problem for a Navier-Stokes-Voigt fluid. *Int. J. Non-Linear Mech.*, 162, 104712, 2024. doi: 10.1016/j.ijnonlinmec.2024. 104712.
10. Berselli, L.C. and Bisconti, L., On the structural stability of the Euler–Voigt and Navier–Stokes–Voigt models. *Nonlinear Anal. Theory Methods Appl.*, 75, 117–130, 2012. doi: 10.1016/j.na.2011.08.011.
11. Chandrasekhar, S., *Hydrodynamic and Hydromagnetic Stability*, Dover, New York, 1981.
12. Chhabra, R.P. and Richardson, J.F., *Non-Newtonian Flow in the Process Industries: Fundamentals and Engineering Applications*, Butterworth-Heinemann, Oxford, UK, 1999.
13. Choudhary, S., Devi, R., Mahajan, A., Sunil, Stability analysis in a couple-stress fluid layer with variable viscosity heated from below: Different conducting boundaries. *Chin. J. Phys.*, 83, 94–102, 2023. doi: 10.1016/j.cjph.2023.02.007.
14. Di Plinio, F., Giorgini, A., Pata, V., Temam, R., Navier–Stokes–Voigt equations with memory in 3D lacking instantaneous kinematic viscosity. *J. Nonlinear Sci.*, 28, 653–686, 2018.
15. Drazin, P.G. and Reid, W.H., *Hydrodynamic Stability*, Cambridge university press, New York, 2004.
16. Finlayson, B.A., *The Method of Weighted Residuals and Variational Principles*, Society for Industrial and Applied Mathematics, Philadelphia, 2013.
17. Galdi, G.P. and Padula, M., A new approach to energy theory in the stability of fluid motion. *Arch. Ration. Mech. Anal.*, 110, 187–286, 1990.
18. Goel, A., Agrawal, S., Agarawal, G., Hydromagnetic stability of an unbounded couple stress, binary fluid mixture having vertical temperature and concentration gradients with rotation. *Indian J. Pure Appl. Math.*, 30, 10, 991–1001, 1999.
19. Irgens, F., *Rheology and Non-Newtonian Fluids*, Springer, New York, 2014.

20. Joseph, D.D., *Fluid Dynamics of Viscoelastic Liquids*, Springer Science & Business Media, New York, 2013.
21. Kavitha, G.N., Shankar, B.M., Shivakumara, I.S., On the magnetohydrodynamic stability of channel flow of Navier–Stokes–Voigt fluid. *Phys. Fluids*, 36, 043105, 2024. doi: 10.1063/5.0196788.
22. Kim, M.C., Lee, S.B., Kim, S., Chung, B.J., Thermal instability of viscoelastic fluids in porous media. *Int. J. Heat Mass Transfer*, 46, 5065–5072, 2003. doi: 10.1016/S0017-9310(03)00363-6.
23. Kumar, D. and Tyagi, V.K., Hydromagnetic thermosolutal instability of an Oldroyd nanofluids in a porous medium. *J. Nanofluids*, 13, 638–649, 2024. doi: 10.1166/jon.2024.2060.
24. Layton, W.J. and Rebholz, L.G., On relaxation times in the Navier-Stokes-Voigt model. *Int. J. Comput. Fluid Dyn.*, 27, 3, 184–187, 2013.
25. Li, Z. and Khayat, R.E., Finite-amplitude Rayleigh–Bénard convection and pattern selection for viscoelastic fluids. *J. Fluid Mech.*, 529, 221–251, 2005. doi: 10.1017/S0022112005003563.
26. Lin, J.R., Linear stability analysis of rotor-bearing system: couple stress fluid model. *Comput. Struct.*, 79, 801–809, 2001. doi: 10.1016/S0045-7949(00)00189-9.
27. Mahajan, A. and Nandal, R., On the stability of penetrative convection in a couple-stress fluid. *Int. J. Appl. Comput. Math.*, 3, 3745–3758, 2017. doi: 10.1007/s40819-017-0324-6.
28. Malashetty, M.S. and Swamy, M., The onset of double diffusive convection in a viscoelastic fluid layer. *J. Non-Newtonian Fluid Mech.*, 165, 19-20, 1129–1138, 2010. doi: 10.1016/j.jnnfm.2010. 05.011.
29. Muskat, M., The flow of fluids through porous media. *J. Appl. Phys.*, 8, 274–282, 1937. doi: 10.1063/1. 1710292.
30. Nield, D.A. and Bejan, A., *Convection in Porous Media*, Springer, New York, 2013.
31. Oskolkov, A.P., Initial-boundary value problems for equations of motion of Kelvin–Voight fluids and Oldroyd fluids. *Trudy Math. Inst. Imeni Steklov*, 179, 126–164, 1988.
32. Oskolkov, A.P., Nonlocal problems for the equations of motion of Kelvin-Voight fluids. *J. Math. Sci.*, 75, 2058–2078, 1995. doi: 10.1007/BF02362946.
33. Pan, Z., Jia, L., Mao, Y., Wang, Q., Transitions and bifurcations in couple stress fluid saturated porous media using a thermal non-equilibrium model. *Appl. Math. Comput.*, 415, 126727, 2022. doi: 10.1016/j. amc.2021.126727.
34. Qin, Y. and Kaloni, P.N., Nonlinear stability problem of a rotating porous layer. *Q. Appl. Math.*, 53, 129–142, 1995.
35. Rosenblat, S., Thermal convection in a viscoelastic liquid. *J. Non-Newtonian Fluid Mech.*, 21, 2, 201–223, 1986. doi: 10.1016/0377-0257(86)80036-2.
36. Roy, K., Ponalagusamy, R., Murthy, P.V.S.N., The effects of double-diffusion and viscous dissipation on the oscillatory convection in a viscoelastic fluid saturated porous layer. *Phys. Fluids*, 32, 2020. doi: 10.1063/5.0020076.

37. Shankar, B. and Shivakumara, I., Stability of natural convection in a vertical layer of Navier-Stokes-Voigt fluid. *Int. J. Heat Mass Transfer*, 144, 106783, 2023.
38. Shankar, B.M., Shivakumara, I.S., Ng, C.O., Stability of couple stress fluid flow through a horizontal porous layer. *J. Porous Media*, 19, 391–404, 2016. doi: 10.1615/JPorMedia.v19.i5.20.
39. Sharma, R.C., Sunil, Chand, S., Thermosolutal instability of Walters' rotating fluid (model b') in porous medium. *Arch. Mech.*, 51, 181–191, 1999.
40. Sharma, S., Sunil, Sharma, P., Stability analysis of thermosolutal convection in a rotating Navier–Stokes– Voigt fluid. *Z. Naturforsch. A*, 79, 2024. doi: 10.1515/zna-2023-0284.
41. Shivakumara, I.S. and Kumar, S.B.N., Linear and weakly nonlinear triple diffusive convection in a couple stress fluid layer. *Int. J. Heat Mass Transf.*, 68, 542–553, 2014. doi: 10.1016/j.ijheatmasstransfer.2013. 09.051.
42. Stokes, V.K., Couple stresses in fluids. *Phys. Fluids.*, 9, 1709–1715, 1966. doi: 10.1063/1.1761925.
43. Straughan, B., A sharp nonlinear stability threshold in rotating porous convection. *Proc. R. Soc. London, Ser. A*, 457, 87–93, 2001. doi: 10.1098/rspa.2000.0657.
44. Straughan, B., *The Energy Method, Stability, and Nonlinear Convection*, Springer Verlag, New York, 2004.
45. Straughan, B., Competitive double diffusive convection in a Kelvin–Voigt fluid of order one. *Appl. Math. Optim.*, 84, Suppl 1, 631–650, 2021a.
46. Straughan, B., Thermosolutal convection with a Navier–Stokes–Voigt fluid. *Appl. Math. Optim.*, 84, 2587– 2599, 2021b. doi: 10.1007/s00245-020-09719-7.
47. Straughan, B., Continuous dependence and convergence for a Kelvin–Voigt fluid of order one. *Ann. dell'Universita Chim. Ferrara*, 68, 49–61, 2022a. doi: 10.1007/s11565-021-00381-7.
48. Straughan, B., Instability thresholds for thermal convection in a Kelvin–Voigt fluid of variable order. *Rend. Circ. Mat. Palermo Ser. II*, 71, 187–206, 2022b. doi: 10.1007/s12215-020-00588-1.
49. Straughan, B., Nonlinear stability for convection with temperature dependent viscosity in a Navier–Stokes– Voigt fluid. *Eur. Phys. J. Plus*, 138, 438, 2023. doi: 10.1140/epjp/s13360-023-04055-5.
50. Straughan, B., Stability problems with generalized Navier–Stokes–Voigt theories. *Ann. dell'Universita Chim. Ferrara*, 70, 4, 1747–1766, 2024.
51. Sunil, and Devi, R., Global stability for thermal convection in a couple stress fluid: rigid boundaries. *Heat Transfer Res.*, 45, 43–57, 2014. doi: 10.1615/HeatTransRes.2013003569.
52. Sunil, Sharma, R.C., Chandel, R.S., Effect of suspended particles on couple-stress fluid heated and soluted from below in porous medium. *J. Porous Media*, 7, 9–18, 2004. doi: 10.1615/JPorMedia.v7.i1.20.

53. Sunil, Sharma, R.C., Divya, Thermosolutal convection in a ferromagnetic fluid saturating a porous medium. *J. Porous Media*, 8, 2005. doi: 10.1615/ JPorMedia.v8.i4.60.
54. Sunil, Devi, R., Mahajan, A., Global stability for thermal convection in a couple-stress fluid. *Int. Commun. Heat Mass Transf.*, 38, 938–942, 2011. doi: 10.1016/j.icheatmasstransfer.2011.03.030.
55. Sunil, Choudhary, S., Bharti, P.K., Global stability for thermal convection in a couple-stress fluid with temperature and pressure dependent viscosity. *Stud. Geophys. Geod.*, 35, 85–102, 2013. doi: 10.2478/ sgem-2013-0032.
56. Thakur, A., Kumar, S., Devi, R., The effect of rotation on ferroconvection in the presence of couple stress forces in porous medium: a nonlinear analysis. *Eur. Phys. J. Plus*, 139, 1–16, 2024a. doi: 10.1140/epjp/ s13360-024-05048-8.
57. Thakur, A., Sunil, Devi, R., The effect of couple stresses on stability analysis of magnetized ferrofluid saturating a porous medium heated from below. *J. Porous Media*, 27, 85–106, 2024b. doi: 10.1615/ JPorMedia.2023050054.
58. Vafai, K., *Handbook of Porous Media*, CRC Press, Boca Raton, 2015.
59. Weng, H.C., Couple-stress effect on the effective viscosity of magnetic fluids. *Physics Procedia*, 9, 63–67, 2010. doi: 10.1016/j.phpro.2010.11.016.
60. Whitaker, S., Advances in theory of fluid motion in porous media. *Ind. Eng. Chem.*, 61, 14–28, 1969. doi: 10.1021/ie50720a004.
61. Zvyagin, V. and Turbin, M., The study of initial-boundary value problems for mathematical models of the motion of Kelvin–Voigt fluids. *J. Math. Sci.*, 168, 157–308, 2010.

8

Convective Heat Transfer and Subcritical Dynamics in Rotating Ferrofluids with Couple Stresses in Porous Media Under Non-Equilibrium Conditions

Sunil[1]**, Akanksha Thakur**[1*] **and Reeta Devi**[2]

[1]Department of Mathematics and Scientific Computing, National Institute of Technology Hamirpur, Hamirpur, Himachal Pradesh, India
[2]Department of Mathematics, Govt. College Nagrota Bagwan, Kangra, Himachal Pradesh, India

Abstract

In this chapter, the applications of ferrofluids in porous media—such as targeted drug delivery, magnetic filtration, and thermal management—are examined through an investigation of convective heat transfer in rotating ferrofluids with couple stresses under local thermal non-equilibrium. A two-temperature model captures heat exchange between solid and liquid phases, while the Darcy–Brinkman model describes ferrofluid flow within the porous medium. Stability and convection onset are analyzed under free–free and rigid–rigid boundary conditions, utilizing linear and nonlinear approaches with normal mode and energy methods, respectively. The Galerkin method is employed to solve the eigenvalue problems, with Rayleigh number differences revealing a subcritical region where convection can initiate below the critical threshold. Key parameters, including magnetization, couple stresses, permeability, rotation, porosity-modified conductivity ratio, and interface heat transfer coefficient, are analyzed for their effects on convection patterns and subcritical behavior. This chapter provides insights into ferrofluid dynamics in porous structures, supporting advancements in targeted heat transfer, enhanced cooling systems, precision drug delivery, and filtration technologies, as well as broader applications in thermal energy management and environmental engineering.

**Corresponding author*: akankshathakur47@gmail.com; ORCID: 0009-0006-5837-8040
Sunil: ORCID: 0000-0001-6738-6815

Dhananjay Yadav, Mukesh Kumar Awasthi and Harith Mohamed Al-Azri (eds.) Non-Newtonian Fluids for Industrial Applications: Modeling and Simulations, (177–208) © 2026 Scrivener Publishing LLC

Keywords: Ferrofluids, convective heat transfer, porous media, couple stresses, local thermal non-equilibrium, subcritical convection, rotating systems, Darcy–Brinkman model

8.1 Introduction

Convection in ferrofluids, particularly when they saturate porous media, has garnered considerable interest in both theoretical and practical domains due to its multifaceted applications in engineering, biomedical systems, and energy technologies [16]. Ferrofluids, which are colloidal suspensions of ferromagnetic nanoparticles, exhibit unique magnetic properties that allow for enhanced control of heat transfer processes compared to traditional fluids. These properties enable ferrofluids to be effectively utilized in advanced cooling systems, drug delivery devices, and as innovative heat transfer fluids in microelectronics. The investigation of convective behavior in ferrofluids is essential for optimizing these applications, particularly in porous media, where the interactions between fluid dynamics, heat transfer, and solid structures create complex phenomena. A foundational reference for ferrofluid research is the work by Rosensweig [25], which provides a comprehensive analysis of ferrofluid mechanics and applications.

The convective stability of ferrofluids can be analyzed using both linear and nonlinear approaches. Linear analysis identifies the onset of instability, while nonlinear analysis provides insights into thresholds where sustained convection occurs. Chandrasekhar [5] extensively reviewed thermal instability and the influence of external forces, with the generalized energy method later developed by Galdi [11] and Galdi and Straughan [12]. Finlayson [10] conducted pioneering studies on ferrofluid convective stability using linear analysis, sparking a broad array of subsequent research. Nonlinear stability analysis for ferrofluids was initiated by Sunil and Mahajan [36], with recent studies like Mahajan and Parashar [18] examining stability in rotating, anisotropic ferrofluid layers, and Dhiman and Sood [9] focusing on oscillatory convection modes. While extensive research has explored ferrofluid stability under diverse influences [3, 14, 20, 29, 39], studies examining combinations of different boundary surface conditions remain limited. Notable exceptions include Dev and Suthar [7], who analyzed nonlinear penetrative convection with internal heating under mixed rigid and free boundary conditions.

In this chapter, convection in ferrofluids saturating a porous medium is examined using the local thermal non-equilibrium (LTNE) model.

The convective behavior of ferrofluids is significantly influenced by the thermal and boundary conditions, particularly under LTNE scenarios. To account for this, Nield and Bejan [21] provided a comprehensive detail of a two-field model for the energy equation, serving as a basic framework for studying local thermal non-equilibrium conditions. In this LTNE model, separate temperature equations are used for the fluid and solid phases, with coupling terms representing interphase heat transfer. Rees and Pop [24] provided thorough reviews of LTNE models, and recent studies have expanded the LTNE model's application across diverse scenarios. Furthermore, the study of LTNE effects has implications in fields such as geothermal energy extraction and the design of heat exchangers, where the understanding of thermal gradients plays a pivotal role. The thermal non-equilibrium effect has been investigated in various convective flows in saturated porous media by authors such as Rees and Pop [23], Banu and Rees [2], and Rees [22]. The integration of LTNE effects into the study of ferrofluid convection is crucial, as it enhances the accuracy of thermal predictions in engineering applications where precise heat management is critical. Only a few research works like, Sunil, Sharma, and Mahajan [40]; Shivakumara, Lee, Ravisha, and Reddy [39]; Sunil, Sharma, and Mahajan [41]; Hajialiloo and Soltanipour [13]; and Manjunatha, Rehman, Kumar, Varma, and Shatanawi [19] have applied the LTNE model to assess ferrofluid stability in porous media, under various conditions.

The complexities of ferrofluid convection are compounded by the presence of rotational forces and couple stresses. The introduction of rotation generates Coriolis forces that can either stabilize or destabilize the convective flow, depending on their interplay with buoyancy and magnetic forces. This rotational influence can lead to altered stability thresholds and modified convective patterns, which are critical for the effective application of ferrofluids in real-world scenarios [5]. Sunil and Mahajan [37] performed a nonlinear analysis for the stability of rotating magnetized ferrofluid. Sunil and Mahajan [38] performed a nonlinear analysis for the stability of rotating magnetized ferrofluid saturating a porous medium. Additionally, couple stresses—characterized by their microstructural properties—introduce another layer of complexity by impacting flow resistance and heat transfer characteristics. These couple stresses arise due to the interactions at the microlevel, leading to nonclassical fluid behaviors that must be accounted for in modeling approaches [33]. The simultaneous examination of rotational effects and couple stresses is vital for a comprehensive understanding of the convective dynamics in ferrofluids, as neglecting either aspect may lead to significant deviations from observed behaviors. Exploring the effects of couple stresses on ferroconvection has broad relevance, with applications in advanced heat transfer, biomedical applications of ferrofluids, microfluidic devices, mechanical systems, and space technology.

Various studies have analyzed the stability of couple stress fluids in specific setups. For instance, Sunil, Devi, and Mahajan [35] conducted linear and nonlinear stability analyses on couple stress fluids, noting a convergence in stability thresholds. Stability analyses in rigid boundaries were undertaken by Sunil and Devi [34]. Similarly, the effect of couple stresses on triple-diffusive convection was examined by Shivakumara and Kumar [28] through linear and weakly nonlinear methods. Research has also covered cases of natural convection in vertical fluid layers [26], penetrative convection in couple stress fluids [17], and, more recently, the influence of variable viscosity and couple stresses on fluids within different free and rigid boundary configurations [4, 6, 27]. Thakur, Kumar, and Devi [42]; Thakur, Sunil, and Devi [44] explored the nonlinear stability of magnetized ferrofluids in porous media with couple stresses across varied boundary setups. Thakur, Kumar, and Devi [42] recently investigated the combined effects of rotation and couple stresses on ferroconvection in porous media.

The existing literature suggests that the effect of couple stresses on convective heat transfer and subcritical behavior in rotating ferrofluids within porous media, particularly under LTNE conditions, remains largely unexplored. To bridge this gap, this study examines how couple stresses and thermal nonequilibrium influence convective heat transfer and subcritical behavior in rotating ferrofluids within a porous medium under two boundary conditions: free–free and rigid–rigid. The Darcy–Brinkman model is applied to describe fluid flow in porous media, while a two-temperature model captures the thermal interaction between the solid and fluid phases. Additionally, the study evaluates how different boundary surfaces affect convection, identifying the optimal surface type for convection in ferrofluid. Key parameters such as magnetization, couple stresses, medium permeability, rotation, porosity-modified conductivity ratio, and the interface heat transfer coefficient are comprehensively analyzed.

Section 8.2 formulates the governing equations of the problem and nondimensionalizes them. In Section 8.3, the nonlinear analysis is conducted on the perturbed nondimensional equations by introducing a generalized energy functional. Then Section 8.4 introduces the eigenvalue problem by using variational principle. The eigenvalue problem is then solved in Section 8.5 by the use of Galerkin method and obtained Rayleigh numbers are presented. Section 8.6 is devoted to the in-depth discussion on the results, and the effect of each parameter has been analyzed in detail. Furthermore, Section 8.7 presents all the concluding points, and Section 8.8 sheds light on the real-world applications of the results of this study.

To our knowledge, this study is the first to explore these combined effects within this framework, offering novel insights into ferrofluid behavior in porous media.

8.2 Formulation of the Problem

8.2.1 Geometrical Configuration and Governing Equations

Consider a horizontal porous layer of characteristic thickness d, saturated with an incompressible ferrofluid and subject to heating from below. This layer is confined by two impermeable walls that are assumed to extend infinitely in the horizontal direction. The Cartesian coordinate system (x, y, z) is oriented such that the x-axis aligns horizontally, and the z-axis points vertically upward, opposite to the direction of gravitational acceleration $\mathbf{g} = (0, 0, -g)$. A constant temperature gradient is maintained by setting the temperature at the lower surface to T_0 and at the upper surface to T_d, where ($T_d < T_0$), resulting in a uniform gradient of $\beta\left(=\left|\frac{dT}{dz}\right|\right)$ (see Figure 8.1).

This ferrofluid layer is subject to a uniform, vertically applied magnetic field, $\mathbf{H} = (0, 0, H)$, and is influenced by rotational effects due to an angular velocity $\Omega = (0, 0, \Omega)$ about the vertical axis. The system is governed by the Boussinesq approximation, where density variations contribute only to the buoyancy force, and the local thermal nonequilibrium model is used to account for temperature differences between the solid matrix and the fluid phase.

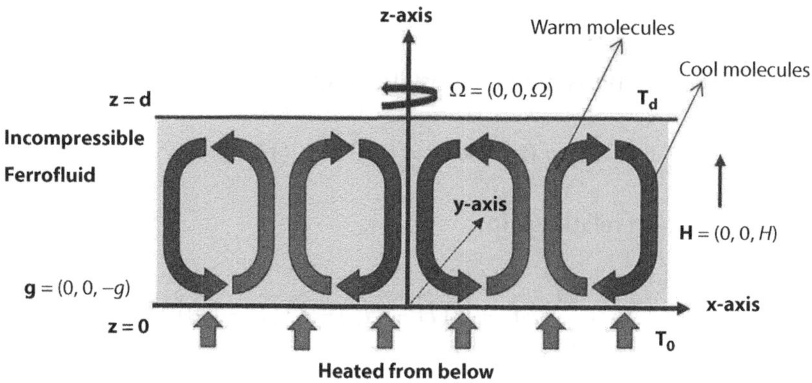

Figure 8.1 Geometrical configuration.

The governing equations for this incompressible, rotating magnetized ferrofluid in a porous medium, incorporating couple stresses and magnetic effects, are derived based on the contributions of prior works on ferrofluid dynamics [25], the stability of ferromagnetic fluids [10], the theory of couple stress fluids [32], and the stability theory in fluid mechanics [15]. The governing equations thus formulated are given as follows.

Mass balance:

$$\nabla \cdot \mathbf{q} = 0. \tag{8.1}$$

Momentum balance:

$$\frac{\rho_f}{\epsilon}\frac{\partial \mathbf{q}}{\partial t} = -\nabla p + \tilde{\mu}\nabla^2\mathbf{q} + \rho_f[1-\alpha(T-T_a)]\,\mathbf{g} + \mu_0(\mathbf{M}\cdot\nabla)\mathbf{H}$$
$$-\frac{1}{K}(\mu-\mu'\nabla^2)\mathbf{q} + \frac{2\rho_f}{\epsilon}(\mathbf{q}\times\boldsymbol{\Omega}). \tag{8.2}$$

Temperature equation (for fluid):

$$\epsilon(\rho C)_f\frac{\partial T_f}{\partial t} + (\rho C)_f\mathbf{q}\cdot\nabla \boldsymbol{T}_f = \epsilon k_f\nabla^2 T_f - h(T_s - T_f). \tag{8.3}$$

Temperature equation (for solid):

$$(1-\epsilon)(\rho C)_s\frac{\partial T_s}{\partial t} = (1-\epsilon)k_s\nabla^2 T_s + h(T_s - T_f). \tag{8.4}$$

Maxwell's equations in magnetostatic limit:

$$\nabla \cdot \mathbf{B} = 0,\ \nabla \times \mathbf{H} = 0,\ \mathbf{B} = \mu_0(\mathbf{H} + \mathbf{M}). \tag{8.5}$$

Magnetization relationship:

$$\mathbf{M} = \frac{\mathbf{H}}{H}[M_0 + \chi(H - H_0) - K_1(T - T_a)]. \tag{8.6}$$

Here, the subscript m denotes the mixture of solid and fluid, while f represents the fluid. $\tilde{\mu}$ refers to the effective viscosity, and **B** and B indicate the magnetic induction and its magnitude, respectively. C_0 is the specific heat at constant pressure, and K_1 denotes the pyromagnetic coefficient. Magnetization is represented by **M**, with M as its magnitude and M_0 as magnetization at magnetic field H_0. Fluid pressure is indicated by p, velocity of the ferromagnetic fluid by **q**, and time by t. Temperature is denoted by T, with T_a as the average temperature and T_0 and T_d as the constant average temperatures at the bottom and upper surfaces, respectively. The coefficient of thermal expansion is represented by α, medium porosity by ϵ, and the del operator by ∇. Thermal diffusivity is denoted by κ, and the thermal conductivity is denoted by κ_m. Dynamic viscosity (constant) is represented by μ', magnetic permeability of vacuum by μ_0, and couple stress viscosity by μ'. Kinematic viscosity is denoted by v, density by ρ, and reference density by ρ_0. Magnetic susceptibility is represented by χ. Finally, K is the permeability of the porous medium, k represents the thermal conductivity, and h is the interface heat transfer coefficient.

8.2.2 Basic State

The basic state defines a stable configuration with no flow, no rotation, and a steady magnetic field and temperature gradient across the layer. This configuration is essential for analyzing how the system responds to disturbances and for determining the onset of convective instability. The basic state is represented by

$$\mathbf{H_b}(z) = \left(H_0 - \frac{K_1\beta z}{1+\chi}\right)\hat{\mathbf{k}}, \quad \mathbf{M_b}(z) = \left(M_0 + \frac{K_1\beta z}{1+\chi}\right)\hat{\mathbf{k}},$$
$$\mathbf{q} = \mathbf{q}_b = (0,0,0),$$
$$\mathrm{p} = \mathrm{p}_b(z), \quad \beta = \frac{T_0 - T_d}{d}, \quad \rho = \rho_b(z) = \rho_f(1+\alpha\beta z),$$
$$T_f = T_s = -bz + T_a. \tag{8.7}$$

Here, the subscript "b" is used to denote the basic state.

8.2.3 Nondimensionalized Perturbation Equations

Now, small perturbations are introduced in the steady-state solutions. Perturbing the system is a fundamental approach in stability analysis that allows us to examine the behavior of a system under small disturbances or fluctuations. The following set of perturbations is introduced in the system:

$$\mathbf{q} = \mathbf{q}_b + \mathbf{q}', \quad p = p_b(z) + p', \quad \rho = \rho_b + \rho', \quad T_f = T_{bf}(z) + \theta,$$
$$T_s = T_{bs}(z) + \psi,$$
$$\mathbf{H} = \mathbf{H}_b(z) + \mathbf{H}', \quad \mathbf{H}' = \nabla\phi', \quad \mathbf{M} = \mathbf{M}_b(z) + \mathbf{M}',$$
$$\mathbf{M}' = (M_1', M_2', M_3'). \tag{8.8}$$

Here, perturbations in temperature for the solid phase, fluid phase, and in magnetic potential are represented by ψ, θ, and ϕ, respectively. The obtained set of perturbation equations is then nondimensionalized using the following nondimensional quantities (the notation * is used for nondimensional representation):

$$\phi^* = \frac{(1+\chi)R^{1/2}}{K_1\beta d^2}\phi', \; t^* = \frac{k_f}{d^2(\rho C)_f}t, \; \mathbf{q}^* = \frac{(\rho C)_f d}{\epsilon k_f}\mathbf{q}' \; \theta^* = \frac{R^{1/2}}{\beta d}\theta',$$
$$p^* = \frac{(\rho C)_f K}{\mu\epsilon k_f}p', \; \psi^* = \frac{R^{1/2}}{\beta d}\psi'.$$

Nondimensionalizing equations simplify mathematical models by consolidating parameters into dimensionless groups, which clarifies dominant physical effects, reduces complexity, and makes models applicable across different scales. This process highlights dominant forces, allows comparison across systems of different scales, and provides clearer physical insight. The perturbed equations in nondimensional form can be expressed as follows (removing asterisks for convenience):

$$\nabla \cdot \mathbf{q} = 0, \tag{8.9}$$

$$\begin{aligned} \frac{1}{V_a}\left(\frac{\partial \mathbf{q}}{\partial t}\right) &= -\nabla p + (F + \tilde{D}_a)\nabla^2\mathbf{q} - \mathbf{q} + R^{1/2}(1+M_1)\theta\hat{\mathbf{k}} \\ &- R^{1/2}M_1\phi_z\hat{\mathbf{k}} - M_1\theta\nabla\phi_z \\ &+ M_1\left(M_3 - \frac{1}{1+\chi}\right)\phi_x\nabla\phi_x + \left(\frac{\chi}{1+\chi}\right)M_1\phi_z\nabla\phi_z \\ &+ M_1\left(M_3 - \frac{1}{1+\chi}\right)\phi_y\nabla\phi_y + V_r\nabla^2\mathbf{q} + T_a^{1/2}(\mathbf{q}\times\hat{\mathbf{k}}), \end{aligned} \tag{8.10}$$

$$\frac{\partial \theta}{\partial t} + \mathbf{q}\cdot\nabla\theta = \nabla^2\theta + \mathcal{H}\gamma'(\psi - \theta) + R^{1/2}w, \tag{8.11}$$

$$A'\frac{\partial \psi}{\partial t} = \nabla^2\psi - \mathcal{H}\gamma'(\psi - \theta), \tag{8.12}$$

$$M_3\nabla^2\phi - (M_3 - 1)\phi_{zz} = \theta_z. \tag{8.13}$$

Here, the following well-established nondimensional parameters have been used to nondimensionalize the perturbed equations:

$$F = \frac{\mu'}{\mu d}, \quad R = \frac{g\alpha\beta\rho_f d^2 K(\rho C)_f}{\mu k_{f\epsilon}}, \quad M_1 = \frac{\mu_0 K_1^2 \beta}{\alpha\rho_f g(1+\chi)},$$

$$M_3 = \frac{1 + M_0/H_0}{(1+\chi)},$$

$$Vr = \frac{\tilde{\mu}}{\mu}, \tilde{D}a = \frac{\tilde{\mu}K}{\mu d^2}, Da = \frac{K}{d^2}, \quad V_a = \frac{\epsilon\nu(\rho C)_f d^2}{k_f K},$$

$$\gamma' = \frac{\epsilon k_f}{(1-\epsilon)k_s},$$

$$\mathcal{H} = \frac{hd^2}{\epsilon k_f}, \quad A' = \frac{(\rho C)_s k_f}{(\rho C)_f k_s}, \quad T_a = \left(\frac{2\Omega d^2}{\kappa\epsilon}\right)^2,$$

$$R_{a_{LTE}} = \left(\frac{\gamma'}{1+\gamma'}\right)R_a.$$

Here, the parameters are defined as follows: F represents the couple stress parameter, R the Rayleigh number, M_1 the ratio of magnetic to gravitational forces, and Vr the ratio of effective viscosity to dynamic viscosity. The Darcy number is denoted by $\tilde{D}_a$, while D_a indicates the Darcy–Brinkman number. V_a is the Vadasz number, γ' the porosity-modified conductivity ratio, $\mathcal{H}$ the interface heat transfer coefficient, and A' the diffusivity ratio. Finally, T_a represents the Taylor number, and M3 is the nonbuoyancy magnetization parameter.

The boundary conditions for the system are

$$w = 0, u_z = 0, v_z = 0, \theta = 0, \psi = 0, \phi_z = 0 \text{ at } z = 0, 1. \tag{8.14}$$

The curl of the velocity vector yields the vorticity, which represents the rotational motion within the fluid. By analyzing the third component of the vorticity equation, we gain insight into the behavior of rotational effects and the development of vertical structures, such as convective rolls or cells. Thus, operating $\hat{\mathbf{k}}$ *curl* and $\hat{\mathbf{k}}\cdot$*curl curl* on Equation (8.10), we obtain

$$\frac{1}{Va}\left(\frac{\partial \zeta}{\partial t}\right) = \zeta + (F+\tilde{D}_a)\nabla^2\zeta + M_1(\theta_y\phi_{zx} - \theta_x\phi_{zy}) + T_A^{1/2}w_z, \quad (8.15)$$

$$\begin{aligned}\frac{1}{Va}\frac{\partial \nabla^2 w}{\partial t} &= R^{1/2}(M_1+1)\nabla_1^2\theta + (F+\tilde{D}_a)\nabla^4 w - \nabla^2 w - M_1R^{1/2}\nabla_1^2\phi_z \\ &+M_1\left(-\nabla_1^2\theta\phi_{zz} + \theta_z\nabla_1^2\phi_z - \theta_x\phi_{zzx} + \theta_{zx}\phi_{zx} - \theta_y\phi_{zzy} + \theta_{zy}\phi_{zy}\right) - T_A^{1/2}\zeta_z,\end{aligned} \quad (8.16)$$

From Equations 8.11 and 8.12, we have

$$\frac{\partial \theta_z}{\partial t} + \mathbf{q_z}\cdot\nabla\theta + \mathbf{q}\cdot\nabla\theta_z = \nabla^2\theta_z + \mathcal{H}\gamma'(\psi_z - \theta_z) + R^{1/2}w_z, \quad (8.17)$$

$$A'\frac{\partial \psi_z}{\partial t} = \nabla^2\psi_z - \mathcal{H}\gamma'(\psi_z - \theta_z), \quad (8.18)$$

Equations 8.15–8.18 are of higher order than Equations 8.9–8.13, so the extra boundary conditions required for the system are given as

$$\theta_{zz} = 0,\ \psi_{zz} = 0,\ \zeta_z = 0,\ w_{zz} = 0 \text{ at } z = 0,1. \quad (8.19)$$

8.3 Nonlinear Analysis

The Rayleigh number obtained through nonlinear analysis using the energy method represents the critical threshold below which stability of the solution is ensured. In this nonlinear analysis, the nonlinear terms in Equations 8.15–8.18 and Equation 8.13 are retained, allowing for the consideration of their potential growth. To initiate the nonlinear analysis, an L^2 energy function $E(t)$ is defined, such that

$$E = \frac{1}{2}\|\theta_z\|^2 + \frac{\lambda_1}{2Va}\|\nabla w\|^2 - \frac{\lambda_3}{2Va}\|\zeta\|^2 + \frac{A'}{2}\|\psi_z\|^2, \quad (8.20)$$

where I_0, D_0, and No, are expressed as

$$I_0 = \lambda_1(1+M_1)R^{1/2}\langle\nabla_1 w\nabla_1\theta\rangle - R^{1/2}\lambda_1 M_1\langle\nabla_1 w\nabla_1\phi_z\rangle$$
$$+T_A^{1/2}\lambda_1\langle w\zeta_z\rangle - T_a^{1/2}\lambda_3\langle\zeta w_z\rangle - \lambda_2\langle\phi\theta_z\rangle + R^{1/2}\lambda_1\langle\theta_z w_z\rangle, \quad (8.21)$$

$$D_0 = \|\nabla\theta_z\|^2 + \|\nabla\psi_z\|^2 + \lambda_1(F+D_a)\|\nabla^2 w\|^2 + \lambda_1\|\nabla w\|^2$$
$$-\lambda 3(F+Da)\|\nabla\zeta\|^2$$
$$-\lambda_3\|\zeta\|^2 - \lambda_2 M_3\|\nabla\phi\|^2 - \lambda_2(M_3-1)\|\phi_z\|^2 + \mathcal{H}\|\theta_z - \psi_z\|^2,$$
$$(8.22)$$

and

$$N_0 = \lambda_1 M_1\left(-\left\langle w\phi_{zz}\nabla_1^2\theta\right\rangle + \left\langle w\theta_z\nabla_1^2\phi_z\right\rangle - \langle w\theta_x\phi_{zzx}\rangle\right)$$
$$+\lambda_1 M_1(\langle w\theta_{zx}\phi_{zx}\rangle + \langle w\theta_y\phi_{zzy}\rangle + \langle w\theta_{zy}\phi_{zy}\rangle) + \langle\mathbf{q}.\theta_{zz}\nabla\theta\rangle$$
$$-\lambda_3 M_1(\langle\zeta\theta_y\phi_{zx}\rangle - \langle\zeta\theta_x\phi_{zy}\rangle). \quad (8.23)$$

Here, λ_1, λ_2, and λ_3 are three positive coupling parameters, $\langle\cdot\rangle$ represents integration over volume V, $\|\cdot\|$ represents the $L^2(V)$ norm, and V represents the standard periodicity cell. I_0, D_0, and N_0 represent the nondefinite production, positive-definite dissipation terms, and nonlinear terms, respectively. The variation in $E(t)$ is expressed as follows:

$$\frac{dE}{dt} = I_0 - D_0 + N_0.$$

Now, we define a positive constant m, such that

$$m = \max_H \frac{I_0}{D_0}, \quad (8.24)$$

where H denotes the space that contains all the possible solutions, and take

$$\frac{dE}{dt} \leq -a_0 D_0 + N_0,$$

where $a_0 = 1 - m$.

8.3.1 Generalized Energy Functional

To control the cubic nonlinear terms in N_0, dissipation terms of higher order need to be introduced. So, the second component of energy is defined as

$$E_1(t) = \frac{1}{2\mathcal{H}}\|\nabla\theta\|^2 + \frac{1}{2Va}\|\nabla\mathbf{q}\|^2 + \frac{A'}{2\mathcal{H}\gamma'}\|\nabla\psi\|^2.$$

Now, taking

$$I_1 = R^{1/2}\left(1 + M_1 + \frac{1}{\mathcal{H}}\right)\langle\nabla w\cdot\nabla\theta\rangle - R^{1/2}M_1\langle\nabla w\cdot\nabla\phi_z\rangle + 2\langle\nabla\psi\cdot\nabla\theta\rangle, \tag{8.25}$$

$$D_1 = (F + Da)\|\nabla^2\mathbf{q}\|^2 + \frac{1}{\mathcal{H}}\|\nabla^2\theta\|^2 + \|\nabla\theta\| + \frac{1}{\mathcal{H}}\|\nabla^2\psi\|^2 + \|\nabla\psi\| \tag{8.26}$$

and

$$\begin{aligned} N_1 = {} & \frac{1}{\mathcal{H}}\langle\nabla^2\theta\mathbf{q}\cdot\nabla\theta\rangle + M_1\langle\theta\nabla^2\mathbf{q}\cdot\nabla\phi_z\rangle - M_1\left(M_3 - \frac{1}{1+\chi}\right) \\ & \langle\phi_x\nabla^2\mathbf{q}\cdot\nabla\phi_x\rangle \\ & - M_1\left(M_3 - \frac{1}{1+\chi}\right)\langle\phi_y\nabla^2\mathbf{q}\cdot\nabla\phi_y\rangle - M_1\left(\frac{\chi}{1+\chi}\right)\langle\phi_z\nabla^2\mathbf{q}\cdot\nabla\phi_z\rangle, \end{aligned} \tag{8.27}$$

an evolution equation for E_1 is found as

$$\frac{dE_1}{dt} \le I_1 - D_1 + N_1.$$

Then, a generalized energy functional is introduced, having the following form [12]:

$$V_g(t) = E(t) + b_0 E_1(t), \tag{8.28}$$

where b_0 is a positive parameter used for coupling the two energy terms. The change in the generalized energy functional $V_g(t)$ is given by

$$\frac{dV_g}{dt} \leq -a_0 D_0 + N_0 + b_0 I_1 - b_0 D_1 + b_0 N_1. \tag{8.29}$$

Now, by using Equation 8.10 and embedding theorems, the following results are derived:

$$\begin{aligned}
&\|\nabla\phi\| \leq \|\theta\|, \|\nabla\phi_z\| \leq \|\nabla^2\phi\| \leq \|\nabla\theta\|, \|\nabla^2\phi_z\| \leq \|\nabla^3\phi\| \leq \|\nabla\theta_z\| \leq \|\nabla^2\theta\|,\\
&\|\nabla^2\phi_x\| \leq \|\nabla^2\theta\|, \|\nabla^2\phi_y\| \leq \|\nabla^2\theta\|, \|\nabla w\| \leq \|\nabla q\|,\\
&\sup|G| \leq C^* \|\nabla^2 G\|, G \in \{\mathbf{q},\theta,\phi_x,\phi_y.\phi_z\}.
\end{aligned}$$

Here, $C^* > 0$ is a constant, which depends entirely on V, and it was determined by Galdi and Straughan [12], and proof of the statement was given by Adams [1]. Now, the following inequality can be obtained from Equation 8.25 using Equations 8.26 and 8.22, embedding theorems, and some standard inequalities:

$$b_0 I_1 \leq 2b_0\varepsilon_0^2 D_1 + \frac{b_0 R}{2\varepsilon_0^2}\left\{\left(1+M_1+\frac{1}{\mathcal{H}_1}\right)^2 + M_1^2\right\} + 2\frac{b_0 R}{2\varepsilon_0^2} D_0. \tag{8.30}$$

Here, $\varepsilon_0^2 = \frac{1}{4}, b_0 = \dfrac{a_0}{4R\left\{\left(1+M_1+\frac{1}{\mathcal{H}_1}\right)^2 + M_1^2\right\}+8}$. and

$$D_2 = \frac{a_0}{2} D_0 + \frac{b_0}{2} D_1. \tag{8.31}$$

Using Equation 8.31 in Inequality 8.30, we get

$$b_0 I_1 < D_2. \tag{8.32}$$

The estimation of the nonlinear terms N_0 and N_1, employing Equations 8.28 and 8.31, and embedding theorems is given as

$$N_1 \leq C^* \left(\frac{2}{b_0}\right)^{1/2} \left(\frac{2}{b_0}\right) \left[\frac{1}{\mathrm{Pr}^{1/2}} + \left\{\frac{1}{\mathcal{H}} + M_1 \left(2M_3 + \frac{2\chi - 1}{1+\chi}\right)\right\}\right] D_2 V_g^{1/2} \tag{8.33}$$

and

$$N_0 \leq 2C^* \frac{\lambda_1 M_1}{b_0} \left(\frac{2}{Da}\right)^{1/2} \left[\left(2M_3 + \frac{2\chi - 1}{1+\chi}\right)\right] D_2 V_g^{1/2}. \tag{8.34}$$

The change in $V_g(t)$ given by Inequality 8.29 can now be represented as (by the virtue of Equation 8.31 and Inequality 8.32)

$$\dot{V}_g(t) \leq -D_2(1 - H' V_g^{1/2}), \tag{8.35}$$

where

$$H' = 2C^* \left(\frac{2}{b_0 Da}\right)^{3/2} \left[\left\{1 + M_1 \left(2M_3 + \frac{2\chi - 1}{1+\chi}\right)\right\} \left(\mathcal{H} + \frac{\lambda_1}{b_0^{1/2}}\right)\right]. \tag{8.36}$$

Theorem 1. *If* $0 < m < 1$ *and* $V_g(0) < H'^{-2}$, *then* $\exists\, K^* > 0$, *such that*

$$V_g(t) \leq V_g(0)\exp(-K^*(1 - AH' V_g^{1/2}(0))t), \quad t \geq 0, \tag{8.37}$$

with H' given by Equation 8.36.

Proof. From Inequality 8.35, it can be obtained that

$$\dot{V}_g(0) < 0. \tag{8.38}$$

Therefore, by following the procedure of Thakur, Sunil, and Devi [44], we can get

$$V_g(t) \leq V_g(0)\exp(-K^*(1 - AH' V_g^{1/2}(0))t), \quad t \geq 0, \tag{8.39}$$

where K^* is given by

$$K^* = \frac{1}{\left(1 + A' + \frac{2}{\epsilon V_a Da}\right)(1 + k_0)}, \tag{8.40}$$

which gives the required theorem.

It is clear from Equation 8.28 that $V_g(t)$ does not contain $\|\nabla\phi\|^2$ (kinetic energy term); so, to observe its behavior, embedding theorems are used and we obtain

$$\|\nabla\phi\|^2 \le \|\theta\|^2. \tag{8.41}$$

Thus, Equation 8.41 and Inequality 8.39 together establish the decay of the kinetic energy.

8.4 Variational Principle

The variational principle is used to find an optimal solution of a system by minimizing or maximizing the functional defined in Equation 8.24, which is an expression for the total energy of the system. This principle underlies many physical theories, as it enables the derivation of the eigenvalue problem by seeking stationary points (minima, maxima, or saddle points) of the functional—meaning the point where small variations of the system's state produce no first-order change in the functional's value. Equation 8.24 is solved for the maximization problem, and the transformations $\hat{\mathbf{q}} = \sqrt{\lambda_1}\mathbf{q}$, $\hat{\phi} = \sqrt{\lambda_2}\phi$ **and** $\hat{\zeta} = \sqrt{\lambda_3}\zeta$ are applied. Thus, the obtained Euler–Lagrange equations can be expressed as follows (omitting the hats):

$$\begin{aligned} &R^{1/2}\lambda_1^{1/2}(1+M_1)\nabla_1^2\theta - \frac{\lambda_1^{1/2}}{\lambda_2^{1/2}}R^{1/2}M_1\nabla_1^2\phi_z - \frac{\lambda_1^{1/2}}{\lambda_3^{1/2}}T_a^{1/2}\zeta_z \\ &\quad - \frac{\lambda_3^{1/2}}{\lambda_1^{1/2}}T_A^{1/2}\zeta_z + \frac{R^{1/2}}{\lambda_1^{1/2}}\theta_{zz} \\ &\quad + 2(F + D_a)\nabla^4 w - 2\nabla^2 w = p, \end{aligned} \tag{8.42}$$

$$(1 + \lambda_1(1 + M_1))\frac{R^{1/2}}{\lambda_1^{1/2}}w + \lambda_2^{1/2}\phi_z + 2\nabla^2\theta_{zz} - 2\mathcal{H}(\theta_{zz} - \psi_{zz}) = 0, \tag{8.43}$$

$$\frac{2}{\gamma'}\nabla^2\psi_{zz} - 2\mathcal{H}\left(\psi_{zz} - \theta_{zz}\right) = 0, \tag{8.44}$$

$$\frac{\lambda_1^{1/2}}{\lambda_2^{1/2}} M_1 R^{1/2} w_z - \lambda_2^{1/2}\theta_z + 2M_3\nabla^2\phi - 2\left(M_3 - 1\right)\phi_{zz} = 0. \tag{8.45}$$

$$2\left(F + D_a\right)\nabla^2\zeta - 2\zeta - \frac{\lambda_1^{1/2}}{\lambda_3^{1/2}} T_a^{1/2} w_z - \frac{\lambda_3^{1/2}}{\lambda_1^{1/2}} T_A^{1/2} w_z = 0 \tag{8.46}$$

Here, the symbol p represents Lagrange's multiplier. A plane covering form

$$(w, \theta, \phi, \psi, \zeta) = \left[W(z), \Theta(z), \Phi(z), \Psi(z), \zeta(z)\right] f(x, y), \tag{8.47}$$

is considered where

$$\nabla_1^2 f + a^2 f = 0,$$

and a is the wave number.

Using Equation 8.47 in Equations 8.42–8.46, we get

$$2\left(F + \tilde{D}a\right)\left(D^2 - a^2\right)^2 W - 2\left(D^2 - a^2\right) W - \frac{R^{1/2}}{\lambda_1^{1/2}}\left(1 + \lambda_1\left(1 + M_1\right)\right) a^2\Theta$$
$$+ \frac{\lambda_1^{1/2}}{\lambda_2^{1/2}} R^{1/2} M_1 a^2 D\Phi = 0, \tag{8.48}$$

$$\left(1 + \lambda_1\left(1 + M_1\right)\right)\frac{R^{1/2}}{\lambda_1^{1/2}} W + \lambda_2^{1/2} D\Phi + 2\left(D^2 - a^2\right)\Theta - 2\mathcal{H}\Theta + 2\mathcal{H}\Psi = 0, \tag{8.49}$$

$$\frac{2}{\gamma'}\left(D^2 - a^2\right)\Psi - 2\mathcal{H}\Psi + 2\mathcal{H}\Theta = 0, \tag{8.50}$$

$$\frac{\lambda_1^{1/2}}{\lambda_2^{1/2}} M_1 R^{1/2} DW - \lambda_2^{1/2} D\Theta + 2M_3\left(D^2 - a^2\right)\Phi - 2\left(M_3 - 1\right) D^2\Phi = 0, \tag{8.51}$$

$$2\left(F+D_a\right)\left(D^2-a^2\right)\Gamma-2\Gamma-\frac{\lambda_1^{1/2}}{\lambda_3^{1/2}}T_a^{1/2}DW-\frac{\lambda_3^{1/2}}{\lambda_1^{1/2}}T_A^{1/2}DW=0 \quad (8.52)$$

with boundary conditions

1. Free–free bounding surfaces:

$$W = D^2W = \Theta = \Psi = D\Phi = D\Gamma = 0,\ \text{at}\ z = 0,1. \quad (8.53)$$

2. Rigid–rigid bounding surfaces:

$$W = DW = \Theta = \Psi = D\Phi = \Gamma = 0,\ \text{at}\ z = 0,1. \quad (8.54)$$

Free bounding surfaces imply that the fluid can move freely at the boundaries, meaning no tangential forces or shear stresses act on these surfaces, allowing slip conditions. In contrast, rigid bounding surfaces allow for tangential forces to act on the system, permitting some degree of fluid motion along the surfaces. While the free–free boundary case is theoretically useful for qualitative insights [31], practical systems generally involve rigid boundary conditions, which are more physically relevant.

To perform the linear analysis, we begin by taking Equations 8.9 through 8.13 and removing the nonlinear terms. Following this, we apply normal mode analysis to formulate the corresponding eigenvalue problem, maintaining the same boundary conditions.

8.5 Method of Solution

Now, following the single-term weighted Galerkin method, as suggested by Finlayson [10] on Equations 8.48–8.51 with boundary conditions 8.53 and 8.54 for respective boundary conditions. The key idea behind the Galerkin method is to reduce a continuous problem to a discrete one by expressing the solution as a series of basis functions. These functions, chosen based on boundary conditions and the problem's specific features, approximate the actual solution. In the Galerkin method, the governing differential equation is multiplied by a set of test functions (often the same as the basis functions) and then integrated over the domain. This integration yields a system of algebraic equations for the coefficients of the basis functions. By solving this system, an approximate solution to the original problem is obtained. The Galerkin method is advantageous because it ensures that the residuals—the errors between the approximate and actual solutions—are orthogonal to the chosen basis functions.

8.5.1 Free–Free Boundaries

The suitable trial functions satisfying the boundary condition 8.53 are

$$W = \sin i\pi z,\ \Theta = \sin i\pi z,\ \Psi = \sin i\pi z,\ \Phi = \cos i\pi z. \tag{8.55}$$

Using solution 8.55 in Equations 8.48–8.51 and opting the same procedure as done in Thakur, Sunil, and Devi [44], we obtain

$$N_e = M_1 R_e = \frac{(1+x)(1+x+(1+\gamma)\mathcal{H}_1)\left((1+(1+x)(F_1+\widehat{D}a))^2(1+x)+T_1\right)(3+4xM_3)}{2x(1+x+\gamma\mathcal{H}_1)\left(1+(1+x)(F_1+\widehat{D}a)\right)(1+2xM_3)}. \tag{8.56}$$

for large values of M_1. Here,

$$\lambda_1 = \lambda_3 = \frac{1}{(1+M_1)} \text{ and } \lambda_2' = \frac{(1+x)\, M_1 \left[1+x+\mathcal{H}'(1+\gamma')\right]}{1+M_1\,[1+x+\mathcal{H}'\gamma']}.$$

and

$$Re = \frac{R}{\pi^2},\ x = \frac{a^2}{\pi^2},\ F_1 = \pi^2 F,\ \widehat{D}a = \pi^2 \tilde{D}_a,\ T_1 = \pi^2 T_a,\ H_1 = \frac{\pi^2}{H},\ \text{and } \lambda_2' = \frac{\lambda_2}{\pi^2}.$$

When $F_1 = 0$, this result (Equation 8.56) aligns entirely with the findings presented by Sunil *et al.* [40]. For linear analysis,

$$N_l = M_1 Re = \frac{(1+x)\,(1+x+(1+\gamma)\mathcal{H}_1)\left((1+(1+x)(F_1+\widehat{D}_a))^2(1+x)+T_1\right)(1+xM_3}{x^2(1+x+\gamma\mathcal{H}_1)\left(1+(1+x)(F_1+\widehat{D}_a)\right)M_3} \tag{8.57}$$

8.5.2 Rigid–Rigid Boundaries

The suitable trial functions satisfying the boundary condition 8.54 are

$$W_i(z) = z^{2i+2} - 2z^{2i+1} + z^{2i},\ \Theta_i(z) = z^{2i} - z^i,\ \Psi_i(z) = z^{2i} - z^i,\ \Phi_i(z) = 3z^{2i} - 2z^{2i+1}. \tag{8.58}$$

Using the abovementioned solution in Equations 8.48–8.51 and employing the same series of steps as for free–free bounding surfaces, we obtain

$$\begin{aligned}N_e = {} & (20999+42334x+21335x^2+520x(168+338x+17x^2)M_3 \\ & +H_1(21335+20999\gamma+21335x(1+\gamma)+1040x(85+84\gamma+85x(1+\gamma))M_3)) \\ & (2(1764+2100x+950x^2+25x^3)(F_1+\widehat{D}_a)^2+4(F_1+\widehat{D}_a)(840+625x+250x^2) \\ & +25(24+20x+3T_1^2))/(15139x(1+x+\gamma H_1)(5+(7+5x)(F_1+\widehat{D}_a))(93+572xM_3))\end{aligned} \tag{8.59}$$

and

$$\begin{aligned}N_l = {} & ((84+85x)(84+85x+85(1+\gamma)\mathcal{H}_1)(42+130xM_3)(0.14(4+(1+4x)(F_1+\widehat{D}_a)) \\ & (540+475x+(1134+1080x+475x^2)(F_1+\widehat{D}_a))+336T_1))/1300x(84 \\ & +85x+85\gamma\mathcal{H}_1)(4+(1+4x)(F_1+\widehat{D}_a)6+6175xM_3.\end{aligned} \tag{8.60}$$

8.6 Results and Discussion

The primary objective of this study is to examine how the couple stress parameter F_1, magnetic parameter M_3, medium permeability Da, rotation T_1, porosity-modified conductivity ratio γ', and heat transfer coefficient $\mathcal{H}_1$ influence the stability of ferroconvection in porous media. Also, we want to see the possibility of a subcritical region and how it is affected by different parameters. In this section, we derived expressions for both linear and non-linear analyses under two different boundary conditions. These expressions demonstrate that the stability thresholds, characterized by N_{cl} and N_{ce}, are directly dependent on the parameters F_1, M_3, Da, T_1, γ', and $\mathcal{H}_1$. Variations in these parameters significantly alter the values of N_{cl} and N_{ce}, thus determining the stability of the ferroconvection system. The effects of these parameters are illustrated in Figures 8.2–8.8. The effect of each of these parameters and the significance of these parameters are discussed below.

8.6.1 Effect of Couple Stresses

Couple stresses present in the system influence the stability of the system. The range of couple stress parameter F_1 is taken from 0 to 1, with 0

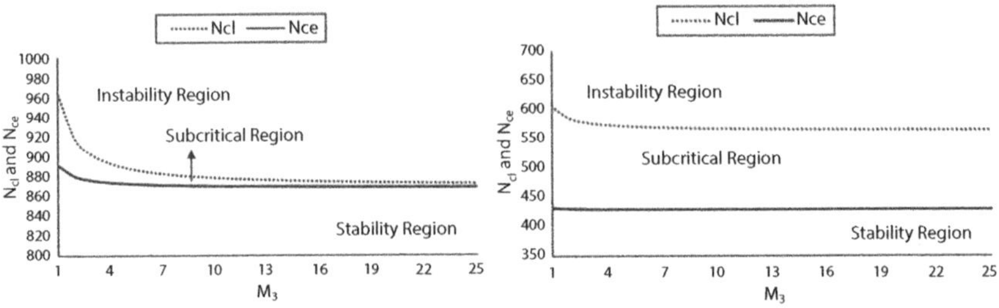

(**a**) Free-free boundaries (**b**) Rigid-rigid boundaries

Figure 8.2 Variations of N_{cl} and N_{ce} with respect to magnetic parameter M_3 for two distinct boundary surfaces for $F_1 = 0.4$, $T_1 = 100$, $D_a = 0.03$, $\gamma' = 0.001$, and $\mathcal{H}_1 = 100$.

indicating the absence of couple stresses [32]. The range can be increased, but the trends observed are the same for larger values too; so we focus on the smaller range for clarity. Figure 8.3 shows the variations in N_{cl} and N_{ce} when F_1 is varied, while keeping the parameters M_3, T_1, Da, γ', and $\mathcal{H}_1$ constant. M_3 is selected from the range specified by Finlayson [10], and the values of M_3, T_1, Da, and $\mathcal{H}_1$ are chosen to be sufficiently small to ensure that they are present in the system while minimizing their impact on the stability region. The range of couple stress parameter F_1 is taken from 0 to 1, with 0 indicating the absence of couple stresses [32]. The results obtained by varying F_1 from 0 to 1 in the expressions obtained in Section 8.5 are presented in Figure 8.3. Figure 8.3a represents the variations of N_{cl} and N_{ce} concerning F_1 for free–free bounding surfaces and Figure 8.3b for rigid–rigid bounding surfaces. Figure 8.3a shows that N_{cl} and N_{ce} increase for increasing F_1, indicating that F_1 stabilizes the system. Also, Figure 8.3a makes it evident that N_{cl} and N_{ce} do not match. The difference between

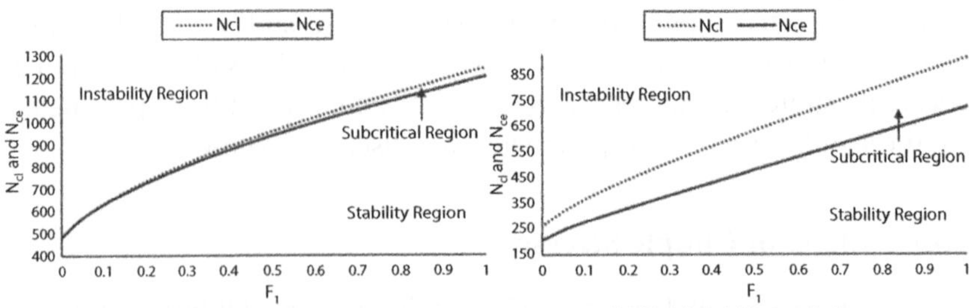

(**a**) Free-free boundaries (**b**) Rigid-rigid boundaries

Figure 8.3 Variations of N_{cl} and N_{ce} with respect to couple stress parameter F_1 for two distinct boundary surfaces for $M_3 = 5$, $D_a = 0.03$, $T_1 = 100$, $\gamma = 0.01$, and $\mathcal{H}_1 = 100$.

these thresholds shows that there is a subcritical region in Figure 8.3a that widens as F_1 increases. For the other bounding surface, Figure 8.3b shows similar trends. For both of these surfaces, F_1 stabilizes the system, and the subcritical region is present. The size of the subcritical region changes when there is a change in the value of F_1.

As vividly depicted in Figures 8.3a and b, an increase in F_1 results in the enlargement of the subcritical region. This effect of the increase in the parameter F_1 is the same for all bounding surfaces, but for rigid-bounding surfaces, the increase in the subcritical region is the largest. As the bounding surfaces are changed to free, this increase in the subcritical region reduces. This implies that free bounding surfaces reduce the difference between linear and nonlinear thresholds and thus shrink the size of the subcritical region. Furthermore, the increase in Rayleigh numbers with higher F_1 values is notably more prominent for free boundary surfaces, suggesting that these boundaries offer greater thermal stability for convection. Moreover, the stabilizing influence of F_1 is apparent throughout Figures 8.3a and b, remaining consistent regardless of the characteristics of the boundary surfaces. Thus, it can be concluded that couple stresses increase the stability of the system.

This behavior implies that manipulating the couple stress parameter F_1 and the rigidity of boundary surfaces can be a strategic approach to enhancing the thermal stability of ferroconvection systems. The pronounced increase in Rayleigh numbers with higher F_1 values, especially for free boundaries, underscores the potential of these findings for practical applications where greater thermal stability is desired.

The observed stabilizing influence of F_1 aligns with the findings of Devi, Sunil, and Paul [8], yet the insights regarding the interaction between boundary conditions and F_1 present a novel avenue for optimizing system stability. These results suggest that by carefully selecting the boundary conditions and tuning F_1, one can achieve a more stable ferroconvection system, which could be of considerable importance in the design and control of such systems in practical applications.

8.6.2 Effect of Magnetization

The parameter M_3 represents the effect of magnetization and signifies the deviation of the magnetic equation of state from linearity, and M_3 is assumed to range between 1 and 25 [10]. Typically, M_3 is approximately equal to 1 (but not exactly), and thus, larger values of M_3 are often considered in this context. To study the effect of magnetization, M_3 was varied from 1 to 25 in the expressions for N_{cl} and N_{ce} obtained in Section 8.5, while keeping the parameters F_1, Da, γ', T_1, and $\mathcal{H}_1$ constant to isolate the influence of M_3. The numerical results are presented graphically in Figure 8.2.

Figure 8.2a shows the variation of N_{cl} and N_{ce} with respect to M_3 for free–free bounding surfaces. It is evident that increasing M_3 decreases both N_{cl} and N_{ce}, indicating a destabilizing effect. The instability region, marked by N_{cl} from linear theory, does not coincide with the stability region defined by N_{ce} from nonlinear theory, suggesting the presence of a subcritical region. Moreover, as M_3 increases, this subcritical region shrinks.

Similarly, Figure 8.2b depicts the scenario for rigid–rigid bounding surfaces, showing that both N_{cl} and N_{ce} decrease with increasing M_3, further destabilizing the system. Again, a subcritical region is present, shrinking as M_3 increases.

In conclusion, M_3 consistently shows a destabilizing effect across both boundary conditions, as illustrated in Figures 8.2a and b. The subcritical region is more pronounced at lower values of M_3 and diminishes as M_3 increases. This reduction in N_{cl} and N_{ce} is particularly significant in free–free boundary surfaces, indicating an earlier onset of convection in these cases.

The destabilizing influence of M_3 aligns with the findings of Sunil and Mahajan [36], confirming that magnetization induces destabilization regardless of boundary conditions, including cases involving couple stresses and porous media. However, Sunil and Mahajan [36] did not explore the effect of M_3 on ferroconvection with rigid–rigid boundaries. The present analysis, therefore, provides important insights into ferroconvection under two different boundary conditions.

8.6.3 Effect of Medium Permeability

The Darcy number Da quantifies how the medium's pore structure influences fluid flow. A higher permeability indicates that the medium allows fluid to pass through more easily, while a lower permeability means that the medium is more resistant to fluid flow. The medium permeability parameter Da is chosen within the range of 0 to 1, as per Nield and Bejan [21]. Da is varied from 0 to 1 in the expressions of N_{cl} and N_{ce} while keeping the parameters M_3, F_1, T_1, γ', and $\mathcal{H}_1$ constant. The results are depicted in Figure 8.4 for all two bounding surfaces. Figure 8.4a shows that when the medium permeability of the system increases, the corresponding thresholds for stability increase. This suggests that the system is stabilized by increasing Da. In Figure 8.4a, a subcritical region is observed as there is a significant difference between N_{cl} and N_{ce}. This subcritical region widens

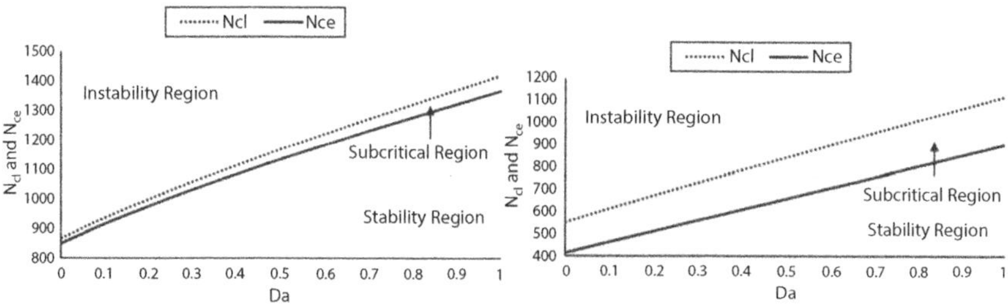

(**a**) Free-free boundaries (**b**) Rigid-rigid boundaries

Figure 8.4 Variations of N_{cl} and N_{ce} with respect to D_a for two distinct boundary surfaces for $M_3 = 5$, $F_1 = 0.4$, $T_1 = 100$, $\gamma = 0.01$, and $\mathcal{H}_1 = 100$.

as the permeability of the system increases. The stabilizing nature of *Da* is seen in Figure 8.4b also. For all considered bounding surfaces, shown in Figures 8.4a and b, as N_{cl} and N_{ce} always have significant difference for both the cases. Figures 8.4a and b also illustrates changes observed in the subcritical region with variations in the values of *Da*, showing that the difference between N_{cl} and N_{ce} increases as *Da* increases, implying a larger subcritical region. This increase is more prominent for free–free bounding surfaces. Notably, free–free surfaces demonstrate greater stability for both lower and higher values of medium permeability. Additionally, the stabilizing effect of *Da* persists across Figure 8.4a and b, regardless of the boundary surfaces, with a more pronounced increase in Rayleigh numbers observed for free–free boundary surfaces.

These findings underscore the critical role of medium permeability in enhancing system stability, especially in the context of different boundary conditions. The results align well with those reported by Sunil *et al.* [40], further validating the observed trends. The ability of *Da* to stabilize the system and influence the size of the subcritical region offers valuable insights into the control of ferroconvection stability.

8.6.4 Effect of Rotation

The Taylor number T_1 denotes the effect of rotation on the system. To analyze the effect of rotation, T_1 is taken to vary from 1 to 10^5, in the expressions obtained for N_{cl} and N_{ce} [5]. The observations are represented graphically in Figure 8.5a and b for respective boundary conditions. From the figures, it can be determined that rotation has a stabilizing effect on the system as N_{cl} and N_{ce} increase as T_1 increases. Figure 8.5a and b shows that a subcritical region is always present in the system. The subcritical region is, however, more prominent in the case of rigid surfaces as seen in Figure 8.5b.

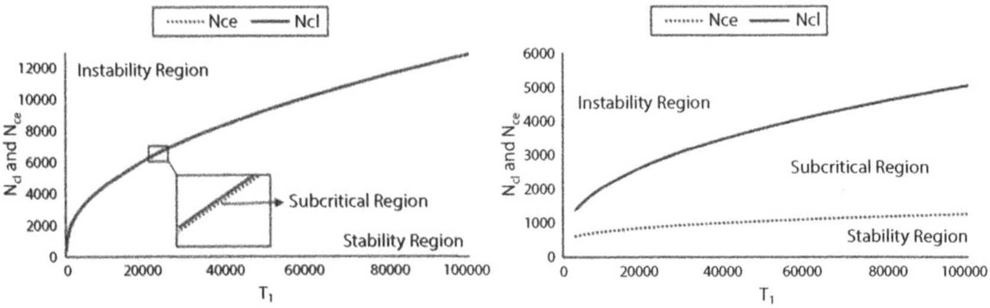

(**a**) Free-free boundaries (**b**) Rigid-rigid boundaries

Figure 8.5 Variations of N_{cl} and N_{ce} with respect to T_1 for two distinct boundary surfaces for $M_3 = 5$, $F_1 = 0.4$, $\gamma' = 0.001$, $D_a = 0.03$, and $\mathcal{H}_1 = 100$.

Moreover, for very small values of rotation, rigid boundaries offer more stability. However, due to the porosity of the medium for high values of rotation, free–free surfaces become more stable. This effect of rotation on boundaries has also been observed by Thakur, Kumar, and Devi [43]. Overall, these results emphasize the importance of rotational effects in stabilizing ferroconvection, particularly under different boundary conditions. The increased stability at higher T_1 values provides key insights into the rotational control of stability in porous media.

8.6.5 Effect of Porosity–Modified Conductivity Ratio

The porosity–muddied conductivity ratio, often denoted by γ', accounts for the influence of the porous medium's structure on heat transfer within the system. The porosity–modified conductivity ratio γ' varies between 0 and 0.1, as per Nield and Bejan [21]. In Figure 8.6, different values of γ' yield corresponding N_{cl} and N_{ce} while keeping the parameters M_3, F_1, T_1, Da, and $\mathcal{H}_1$ constant. Once again, the figures clearly demonstrate a significant

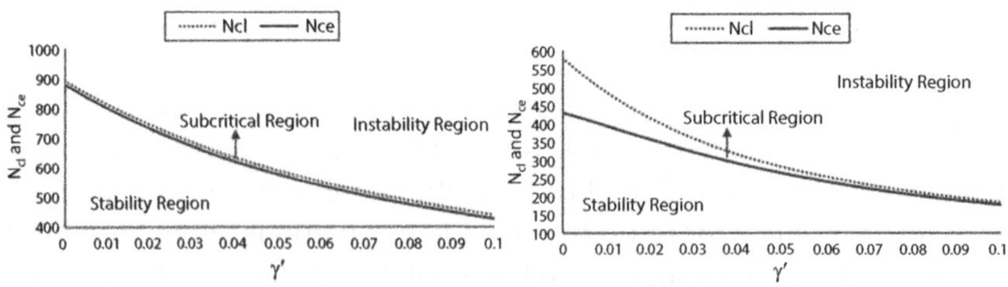

(**a**) Free-free boundaries (**b**) Rigid-rigid boundaries

Figure 8.6 Variations of N_{cl} and N_{ce} with respect to γ' for two distinct boundary surfaces for $M_3 = 5$, $F_1 = 0.4$, $T_1 = 100$, $D_a = 0.03$, and $\mathcal{H}_1 = 100$.

difference between N_{cl} and N_{ce} across all bounding surfaces, indicating the subcritical region is present in all two cases, as shown in Figures 8.6a and b.

Figures 8.6a and b also showcases the variations in the subcritical region with respect to the boundary surfaces. The subcritical region shrinks as boundaries become rigid. Thus, for the rigid–rigid boundary surfaces, the increase in the subcritical region is the largest, as clearly seen in Figure 8.6b. Another important observation made by Figures 8.6a and b is that for an increasing value of γ', the values of Rayleigh numbers N_{cl} and N_{ce} decrease. Thus, the destabilizing nature of γ' remains the same for both boundary surfaces. However, the decrease in N_{cl} and N_{ce} is largest in the case of free–free boundaries.

This analysis highlights the critical role of γ' in determining the stability of ferroconvection in porous media, particularly in scenarios involving rigid boundary surfaces, where the destabilizing influence of γ' is most pronounced. The findings align with and extend the understanding of the thermal behavior of ferroconvection in porous structures, providing insights that are consistent across different boundary conditions.

8.6.6 Effect of Heat Transfer Coefficient

Heat transfer coefficient $\mathcal{H}_1$ has been taken to vary from 0 to 10^6 to cover a wide range of porous materials [21]. Figures 8.7 and 8.8 show the effect of heat transfer coefficient $\mathcal{H}_1$ on the stability threshold (N_{ce}) and instability thresholds (N_{cl}). These effects are analyzed by keeping the other variables of the system M_3, F_1, T_1, Da, and γ' constant. Figure 8.7 presents the results of the analysis for small values of $\mathcal{H}_1$, while Figure 8.8 displays the corresponding results for large values of $\mathcal{H}_1$. This analysis has been done for

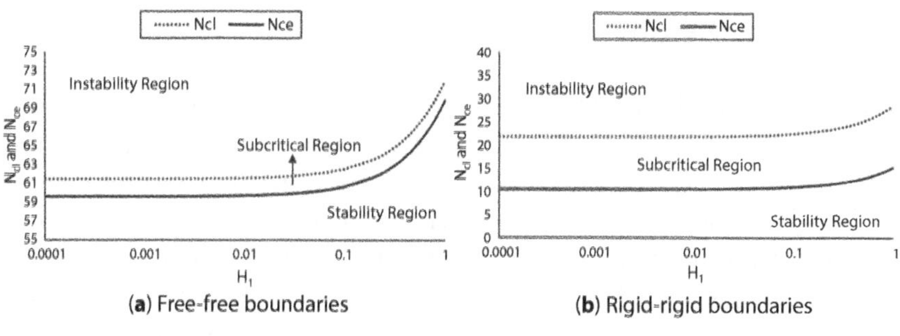

Figure 8.7 Variations of N_{cl} and N_{ce} with respect to small values of $\mathcal{H}_1$ for two distinct boundary surfaces for $M_3 = 5$, $F_1 = 0.4$, $T_1 = 100$, $D_a = 0.03$, and $\gamma' = 0.001$.

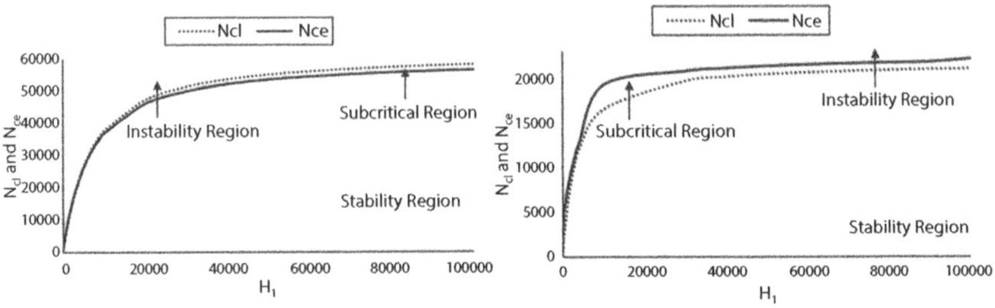

(**a**) Free-free boundaries (**b**) Rigid-rigid boundaries

Figure 8.8 Variations of N_{cl} and N_{ce} with respect to big values of $\mathcal{H}_1$ for two distinct boundary surfaces for $M_3 = 5$, $F_1 = 0.4$, $T_1 = 100$, $D_a = 0.03$, and $\gamma' = 0.001$.

two different scales of $\mathcal{H}_1$ because the nature of $\mathcal{H}_1$ is not found to be consistent.

For both scales, $\mathcal{H}_1$ demonstrates a stabilizing effect, evident in the increase in values of N_{cl} and N_{ce}. Although the difference between N_{cl} and N_{ce} may be small, it consistently indicates the presence of a subcritical region within the system. Notably, the behavior of the subcritical region differs based on the scale of $\mathcal{H}_1$: for smaller values of $\mathcal{H}_1$, the subcritical region decreases as $\mathcal{H}_1$ increases; however, for larger values, the subcritical region expands with increasing $\mathcal{H}_1$.

Boundary conditions exert a consistent influence across both scales, with rigid boundaries offering enhanced thermal stability, as seen in the higher values of N_{cl} and N_{ce}. This analysis underscores the dual role of $\mathcal{H}_1$ in system stability, with the subcritical region being highly sensitive to the scale of $\mathcal{H}_1$. These findings highlight the significance of $\mathcal{H}_1$ in ferroconvection stability within porous media, emphasizing that the subcritical region is a defining characteristic in such systems.

8.7 Conclusions

This study utilized the LTNE model to analyze how couple stresses influence heat transfer in a rotating ferrofluid within a porous medium, considering two types of boundary conditions. The normal mode method was applied for linear analysis, while the energy method was used for nonlinear analysis, resulting in a set of eigenvalue problems. These were solved using the single-term Galerkin method. The findings, presented graphically, were accompanied by an in-depth discussion. Key conclusions are summarized as follows:

- A nonlinear energy analysis confirms the decay of energy over time.
- Across varying parameters and boundary types, a consistent difference was observed between linear instability and nonlinear stability thresholds, indicating a subcritical region.
- The difference between linear instability and nonlinear stability thresholds reduces significantly when free surfaces are replaced by rigid ones, suggesting that rigid boundaries enhance the subcritical region.
- Magnetization and porosity-modified conductivity ratio consistently promote heat transfer, destabilizing the system. In contrast, couple stresses, heat transfer coefficient, and permeability restrict heat transfer, delaying convection onset.
- For lower rotational values, rigid boundaries provide more stability, whereas at higher values, free surfaces become more stable.
- Notably increasing magnetization and the porosity-modified conductivity ratio reduce the subcritical region, while higher medium permeability and couple stresses expand it. The subcritical region decreases with a low heat transfer coefficient but expands with higher values.
- In the absence of couple stresses, the findings align well with existing literature.

8.8 Applications

The parametric analysis—covering magnetization, couple stresses, medium permeability, porosity-modified conductivity ratio, and the heat transfer coefficient—reveals their substantial impact on system stability. By carefully adjusting these parameters, the extent of the subcritical region can be minimized, thus enhancing overall system stability. The conclusions of this research on convective heat transfer in rotating ferrofluids within porous media highlight several key applications and underscore its importance across a range of industries.

- **Thermal management in microelectronics:** The ability to control and optimize heat transfer through ferrofluids makes this study valuable in the development of cooling systems for high-performance electronics, where precise control of thermal gradients is critical for efficient energy dissipation. The findings on how boundary conditions affect stability thresholds can help improve cooling technology designs, extending the operational lifespans of electronic devices.

- **Geothermal and energy systems:** Understanding how parameters like magnetization, rotation, and permeability influence convective behavior aids in the efficient management of heat transfer in geothermal energy systems and other porous-media applications. Improved stability control could enhance geothermal extraction processes and other energy-related heat transfer systems by allowing operators to manage convective instabilities effectively.
- **Biomedical devices and drug delivery:** Ferrofluids are used in various biomedical applications, including targeted drug delivery systems. Insights into subcritical stability regions and how parameters like magnetization impact heat transfer could aid in fine-tuning these applications, where controlled fluid dynamics is essential for delivering medications to within the target body part, safely and effectively.
- **Advanced materials and manufacturing:** This research contributes to materials science by enhancing the understanding of ferrofluid behavior in porous media. The insights on couple stresses and boundary effects are particularly useful for developing advanced manufacturing processes involving fluid flow through porous structures, such as filtration, 3D printing, and enhanced oil recovery.
- **Magnetic and rotating machinery:** The study provides a foundation for optimizing rotating machinery that employs ferrofluids for lubrication, cooling, or damping. The knowledge of stability thresholds under different conditions could improve machine efficiency and stability, especially in high-stress environments where thermal stability is critical.

AI disclosure statement: The author confirms that no artificial intelligence (AI)-generated content or AI-assisted writing tools were used in the preparation of this chapter (including summaries and the abstract). All text, analysis, and conclusions are the result of the author's original work.

References

1. Adams, R., *Sobolev spaces*, Academic Press, New York, 1975.
2. Banu, N. and Rees, D.A.S., Onset of Darcy-Benard convection using a thermal non-equilibrium model. *Int. J. Heat Mass Transfer*, *45*, 11, 2221–2228, 2002. https://doi.org/10.1016/S0017-9310(01)00331-3.
3. Canu, R. and Renoult, M.C., Linear stability analysis of a Newtonian ferrofluid cylinder under a magnetic field. *J. Fluid Mech.*, *915*, A137, 2021. https://doi.org/10.1017/jfm.2021.171.
4. Chandel, V., Sunil, Sharma, P., Study of global stability of rotating partially-ionized plasma saturating a porous medium. *Spec. Top. Rev. Porous Media: Int. J.*, *15*, 6, 27–46, 2024. https://doi.org/10.1615/SpecialTopicsRevPorousMedia.2024050815.
5. Chandrasekhar, S., *Hydrodynamic and Hydromagnetic Stability*, Dover Publications, New York, 1981.
6. Choudhary, S., Devi, R., Mahajan, A., Sunil, Stability analysis in a couple-stress fluid layer with variable viscosity heated from below: Different conducting boundaries. *Chin. J. Phys.*, *83*, 94–102, 2023. https://doi.org/10.1016/j.cjph.2023.02.007.
7. Dev, K. and Suthar, O.P., Nonlinear stability analysis of penetrative convection in ferrofluids *via* internal heating. *J. Magn. Magn. Mater.*, *564*, 2, 170064, 2022. https://doi.org/10.1016/j.jmmm.2022.170064.
8. Devi, R., Sunil, Paul, R., The effect of rotation on thermal convection in a couple-stress fluid saturating a porous medium using Galerkin method: A nonlinear stability analysis. *Res. J. Sci. Tech.*, *5*, 1, 10–19, 2013. https://doi.org/https://rjstonline.com/AbstractView.aspx?PID=2013-5-1-2.
9. Dhiman, J.S. and Sood, S., Linear and weakly non-linear stability analysis of oscillatory convection in rotating ferrofluid layer. *Appl. Math. Comput.*, *430*, 127239, 2022. https://doi.org/10.1016/j.amc.2022.127239.
10. Finlayson, B.A., Convective instability of ferromagnetic fluids. *J. Fluid Mech.*, *40*, 4, 753–767, 1970. https://doi.org/10.1017/S0022112070000423.
11. Galdi, G.P., Nonlinear stability of the magnetic Bénard problem *via* a generalized energy method. *Arch. Ration. Mech. Anal.*, *87*, 167–186, 1985. https://doi.org/10.1007/BF00280699.
12. Galdi, G.P. and Straughan, B., A nonlinear analysis of the stabilizing effect of rotation in the Bénard problem. *Proc. R. Soc. London, Ser. A*, *402*, 257–283, 1985. https://doi.org/10.1098/rspa.1985.0118.
13. Hajialiloo, H. and Soltanipour, H., Numerical assessment of thermomagnetic convection of ferrofluid in a porous cavity employing a local thermal nonequilibrium model. *Therm. Sci. Eng. Prog.*, *53*, 102693, 2024.
14. Javed, T., Mehmood, Z., Abbas, Z., Natural convection in square cavity filled with ferrofluid saturated porous medium in the presence of uniform magnetic field. *Physica B*, *506*, 122–132, 2017. https://doi.org/10.1016/j.physb.2016.11.008.

15. Joseph, D.D., *Stability of Fluid Motions I*, vol. 27, Springer Science & Business Media, New York, 2013.
16. Kole, M. and Khandekar, S., Engineering applications of ferrofluids: A review. *J. Magn. Magn. Mater.*, *537*, 168222, 2021. https://doi.org/10.1016/j.jmmm.2021.168222.
17. Mahajan, A. and Nandal, R., On the stability of penetrative convection in a couple-stress fluid. *Int. J. Appl. Comput. Math.*, *3*, 3745–3758, 2017. https://doi.org/10.1007/s40819-017-0324-6.
18. Mahajan, A. and Parashar, H., Linear and weakly nonlinear stability analysis on a rotating anisotropic ferrofluid layer. *Phys. Fluids*, *32*, 2, 024101, 2020. https://doi.org/10.1063/1.5133102.
19. Manjunatha, S., Rehman, K.U., Kumar, J.S., Varma, S., Shatanawi, W., On thermal non-equilibrium (tne) model for heat transfer in ternary ferro-nanofluid with rectangular porous fn. *Int. J. Thermofluids*, *23*, 100757, 2024.
20. Nanjundappa, C., Shivakumara, I., Prakash, H., Effect of Coriolis force on thermomagnetic convection in a ferrofluid saturating porous medium: A weakly nonlinear stability analysis. *J. Magn. Magn. Mater.*, *370*, 140–149, 2014. https://doi.org/10.1016/j.jmmm.2014.06.035.
21. Nield, D.A. and Bejan, A., *Convection in Porous Media*, vol. 3, Springer, New York, 2006.
22. Rees, D.A.S., Vertical free convective boundary-layer flow in a porous medium using a thermal nonequilibrium model: Elliptical effects. *Z. Angew. Math. Phys.*, *54*, 3, 437–448, 2003. https://doi.org/10.1007/s00033-003-0032-4.
23. Rees, D.A.S. and Pop, I., Vertical free convective boundary-layer flow in a porous medium using a thermal nonequilibrium model. *J. Porous Media*, *3*, 1, 31–44, 2000. https://doi.org/10.1615/JPorMedia.v3 .i1.30.
24. Rees, D.A.S. and Pop, I., Local thermal non-equilibrium in porous medium convection, in: *Transport Phenomena in Porous Media III*, pp. 147–173, Elsevier, Oxford, 2005.
25. Rosensweig, R.E., *Ferrohydrodynamics*, Cambridge University Press, Cambridge, 1985.
26. Shankar, B.M., Kumar, J., Shivakumara, I.S., Stability of natural convection in a vertical couple stress fluid layer. *Int. J. Heat Mass Transfer*, *78*, 447–459, 2014. https://doi.org/10.1016/j.ijheatmasstransfer .2014.06.087.
27. Sharma, S., Sunil, Sharma, P., Stability analysis of thermosolutal convection in a rotating Navier–Stokes–Voigt fluid, in: *Z. Naturforsch. A: Phys. Sci*, 2024, in press, https://doi.org/10.1515/zna-2023-0284.
28. Shivakumara, I. and Kumar, S.B.N., Linear and weakly nonlinear triple diffusive convection in a couple stress fluid layer. *Int. J. Heat Mass Transfer*, *68*, 542–553, 2014. https://doi.org/10.1016/j.ijheatmasstransfer.2013.09.051.
29. Shivakumara, I., Lee, J., Nanjundappa, C., Ravisha, M., Ferromagnetic convection in a rotating ferrofluid saturated porous layer. *Transp. Porous Media*, *87*, 251–273, 2011. https://doi.org/10.1007/s11242-010-9678-5.

30. Shivakumara, I., Lee, J., Ravisha, M., Reddy, R.G., The onset of brinkman ferroconvection using a thermal non-equilibrium model. *Int. J. Heat Mass Transfer*, *54*, 9-10, 2116–2125, 2011.
31. Spiegel, E.A., Convective instability in a compressible atmosphere. *I. Astrophys. J.*, *141*, 1068–1090, 1965. https://doi.org/10.1086/148197.
32. Stokes, V., Couple stress in fluids. *Phys. Fluids*, *9*, 1709–1715, 1966. https://doi.org/10.1063/1.1761925.
33. Stokes, V., Couple stresses in fluids, in: *Theories of Fluids with Microstructure: An Introduction*, pp. 34–80, 1984, https://doi.org/10.1063/1.1761925.
34. Sunil, and Devi, R., Global stability for thermal convection in a couple stress fluid: Rigid boundaries. *Heat Transfer Res.*, *45*, 1, 43–57, 2014. https://doi.org/10.1615/HeatTransRes.2013003569.
35. Sunil, Devi, R., Mahajan, A., Global stability for thermal convection in a couple-stress fluid. *Int. Commun. Heat Mass Transfer*, *38*, 7, 938–942, 2011. https://doi.org/10.1016/j.icheatmasstransfer.2011.03.030.
36. Sunil, and Mahajan, A., A nonlinear stability analysis for magnetized ferrofluid heated from below. *Proc. R. Soc. A*, *464*, 2089, 83–98, 2008a. https://doi.org/10.1098/rspa.2007.1906.
37. Sunil, and Mahajan, A., A nonlinear stability analysis for rotating magnetized ferrofluid heated from below. *Appl. Math. Comput.*, *204*, 1, 299–310, 2008b. https://doi.org/10.1016/j.amc.2008.06.043.
38. Sunil, and Mahajan, A., A nonlinear stability analysis for rotating magnetized ferrofluid heated from below saturating a porous medium. *Z. Angew. Math. Phys.*, *60*, 2, 344–362, 2009a. https://doi.org/10.1007/s00033-008-7133-z.
39. Sunil, and Mahajan, A., A nonlinear stability analysis for thermoconvective magnetized ferrofluid saturating a porous medium. *Transp. Porous Media*, *76*, 327–343, 2009b. https://doi.org/10.1007/s11242-008-9249-1.
40. Sunil, Sharma, P., Mahajan, A., Nonlinear ferroconvection in a porous layer using a thermal nonequilibrium. *Spec. Top. Rev. Porous Media: Int. J.*, *1*, 2, 105–121, 2010. https://doi.org/10.1615/SpecialTopicsRevPorousMedia.v1.i2.30.
41. Sunil, Sharma, P., Mahajan, A., Onset of Darcy–Brinkman ferroconvection in a rotating porous layer using a thermal non-equilibrium model: a nonlinear stability analysis. *Spec. Top. Rev. Porous Media: Int. J.*, *88*, 2, 421–439, 2011. https://doi.org/10.1615/SpecialTopicsRevPorousMedia.v1.i2.30.
42. Thakur, A., Kumar, S., Devi, R., A nonlinear stability analysis for magnetized ferrofluid heated from below in the presence of couple stresses for combination of different bounding surfaces. *Numer. Heat Transf. B Fundam.*, 2023. https://doi.org/10.1080/10407790.2023.2275728.
43. Thakur, A., Kumar, S., Devi, R., The effect of rotation on ferroconvection in the presence of couple stress forces in porous medium: A nonlinear analysis. *Eur. Phys. J. Plus*, *139*, 3, 236, 2024. https://doi.org/10.1140/epjp/s13360-024-05048-8.

44. Thakur, A., Sunil, Devi, R., The effect of couple stresses on stability analysis of magnetized ferrofluid saturating a porous medium heated from below. *J. Porous Media*, *27*, 5, 85–106, 2024. https://doi.org/10.1615/JPorMedia.2023050054.

9

Non-Newtonian Casson Fluid through a Porous Rotating Channel with Seepage Flow

Abdul Faiz Ansari[1*], Sameera Iqram[1], Vinod Y.[2], Mohd. Asif[3] and Piyush Jaiswal[4]

[1]Department of Basic Science and Humanities, Pranveer Singh Institute of Technology, Kanpur, India
[2]Department of Mathematics, Davangere University, Davangere, India
[3]Department of Mathematics & Statistics, Integral University, Lucknow, India
[4]Department of Mathematics & Astronomy, University of Lucknow, Lucknow, India

Abstract

Rotating channels represent complex geometries that play a crucial role in various engineering applications, including the optimization of blood flow simulators, filtration systems, and chemical reactors. In these systems, non-Newtonian fluids, such as Casson fluid, exhibit unique flow characteristics, especially when subjected to rotational and porous conditions. This study focuses on the steady, laminar flow of incompressible Casson fluid through a rotating channel filled with porous material, where a uniform angular velocity is applied and seepage occurs at the upper wall of the channel. The governing equations for this flow are derived from the Darcy–Brinkman model, and through the use of similarity transformations, these equations are converted into a sixth-order ordinary differential equation (ODE). The analytical solution to this ODE is obtained, providing insights into the behavior of the fluid under the given conditions. The impact of the Casson parameter on the primary, secondary, and seepage velocities is thoroughly analyzed, with particular attention

**Corresponding author*: imfaizofcial@gmail.com; ORCID: https://orcid.org/0000-0001-8373-2722
Sameera Iqram: ORCID: https://orcid.org/0000-0002-4977-5953
Vinod Y.: ORCID: https://orcid.org/0000-0001-6345-2694
Mohd. Asif: ORCID: https://orcid.org/0000-0003-1383-1973
Piyush Jaiswal: ORCID: https://orcid.org/0009-0003-5146-8279

Dhananjay Yadav, Mukesh Kumar Awasthi and Harith Mohamed Al-Azri (eds.) Non-Newtonian Fluids for Industrial Applications: Modeling and Simulations, (209–222) © 2026 Scrivener Publishing LLC

paid to the influence of rotational effects in a porous medium. Graphical representations are employed to illustrate the flow behavior and highlight the significant changes in velocity profiles due to variations in the Casson parameter. This comprehensive analysis contributes to a deeper understanding of non-Newtonian fluid dynamics in rotating porous channels, with potential applications in engineering fields such as biomedicine, environmental engineering, and industrial fluid systems.

***Keywords*:** Casson fluid, Non-Newtonian fluid, rotating channel, porous medium, seepage flow, Darcy–Brinkman model, similarity transformation

9.1 Introduction

This paper holds significant applications in various engineering and scientific fields where fluid mechanics plays a crucial role. The study's insights into the behavior of Casson fluids—common in biological systems, industrial processes, and food technology—when influenced by rotation, porous media, and seepage effects, can enhance the design and optimization of filtration systems, chemical reactors, and biomedical devices like blood flow simulators. Additionally, the research is valuable for environmental engineering, particularly in groundwater flow analysis and pollution control in porous media, where such non-Newtonian fluids are encountered.

The Casson fluid model was introduced in 1959 by British rheologist N.W. Casson [1] in his study of colloidal suspension flows. Casson proposed a mathematical framework to characterize materials with yield stress and shear-thinning properties, serving as a bridge between Newtonian and non-Newtonian fluid mechanics. His contributions were particularly impactful in industries such as printing and ink manufacturing, where understanding the flow behavior of complex fluids like printing inks is crucial. Casson fluids are characterized by two key flow parameters: yield stress, arising from attractive forces between fluid particles, and plastic viscosity, linked to particle asymmetry [10]. Micropolar fluids [14], which incorporate microstructure and couple stress effects, exhibit nonsymmetric stress components [15]. Rott [16] presented an analytical solution for viscous flow at a stagnation point on a vertical plane. Bluman and Cole's seminal work, *Similarity Methods for Differential Equations*, highlights the relationship between dimensional analysis and similarity solutions [17]. Chakraborti and Gupta [18] proposed a similarity solution for flow with constant suction velocity along a smooth surface, while Vajravelu and Rollins [19] explored heat transfer in a conducting fluid over a fat plane. Studies have also examined the influence of chemical reactions [20] on convective heat and mass transfer over a porous stretching surface. Arthur *et al.* [21] investigated Casson fluid flow over a porous surface under a uniform magnetic field, and Rasool *et al.* [22] analyzed the flow of incompressible viscous MHD nanofluid over a nonlinear stretching surface.

Rotating channels in porous media play a vital role in engineering applications such as geothermal energy extraction, filtration systems, and enhanced oil recovery, where fluid flow dynamics are significantly influenced by both rotation and porous structures. Recently, Wang [23] examined the flow of Newtonian fluid through a porous rotating channel with seepage flow, analyzing the effects of permeability and rotation on primary, secondary, and seepage flow using streamlines. He also performed asymptotic analysis for large parameter values. Vidyanidhi and Nigam [2] pioneered the study of flow behavior in rotating channels, identifying secondary velocity effects due to rotation. Later, Vidyanidhi and Rao [3] extended this research to porous media with heat transfer, analyzing the impact of permeability on primary and secondary flows in rotating channels. Verma and Ansari [9] investigated MHD flow in rotating porous channels, focusing on the influence of magnetic fields on primary and secondary velocities; they [6] also investigated slip conditions effect on these MHD flow with effect of viscous dissipation on heat transfer. Wang [4] addressed unsteady flow in isotropic rotating porous channels and later expanded this work to steady flow in anisotropic porous channels [13], studying the effect of varying permeabilities in different directions. Wang [5] discussed the oscillatory behavior of flow in a rotating porous channel. Verma and Ansari [8] further explored the role of magnetic fields in these systems. They [7] also investigated the Couette flow through a porous rotating channel where permeability is taken as anisotropic (for more anisotropic types, see [24, 25]).

These present studies primarily focus on Newtonian fluids, providing the motivation to extend the analysis to non-Newtonian fluids in rotating porous channels with seepage velocity. The present work considers a porous channel rotating with a constant angular velocity. An analytical solution is derived by solving the governing Darcy–Brinkman equations of motion, with graphical representations illustrating the effects of the Casson parameter on the flow for both rotating and nonrotating cases.

Blood simulator can be a direct application of this paper because blood exhibits non-Newtonian behavior similar to a Casson fluid, where its viscosity decreases with an increasing shear rate. This study's analysis of Casson fluid flow in a porous rotating channel with seepage directly aligns with modeling blood flow through biological systems, such as capillaries and porous tissue. The rotating and seepage effects could simulate physiological conditions like blood flow influenced by rotational forces (e.g., in the heart or centrifugation) and permeation through porous biological membranes. This makes these findings particularly relevant for developing realistic and accurate blood flow simulators, enhancing medical diagnostics and treatment planning.

9.2 Problem Formulation

The following assumptions are considered in this study:

1. The flow is steady and laminar, and the non-Newtonian Casson fluid is incompressible.
2. Fluid with velocity U is driven through the top wall.
3. The primary flow $u'(z')$ is driven by applied pressure gradient, while the secondary flow $v'(z')$ is generated due to rotation of the channel about normal to the z'-direction with uniform velocity Ω (as shown in Figure 9.1).
4. u', v', and w' denote velocity components of the fluid in the x'-, y'-, and z'-directions, respectively.
5. The fluid adheres to the no-slip condition at the walls of the channel, that is, $u' = v' = w' = 0$ at $z' = 0$ and $u' = v' = 0$, $w' = U$ at $z' = h$.
6. The rheological equations of stress tensor τ_{ij} for an incompressible Casson fluid motion [10–12] are

$$\tau_{ij} = \begin{cases} 2(S_Y/\sqrt{2\Pi} + \mu_{PD})e_{ij}, & \text{if } \Pi > \Pi_c \\ 2(S_Y/\sqrt{2\Pi_c} + \mu_{PD})e_{ij}, & \text{if } \Pi < \Pi_c. \end{cases} \tag{9.1}$$

where μ_{PD}, e_{ij}, and S_Y are the coefficient of plastic dynamic viscosity, the $(i, j)^{th}$ component of rate of deformation, and yield stress of Casson fluid, respectively. Also, Π_c is the critical value of $\Pi = e_{ij}e_{ij}$.

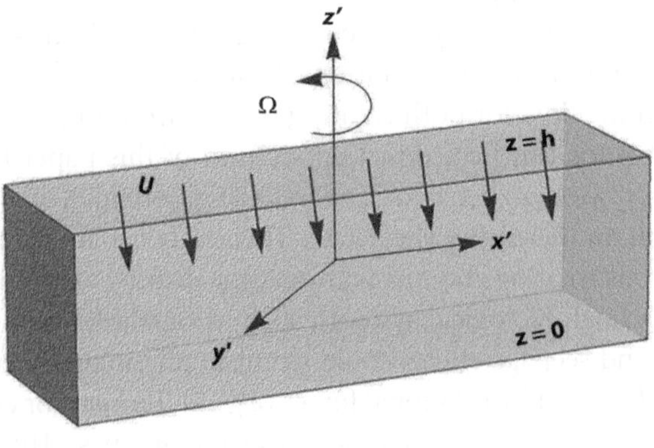

Figure 9.1 Schematic diagram of considered problem.

Based on the above assumptions, the modified Darcy–Brinkman equations for motion are [6–9, 13, 23]

$$\frac{\partial p'}{\partial x'} = \mu_{PD}(1+\frac{1}{\beta})\nabla^2 u' - \frac{\mu}{K}u' + 2\rho\Omega v', \tag{9.2}$$

$$\frac{\partial p'}{\partial y'} = \mu_{PD}(1+\frac{1}{\beta})\nabla^2 v' - \frac{\mu}{K}v' - 2\rho\Omega u', \tag{9.3}$$

$$\frac{\partial p'}{\partial z'} = \mu_{PD}(1+\frac{1}{\beta})\nabla^2 w' - \frac{\mu}{K}w'. \tag{9.4}$$

where β is the nondimensional Casson parameter, μ is the viscosity, ρ is the density, ρ' is the pressure, and K is the isotropic permeability of the porous channel. Introducing nondimensional quantities:

$$x = \frac{x'}{h},\ y = \frac{y'}{h},\ z = \frac{z'}{h},\ u = \frac{u'}{U},\ v = \frac{v'}{U},\ w = \frac{w'}{U},\ \text{and}\ p = \frac{hp'}{\mu_{PD}U}.$$

$$\frac{\partial p}{\partial x} = (1+\frac{1}{\beta})\nabla^2 u - \alpha u + \gamma v, \tag{9.5}$$

$$\frac{\partial p}{\partial y} = (1+\frac{1}{\beta})\nabla^2 v - \alpha v - \gamma u, \tag{9.6}$$

$$\frac{\partial p}{\partial z} = (1+\frac{1}{\beta})\nabla^2 w - \alpha w. \tag{9.7}$$

With boundary conditions: $u = 0 = v = w$ at $z = 0$, $u = 0 = v$ and $w = 1$ at $z = 1$, where $\alpha = \frac{\mu h^2}{\mu_{PD}K}$ is the permeability parameter and $\gamma = \frac{2\rho\Omega h^2}{\mu_{PD}}$ is the rotation parameter. Both α and γ are nondimensional parameters.

9.3 Solution of Problem

Using similarity transformation, $u = xf'(z)$, $v = -x\gamma g(z)$, and $w = -f(z)$, and consider the applied pressure gradient in the x-direction as follows:

$$p = p_s + Cx^2 - p^*(z)$$

$$(1+\frac{1}{\beta})f'''' - \alpha f'' - \gamma^2 g' = 0, \tag{9.8}$$

$$(1+\frac{1}{\beta})g'' - \alpha g + f' = 0, \tag{9.9}$$

$$(1+\frac{1}{\beta})f'' - \alpha f + p^{*\prime} = 0. \tag{9.10}$$

With boundary condition $f(0) = 0 = f'(0) = f'(1) = g(0) = g'(0)$ and $f(1) = 1$.

In order to evaluate solutions analytically of these equations, we can eliminate g from Equation 9.8, and we use Equation 9.9 to transform into a sixth-order differential equation in terms of function f as follows:

$$(1+\frac{1}{\beta})f'''''' - 2\alpha(1+\frac{1}{\beta})f'''' + (\alpha^2+\gamma^2)f'' = 0. \tag{9.11}$$

Again using Equations 9.8 and 9.9 for two more boundary conditions, $f''(0) = f''(1)$ and $(1+\frac{1}{\beta})f'''''(0) = \alpha f'''(0)$. Since the independent solution of ODE 9.11 will be proportional to $f(z) = \exp(\eta z)$, using Equation 9.11, we have

$$(1+\frac{1}{\beta})\eta^4 - 2\alpha(1+\frac{1}{\beta})\eta^2 + (\alpha^2+\gamma^2) = 0 \Rightarrow \eta = \pm\sqrt{\frac{(\alpha^2+\gamma^2)}{(1+\frac{1}{\beta})}}\exp(\pm i\theta/2), \tag{9.12}$$

where $\theta = \tan^{-1}(\frac{\gamma}{\alpha})$. Therefore, consider $\eta_1 = \sqrt{\frac{(\alpha^2+\gamma^2)}{(1+\frac{1}{\beta})}}\cos(\theta/2)$ and $\eta_2 = \sqrt{\frac{(\alpha^2+\gamma^2)}{(1+\frac{1}{\beta})}}\sin(\theta/2)$ so that following the solution satisfies all boundary conditions given as

$$f(z) = A_1(1-\cosh(\eta_1 z)\cos(\eta_2 z)) + A_2 z + A_3 \cosh(\eta_1 z)\sin(\eta_2 z)$$
$$+A_4 \sinh(\eta_1 z)\cos(\eta_2 z) + A_5 \sinh(\eta_1 z)\sin(\eta_2 z). \tag{9.13}$$

where A_i are arbitrary constants that can be evaluated using boundary conditions $f(0) = 0 = f'(0) = f'(1), f(1) = 1$, and $f''(0) = f''(1)$. Due to burdensome expressions of A_i's, these are not mentioned here. Now using Equation 9.8 to evaluate the function $g(z)$ as

$$g(z) = \frac{1}{\gamma^2}((1+\frac{1}{\beta})(f'''(z) - f'''(0)) - \alpha f'(z)) \tag{9.14}$$

Also from Equation 9.10, the pressure function $p^*(z)$ is given as

$$p^*(z) = \alpha \int_0^z f dz - (1+\frac{1}{\beta})f' \tag{9.15}$$

The solutions $f(z), f'(z)$, and $g(z)$ are precise yet too complex to allow for a straightforward physical interpretation.

9.4 Results and Discussion

This section presents the analysis of velocity profiles, represented by the functions $f'(z), f(z)$, and $g(z)$, across the rotating channel, highlighting the influence of the Casson parameter while keeping the permeability parameter constant.

Figures 9.2 and 9.3 provide a detailed illustration of the impact of the Casson parameter β on the velocity profile $f'(z)$ across the channel under two distinct conditions: when the channel is not rotating (i.e., rotation parameter $\gamma = 0$) and when the channel is rotating (i.e., rotation parameter $\gamma = 100$). As the Casson parameter increases, a noticeable decrease in $f'(z)$ is observed, which is consistent across both rotating and nonrotating cases. However, the most interesting observation is that the differences among the profiles are relatively small in the absence of rotation, indicating that the effect of the Casson parameter is more subtle when no rotational forces are applied. In contrast, when rotation is introduced, particularly at a high rotation parameter of $\gamma = 100$, the profiles of $f'(z)$ show a much more pronounced difference. This significant variation in the profiles, as depicted in Figure 9.3, highlights how the rotation parameter amplifies the effect of the

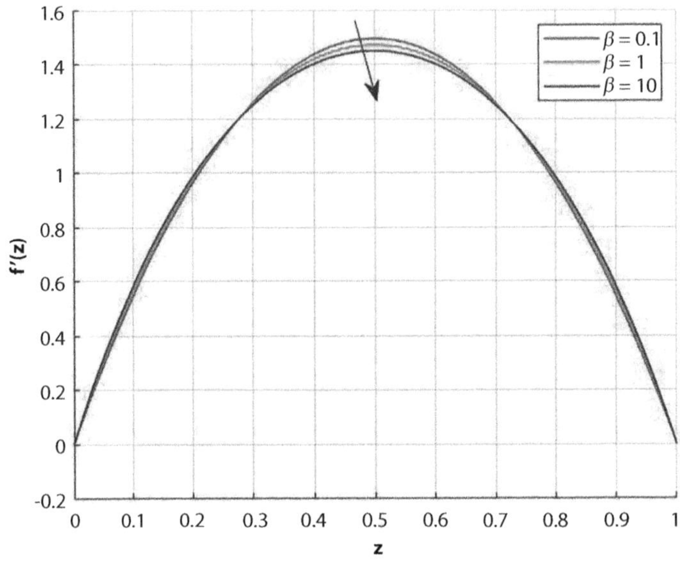

Figure 9.2 Effect of β on profiles of $f'(z)$ across the channel when $\gamma = 0$.

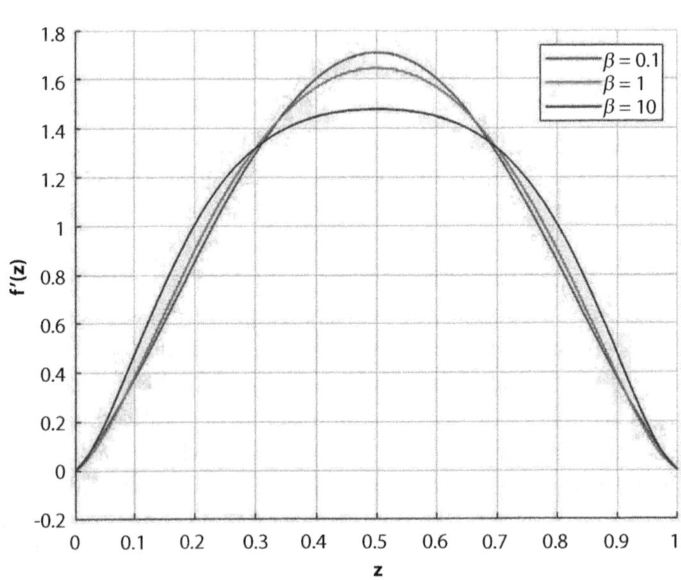

Figure 9.3 Effect of β on profiles of $f'(z)$ across the channel when $\gamma = 100$.

Casson parameter on the flow dynamics. Furthermore, it is observed that the profiles of $f'(z)$ remain consistently parabolic throughout the domain, starting from zero at the lower plate ($z = 0$) and ending at zero at the upper plate ($z = 1$), in accordance with the no-slip boundary conditions applied at both surfaces. This parabolic shape of the velocity profiles is a crucial feature of the flow, indicating the development of a laminar profile under these boundary conditions.

Figures 9.4 and 9.5 present the influence of the Casson parameter β on the velocity profile $f(z)$ across the channel for two distinct conditions: when the channel is not rotating (i.e., rotation parameter $\gamma = 0$) and when the channel is rotating (i.e., rotation parameter $\gamma = 100$). As the Casson parameter increases, there is a noticeable decrease in $f(z)$, which is consistent across both cases. Despite this decrease, it is observed that the differences among the profiles remain minimal in both the non-rotating and rotating scenarios. This suggests that, although the Casson parameter influences the velocity profile, its impact on $f(z)$ does not lead to significant variations in the overall flow structure under the conditions studied. The minimal difference in profiles highlights the relatively small effect of rotation on the velocity profile in this particular case, despite the rotation parameter being increased to $\gamma = 100$.

Figures 9.6 and 9.7 illustrate the effect of the Casson parameter β on the velocity profile $g(z)$ across the channel for two different rotation parameters:

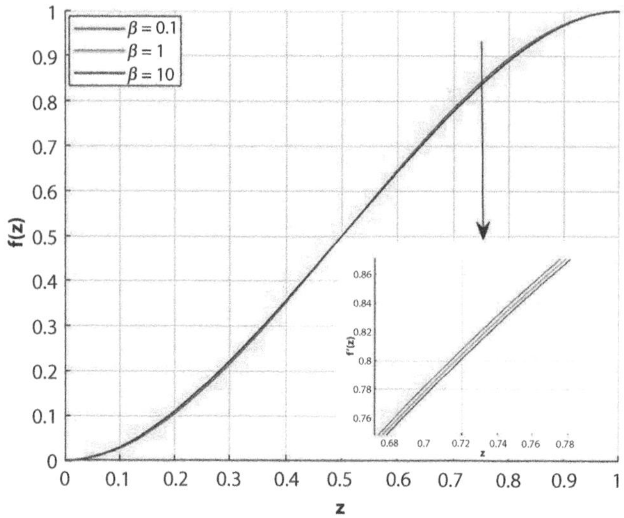

Figure 9.4 Effect of β on profiles of $f(z)$ across the channel when $\gamma = 0$.

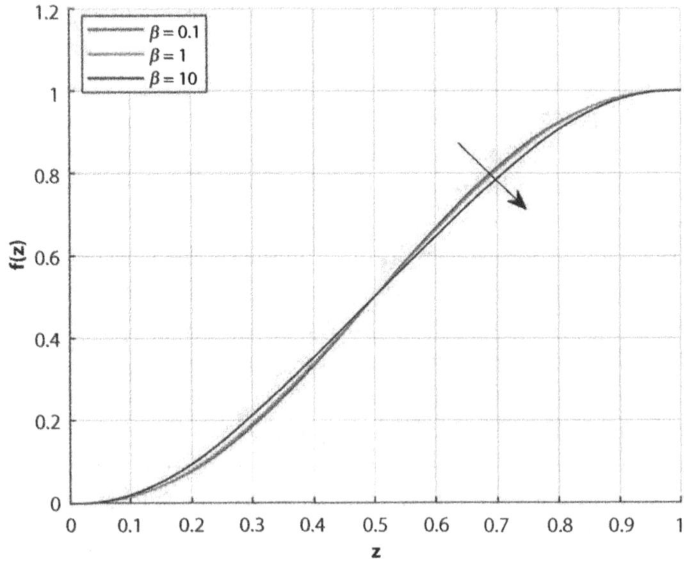

Figure 9.5 Effect of β on profiles of $f(z)$ across the channel when $\gamma = 100$.

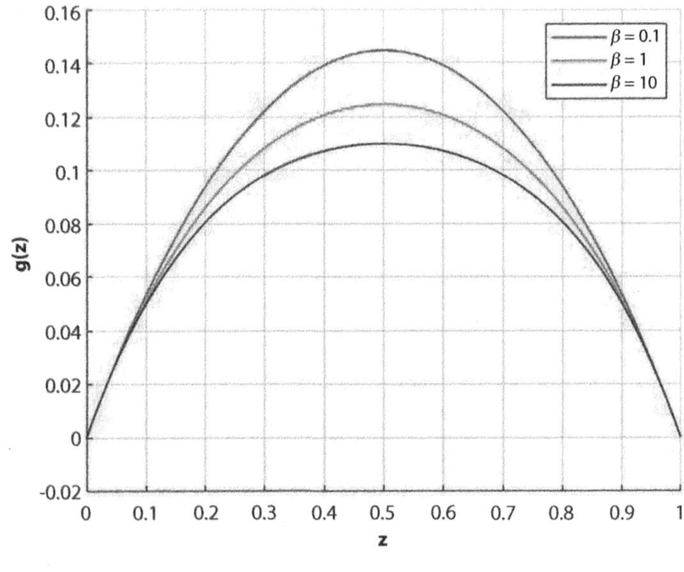

Figure 9.6 Effect of β on profiles of $g(z)$ across the channel when $\gamma = 10$.

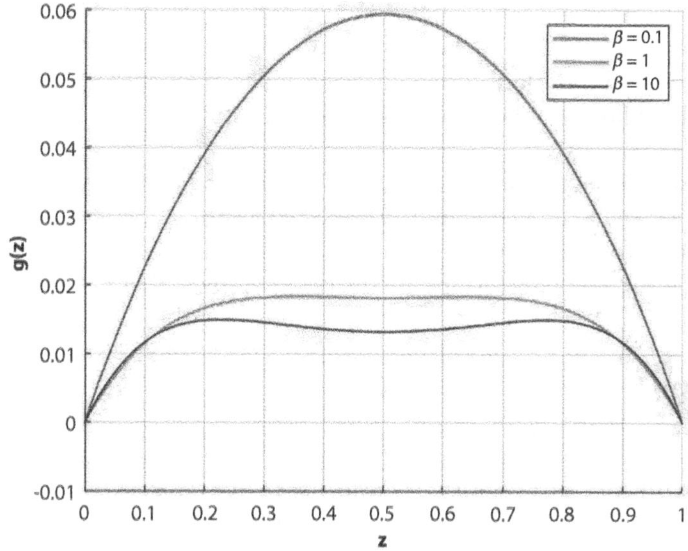

Figure 9.7 Effect of β on profiles of $g(z)$ across the channel when $\gamma = 100$.

$\gamma = 10$ and $\gamma = 100$. As the Casson parameter increases, the velocity profile $g(z)$ consistently decreases. Notably, for the case of $\gamma = 100$, the profile of $g(z)$ becomes nonparabolic when $\beta = 10$, indicating a departure from the typical parabolic flow pattern. In contrast, for all other values of the Casson parameter, the profile of $g(z)$ remains parabolic, even at higher rotation parameters. This highlights the influence of the Casson parameter on the shape of the velocity profile, especially under higher rotation conditions, where the nonparabolic behavior is more pronounced.

9.5 Conclusion

This study investigates the flow dynamics of incompressible Casson fluid through a rotating porous channel with seepage at the upper wall, employing similarity transformations and analytical solutions to the governing Darcy–Brinkman equations. The results reveal that the Casson parameter significantly influences the velocity profiles, with an increase in the parameter leading to a decrease in velocity across all scenarios. While the profiles remain parabolic under no rotation, introducing rotation enhances the impact of the Casson parameter, creating greater differences among the velocity profiles, particularly for higher rotation parameters. Notably, the secondary velocity profile becomes nonparabolic for high Casson parameter values when the rotation parameter is large. These findings provide insights into the complex interplay between non-Newtonian fluid properties, rotation, and porous media, with implications for engineering applications such as filtration systems, blood flow simulators, and chemical reactors.

AI disclosure statement: The author confirms that no artificial intelligence (AI)-generated content or AI-assisted writing tools were used in the preparation of this chapter (including summaries and the abstract). All text, analysis, and conclusions are the result of the author's original work.

References

1. Casson, N., Flow equation for pigment-oil suspensions of the printing ink-type, in: *Rheology of disperse systems*, pp. 84–104, 1959.
2. Vidyanidhi, V. and Nigam, S.D., Secondary flow in a rotating system. *J. Math. Phys. Sci.*, 1, 85–94, 1967.
3. Vidyanidhi, V. and Rao, V.D., Heat transfer through in a rotating channel in porous medium. *Acta Phys. Acad. Sci. Hung.*, 45, 179–189, 1978.
4. Wang, C.Y., Unsteady Poiseuille flow in a porous rotating channel. *J. Porous Media*, 16, 267–275, 2013.
5. Wang, C.Y., Oscillatory channel flow in a rotating system. *Arch. Appl. Mech.*, 29, 1271–1276, 2012.
6. Verma, V.K. and Ansari, A.F., Effect of Magnetic Field and Slip Conditions on Flow in a Rotating Porous Channel With Viscous Dissipation. *Heat Transfer*, 54, 2, 1562–1573, 2025. https://doi.org/10.1002/htj.23241.
7. Verma, V.K. and Ansari, A.F., Couette flow through an anisotropic porous rotating channel: Brinkman approach. *ZAMM-J. Appl. Math. Mechanics/ Zeitschrift für Angew. Math. und Mech.*, 105, 2025, e202400353, 2024. https://doi.org/10.1002/zamm.202400353.
8. Verma, V.K. and Ansari, A.F., Darcy-Brinkman Flow In An Anisotropic Rotating Porous Channel Under The Influence Of Magnetic Field. *J. Porous Media*, 27, 6, 31–43, 2024. https://doi.org/10.1615/JPorMedia.2023050260.
9. Verma, V.K. and Ansari, A.F., Effect of magnetic field on Darcy-Brinkman flow through rotating porous channel system, in: *Special Topics & Reviews in Porous Media*, vol. 16, no. 3, pp. 99–110, 2025. https://doi.org/10.1615/SpecialTopicsRevPorousMedia.2024052434.
10. Abdus, S.S. and Sharif, U., *MHD Boundary Layer Casson Fluid Over a Nonlinear Stretching Sheet*, Lambert Academic Publishing, India, 2015.
11. Maurya, D.K., Deo, S., Maurya, P.K., Hydromagnetic flow of Casson fluid over a stretching plane through porous medium. *Z. Angew. Math. Mech.*, e202300675, 2024. https://doi.org/10.1002/zamm.202300675.

12. Mukhopadhyay, S., Vajravelu, K., Van Gorder, R.A., Casson fluid flow and heat transfer at an exponentially stretching permeable surface. *J. Appl. Mech.*, 80, 5, 054502, 2013. https://doi.org/10.1115/1.4023618.
13. Wang, C.Y., Darcy-Brinkman flow in a rotating channel filled with an anisotropic porous medium. *J. Porous Media*, 25, 1, 31–40, 2022. https://doi.org/10.1615/JPorMedia.2021038567.
14. Lukaszewicz, G., *Micropolar Fluids: Theory and Applications*, Springer, New York, 1999.
15. Stokes, V.K., *Theories of Fluids with Microstructure*, Springer, New York, 1984.
16. Rott, N., Unsteady viscous flow in the vicinity of a stagnation point. *Q. Appl. Math.*, 13, 4, 444–451, 1956.
17. Bluman, G.W. and Cole, J.D., *Similarity Methods for Differential Equations*, Springer, New York, 1974.
18. Chakraborti, A. and Gupta, A.S., Hydromagnetic flow and heat transfer over a stretching sheet. *Q. Appl. Math.*, 37, 1, 73–78, 1979.
19. Vajravelu, K. and Rollins, D., Heat transfer in an electrically conducting fluid over a stretching surface. *Int. J. Non-Linear Mech.*, 27, 2, 265–277, 1992.
20. Chamkha, A.J., Aly, A.M., Mansour, M.A., Similarity solution for unsteady heat and mass transfer from a stretching surface embedded in a porous medium with suction/injection and chemical reaction effects. *Chem. Eng. Commun.*, 197, 6, 846–858, 2010.
21. Arthur, E.M., Seini, I.Y., Bortteir, L.B., Analysis of Casson fluid flow over a vertical porous surface with chemical reaction in the presence of magnetic field. *J. Appl. Math. Phys.*, 3, 713–723, 2015.
22. Rasool, G., Chamkha, A.J., Muhammad, T., Shafq, A., Khan, I., Darcy-Forchheimer relation in Casson type MHD nanofluid flow over nonlinear stretching surface. *Propul. Power Res.*, 9, 2, 159–168, 2020.
23. Wang, C.Y., Fluid Seepage into a Rotating Channel Filled with a Porous Medium. *J. Eng. Mech.*, 151, 2, 04024107, 2025.
24. Verma, V.K. and Ansari, A.F., Darcy-Brinkman Flow of a Micropolar Fluid through an Anisotropic Porous Channel. *J. Comput. Theor. Transp.*, 53, 7, 575–594, 2024. https://doi.org/10.1080/23324309.2024.2413187.
25. Verma, V.K. and Ansari, A.F., Couette flow of micropolar fluid in a channel filled with anisotropic porous medium. *Arch. Mech. Eng.*, 71, 4, 581–596, 2024. https://doi.org/10.24425/ame.2024.151336.

10

Stationary Thermosolutal Convection of a Rotating Walters' (Model B') Nanofluid in a Porous Medium Under Rigid–Rigid and Rigid–Free Boundary Conditions

Pushap Lata Sharma[1], Praveen Lata[1*], Ajit Kumar[1], G.C. Rana[2] and Dhananjay Yadav[3]

[1]Department of Mathematics & Statistics, H.P.U., Shimla, India
[2]Govt. College, Dhaneta, Hamirpur, Himachal Pradesh, India
[3]Department of Mathematical & Physical Sciences, University of Nizwa, Nizwa, Oman

Abstract

The impact of rotational effects on thermohaline convection in Walters' (Model B') nanofluid saturated in a porous medium is investigated in this work under two different boundary conditions. We examine the effects of rotation on the nature and beginning of convective patterns in the nanofluid using both theoretical analysis and numerical simulations. For the rigid–rigid boundary condition, rotational effects are found to stabilize the system by moderating convective instabilities, thereby enhancing both thermal and solutal diffusion within the porous medium. In contrast, the rigid–free boundary condition reveals a more intricate interaction between centrifugal and Coriolis forces induced by rotation, leading to varying impacts on convective stability. Parameters like porosity, Soret parameter, modified diffusivity ratio, thermoparticle Lewis number, and Taylor number have stabilizing effect that delays the start of convection, and solutal Rayleigh number, Dufour parameter, thermo-solutal

**Corresponding author*: praveenlata5@gmail.com; ORCID: https://orcid.org/0009-0002-9726-8814
Pushap Lata Sharma: ORCID: https://orcid.org/0000-0001-5848-9214
Ajit Kumar: ORCID: https://orcid.org/0000-0002-5888-3761
G.C. Rana: ORCID: https://orcid.org/0000-0003-2724-8308
D. Yadav: ORCID: https://orcid.org/0000-0001-8404-2053

Dhananjay Yadav, Mukesh Kumar Awasthi and Harith Mohamed Al-Azri (eds.) Non-Newtonian Fluids for Industrial Applications: Modeling and Simulations, (223–242) © 2026 Scrivener Publishing LLC

Lewis number, and concentration Rayleigh number possess destabilizing effect that enhances the beginning of convection on both the boundary conditions. The above-mentioned parameters are obtained graphically by using MATLAB software.

Keywords: Thermosolutal convection, Brownian motion, porous medium, Walters' (Model B')

10.1 Introduction

Thermosolutal convection in porous media is a vital area of research with significant implications for various fields, including geophysics, soil sciences, bioengineering, cancer therapy, and oceanography. The use of nanofluids, which are engineered mixtures of nanoparticles and fluids, in these porous environments can significantly enhance thermal and solutal properties due to the unique characteristics of nanoparticles, such as high thermal conductivity and large surface area-to-volume ratio. This highlights the importance of understanding the behavior of nanofluids in porous media to drive advancements in these critical fields. Thermosolutal convection is a complex phenomenon where both temperature and concentration gradients induce density variations, leading to convective currents. [25] built the basis for studying thermohaline convection, looking at how the system reacts to small changes. Building upon this, [26] used the stability of thermohaline convection in porous media heated from below and is examined using the Darcy–Maxwell model.

Furthermore, [1, 6, 7, 11, 12, 15, 16, 22, 27, 28] explored that the double diffusive convection starts in a porous layer saturated with a viscoelastic fluid, accounting for anisotropic properties and exploring the transition from stability to instability. These contributions have significantly expanded our knowledge of thermosolutal convection media. These studies have significantly advanced our understanding of thermosolutal convection and its applications. Thermosolutal convection has numerous applications in various fields, including chemical engineering, environmental engineering, aerospace engineering, geothermal systems, and biological and biomedical applications. Understanding the behavior of nanofluids in porous media is crucial for applications like heat transfer enhancement, geothermal energy, and electronic cooling. The research conducted by [3] has probably helped us learn more about thermosolutal convection in Jeffrey nanofluids filled porous media, which has led to new ideas for how it can be used in real life. Later, [13, 20], and [14] investigated how thermosolutal convection in a Jeffrey nanofluid was affected by rotation and magnetic field. Building on this research, [18] and [19] have conducted extensive research on thermosolutal convection in various fluids, exploring its behavior, dynamics, and applications by explaining the onset of convection and the behavior of nanofluids in a porous medium.

[2] suggested that the base fluid rate plus a relative slip rate can be added to determine the absolute velocity of nanoparticles. [5] discussed in detail how a Newtonian fluid would heat up under various hydrodynamic and hydromagnetic assumptions. [8, 9], and [10] used the Brinkman model, which takes into consideration how porosity and viscosity affect fluid flow and heat transfer. [4, 17, 23, 24], and [21] studied the instability of the nanofluid layer through experiments. The findings reveal significant influences of these effects on the system's behavior, addressing a previously existing knowledge gap. This original work, which has not been published elsewhere, provides valuable insights into the stability of such systems under various boundary conditions. A thorough literature review confirms the originality of this research and has not been previously published.

10.2 Mathematical Model

Consider a horizontal layer of fluid enclosed between two planes $z^* = 0$ and $z^* = d$ (refer Figure 10.1). With the help of gravity $g(0,0-g)$, the fluid layer rises after receiving heat from below. The temperature, volumetric fraction, and concentration at the lower wall are given by T_0^*, ϕ_0^*, and C_0^*, while those at the upper wall are given by T_1^*, ϕ_1^*, and C_1^*, respectively.

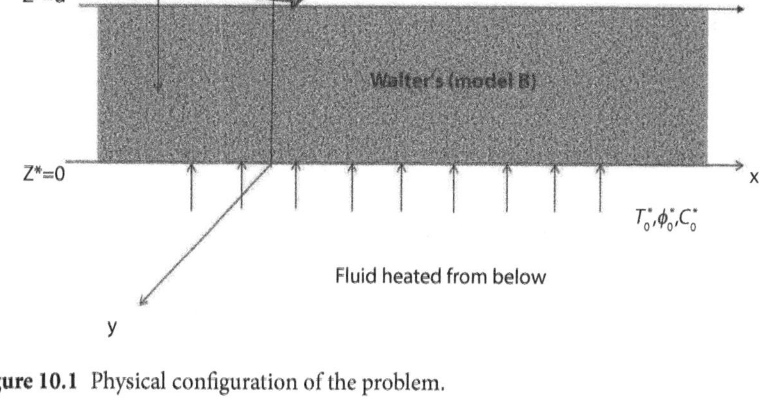

Figure 10.1 Physical configuration of the problem.

10.2.1 Governing Equations

Here, $\rho_f, \varepsilon, , \mathrm{v}^*, p^*, t^*, \mu, \mu', k_3, \Omega, \rho_P, \alpha_T, \alpha_C, D_B, D_T, k_m, \alpha_m, \sigma, D_{SM}$, and D_{CT} are the fluid density, porosity, Darcy velocity, pressure, time, viscosity of the fluid, kinematic viscoelasticity, permeability, angular velocity, nanoparticle mass density, coefficient of thermal expansion, analogous solvent coefficient of expansion, Brownian diffusion coefficient, thermophoresis diffusion coefficient, effective thermal conductivity, thermal diffusivity, heat capacity ratio, solute diffusivity, and diffusivity of the Soret type, respectively.

According to [2, 13, 14, 17, 20], and [21], the governing equations of conservation of mass and momentum

$$\nabla^*.\mathbf{v}^* = 0, \tag{10.1}$$

$$\frac{\rho_f}{\varepsilon}\frac{dv^*}{dt^*} = -\nabla^* p^* + \rho g - \frac{1}{k_3}\left(\mu - \mu'\frac{\partial}{\partial t}\right)v^* + \frac{2\rho_f}{\varepsilon}\left(v^* \times \Omega\right). \tag{10.2}$$

One way to express the nanofluids density is as

$$\rho = \phi^* \rho_P + \left(1 - \phi^*\right)\rho_f. \tag{10.3}$$

We used the base fluids density to determine the nanofluids density; we deliberate $\rho = \rho_f$.

The base fluid's Boussinesq approximation is shown; the specific weight ρg in Equation 10.2 is given as

$$\rho g \approx \left[\phi^* \rho_P + \left(1 - \phi^*\right)\left\{\rho_f\left(1 - \alpha_T\left(T^* - T_1^*\right) - \alpha_C\left(C^* - C_1^*\right)\right)\right\}\right]. \tag{10.4}$$

If one presents a buoyancy force [2, 17], and [21], Equation 10.2 becomes

$$0 = -\nabla^* p^* + \left[\phi^* \rho_P + \left(1 - \phi^*\right)\left\{\rho_f\left(1 - \alpha_T\left(T^* - T_1^*\right) - \alpha_C\left(C^* - C_1^*\right)\right)\right\}\right] g - \frac{1}{k_3}\left(\mu - \mu'\frac{\partial}{\partial t}\right)\mathbf{v}^* + \frac{2\rho_f}{\varepsilon}\left(\mathbf{v}^* \times \Omega\right). \tag{10.5}$$

For nanoparticles, the equation of continuity is

$$\frac{\partial \phi^*}{\partial t^*}+\frac{1}{\varepsilon}\mathrm{v}^*.\nabla^*\phi^* = D_B\nabla^{*2}\phi^* + \left(D_T / T_c^*\right)\nabla^{*2}T^*. \tag{10.6}$$

For nanoparticles, the thermal energy equation is

$$\left(\rho c\right)_f \mathbf{v}^*\cdot\nabla^* T^* + \left(\rho c\right)_m \frac{\partial T^*}{\partial t^*} = k_m\nabla^{*2}T^* + \varepsilon\left(\rho c\right)_p \left[D_B\nabla^*\phi^*\cdot\nabla^*T^* + \left(D_T / T_c^*\right)\nabla^*T\cdot\nabla^*T^*\right]. \tag{10.7}$$

For solute concentration, the conservation equation is

$$\frac{\partial C^*}{\partial t^*}+\frac{1}{\varepsilon}\mathbf{v}^*.\nabla^* C^* = D_{SM}\nabla^{*2}C^* + D_{CT}\nabla^{*2}T^*. \tag{10.8}$$

$$\left.\begin{aligned} &w^*=0,\quad T^*=T_0^*,\quad C^*=C_0^*,\quad \phi^*=\phi_0^*,\quad D_B\frac{\partial \phi^*}{\partial z^*}+\frac{D_T}{T_1}\frac{\partial T^*}{T_1^*}=0 \quad \text{at} \quad z^*=0 \\ &\text{and} \\ &w^*=0,\quad T^*=T_1^*,\quad C^*=C_1^*,\quad \phi^*=\phi_1^*,\quad D_B\frac{\partial \phi^*}{\partial z^*}+\frac{D_T}{T_1}\frac{\partial T^*}{T_1^*}=0 \quad \text{at} \quad z^*=d \end{aligned}\right\}, \tag{10.9}$$

which represents the boundary condition.

We present the following dimensionless quantities as

$$\left(x,y,z\right)=\frac{\left(x^*,y^*,z^*\right)}{d},\quad P=p^*k\big/\mu\alpha_m,\quad \left(u,v,w\right)=\frac{\left(u^*,v^*,w^*\right)}{\alpha_m}d,$$
$$t=t^*\alpha_m\big/\sigma d^2,\quad \phi=\phi^*-\phi_0^*\big/\phi_1^*-\phi_0^*,$$
$$T=\frac{T^*-T_1^*}{T_0^*-T_1^*},\quad C=\frac{C^*-C_1^*}{C_0^*-C_1^*}.$$

We obtain Equations 10.1 and 10.5–10.8 in dimensionless form as follows:

$$\nabla.\mathbf{v}=0, \tag{10.10}$$

$$-\nabla p-\left(1-F\frac{\partial}{\partial t}\right)\mathbf{v}-R_n\phi\hat{e}_z-R_m\hat{e}_z+R_aT\hat{e}_z+\frac{R_S}{L_e}C\hat{e}_z+\sqrt{T_a}\,(\mathbf{v}\times\hat{e}_z)=0, \tag{10.11}$$

$$\frac{1}{\sigma}\frac{\partial\phi}{\partial t}+\frac{1}{\varepsilon}\mathbf{v}.\nabla\phi-\frac{1}{L_n}\nabla^2\phi-\frac{N_A}{L_n}\nabla^2T=0, \tag{10.12}$$

$$\frac{\partial T}{\partial t}+\mathbf{v}\cdot\nabla T-\nabla^2T-\frac{N_B}{L_n}\nabla\phi\cdot\nabla T-\frac{N_AN_B}{L_n}\nabla T\cdot\nabla T-N_{TC}\nabla^2C=0, \tag{10.13}$$

$$\frac{1}{\sigma}\frac{\partial C}{\partial t}+\frac{1}{\varepsilon}\mathbf{v}.\nabla C-\frac{1}{L_e}\nabla^2C-N_{CT}\nabla^2=0. \tag{10.14}$$

$$\left.\begin{array}{l} w=0,\quad \phi=1,\quad T=1,\quad C=1,\quad \dfrac{\partial\phi}{\partial z}+N_A\dfrac{\partial T}{\partial z}=0\quad \text{at}\quad z=0\\ \text{and}\\ w=0,\quad \phi=0,\quad T=0,\quad C=0,\quad \dfrac{\partial\phi}{\partial z}+N_A\dfrac{\partial T}{\partial z}=0\quad \text{at}\quad z=1 \end{array}\right\}. \tag{10.15}$$

Here, $F=\dfrac{\mu'\kappa_m}{\mu\sigma d^2}$ is the kinematic viscoelasticity, $L_e=\dfrac{\alpha_m}{D_B}$ is the Lewis number, $N_B=\dfrac{(\rho c)_p}{(\rho c)_m}(\phi_1^*-\phi_0^*)$ is the modified particle-density increment, $L_n=\dfrac{\alpha_m}{D_B}$ is the thermonanofluid Lewis number, $N_A=\dfrac{D_T(T_0^*-T_1^*)}{D_BT_1^*(\phi_1^*-\phi_0^*)}$ is the modified diffusivity ratio, $N_{CT}=\dfrac{D_{CT}(T_0-T_1)}{\alpha_m(C_0-C_1)}$ is the Soret parameter, $N_{TC}=\dfrac{D_{TC}(C_0-C_1)}{\alpha_m(T_0-T_1)}$ is the Dufour parameter, $R_a=\dfrac{\rho\beta Kdg(T_0^*-T_1^*)}{\mu\alpha_m}$ is the thermal Rayleigh–Darcy number, $R_n=\dfrac{(\rho_p-\rho)(\phi_1^*-\phi_0^*)kdg}{\mu\alpha_m}$ is the concentration Rayleigh number, $R_m=\dfrac{\left[\rho_p\phi_1^*+\rho(1-\phi_1^*)\right]Kdg}{\mu\alpha_m}$ is the basic density Rayleigh number, $R_s = \dfrac{\rho c\left(C_0^* - C_1^*\right)kdg}{\mu D_s}$ is the solutal Rayleigh number, and $T_a = \left(\dfrac{2\Omega H^2 \rho}{\mu}\right)^2$ is the Taylor number.

10.2.2 Basic State Solutions

It is presumed that the nanofluid fundamental state is independent of time by [3, 17, 20] and [21]

$$v(u,v,w) = 0, \quad T = T_b(z), \quad p = p_b(z), \quad \phi = \phi_b(z), \quad C = C_b(z). \tag{10.16}$$

Substituting Equation 10.16 into Equations 10.11–10.14, these equations become

$$-\frac{dp_b}{dz} - R_m\hat{e}_z + R_a T_b \hat{e}_z - R_n \phi_b \hat{e}_z + \frac{R_S}{L_e} C_b \hat{e}_z = 0, \tag{10.17}$$

$$\frac{d^2\phi_b}{dz^2} + N_A \frac{d^2 T_b}{dz^2} = 0, \tag{10.18}$$

$$\frac{d^2 T_b}{dz^2} + \frac{N_A N_B}{L_n}\left(\frac{dT_b}{dz}\right)^2 + \frac{N_B}{L_n}\frac{d\phi_b}{dz} \cdot \frac{dT_b}{dz} + N_{TC}\frac{d^2 C_b}{dz^2} = 0, \tag{10.19}$$

$$\frac{d^2 C_b}{dz^2} + L_e N_{CT}\frac{d^2 T_b}{dz^2} = 0. \tag{10.20}$$

Using boundary conditions 10.15 in Equations 10.18–10.20, the solution is given as

$$T_b = 1 - z, \quad \phi_b = N_A z + \phi, \quad C_b = 1 - z, \tag{10.21}$$

which coincides with the results obtained by [21].

10.2.3 Perturbation Solutions

We applied small perturbations to the fundamental state in order to examine the system's stability; the disturbed state is provided as [17] and [21]

$$\mathbf{v}(u,v,w)=0+\mathbf{v}'(u,v,w),\quad T=T_b+T',\quad p=p_b+p',$$
$$C=C_b+C',\phi=\phi_b+\phi'. \tag{10.22}$$

Using Equation 10.22 into Equations 10.10–10.15, we get the following equations:

$$\nabla.\mathbf{v}'=0, \tag{10.23}$$

$$-\nabla p'-\left(1-F\frac{\partial}{\partial t}\right)\mathbf{v}'-R_n\phi'\hat{e}_z+R_aT'\hat{e}_z+\frac{R_S}{L_e}C'\hat{e}_z+\sqrt{T_a}(\mathrm{v}'\times\hat{e}_z)=0, \tag{10.24}$$

$$\frac{1}{\sigma}\frac{\partial\phi'}{\partial t}+\frac{N_A}{\varepsilon}\mathbf{v}'-\frac{1}{L_n}\nabla^2\phi'-\frac{N_A}{L_n}\nabla^2T'=0, \tag{10.25}$$

$$\frac{\partial T'}{\partial t}-\mathbf{v}'-\nabla^2T'-\frac{N_B}{L_n}\left(N_A\frac{\partial T'}{\partial z}-\frac{\partial\phi'}{\partial z}\right)+\frac{2N_AN_B}{L_n}\frac{\partial T'}{\partial z}-N_{TC}\nabla^2C'=0, \tag{10.26}$$

$$\frac{1}{\sigma}\frac{\partial C'}{\partial t}-\frac{1}{\varepsilon}\mathbf{v}'-\frac{1}{L_e}\nabla^2C'-N_{CT}\nabla^2T'=0. \tag{10.27}$$

$$\left.w'=0,\ T'=0,\ C'=0,\ \phi'=0,\ \frac{\partial\phi'}{\partial z}+N_A\frac{\partial T'}{\partial z}=0\ \text{at}\ z=0,\ z=1\right\}. \tag{10.28}$$

The nonentities u',v',w',T',p',ϕ' and C' can be compact into four nonentities by taking Equation 10.24 with $\hat{e}_z$.*Curl.Curl*, which yields

$$\nabla^2 w' - F\frac{\partial}{\partial t}\nabla^2 w' - R_a \nabla_H^2 T' + R_n \nabla_H^2 \phi' - \frac{R_S}{L_e}\nabla_H^2 C' - T_a \frac{\partial^2 w'}{\partial z^2} = 0. \tag{10.29}$$

Here, $\nabla^2 = \frac{\partial^2}{\partial x^2} + \frac{\partial^2}{\partial y^2} + \frac{\partial^2}{\partial z^2}$ is the Laplace operator, and $\nabla_H^2 = \frac{\partial^2}{\partial x^2} + \frac{\partial^2}{\partial y^2}$ is the Laplace operator in two dimensions.

10.2.4 Normal Mode Analysis

Assuming that the perturbed values are of the following type and using the normal mode approach to analyze the disturbances by [17] and [21]:

$$[w', \phi', T', C'] = [W(z), \Phi(z), \Theta(z), \Sigma(z)]\exp(ilx + imy + st). \tag{10.30}$$

Substituting Equation 10.30 into Equations 10.25–10.27 and 10.29, we get

$$(1 - sF)(D^2 - a^2)W + T_a D^2 W + R_a a^2 \Theta - R_n a^2 \Phi + \frac{R_S}{L_e} a^2 \Sigma = 0, \tag{10.31}$$

$$\frac{W}{\varepsilon} + \left(\frac{s}{\sigma} - \frac{(D^2 - a^2)}{L_e}\right)\Sigma + N_{CT}(D^2 - a^2)\Theta = 0, \tag{10.32}$$

$$W - \left(s - (D^2 - a^2) + \frac{N_A N_B}{L_n}\right)\Theta - \frac{N_B}{L_n} D\Phi + N_{TC}(D^2 - a^2)\Sigma = 0, \tag{10.33}$$

$$\frac{N_A}{\varepsilon} W - \left(\frac{(D^2 - a^2)}{L_n} - \frac{s}{\sigma}\right)\Phi - \frac{N_A}{L_n}(D^2 - a^2)\Theta = 0. \tag{10.34}$$

$$W = 0,\ \Sigma = 0,\ \Theta = 0,\ \Phi = 0,\ D\Phi + N_A D\Theta = 0 \text{ at } z = 0 \text{ and } z = 1. \tag{10.35}$$

where $D = d/dz$ and $a = \sqrt{l^2 + m^2}$ is the dimensionless wave number.

10.3 Linear Stability Analysis

10.3.1 For Rigid–Rigid Boundaries

The boundary conditions for rigid–rigid surfaces are given by [17] and [21]

$$W = DW = \Theta = \Phi = \Sigma = 0 \text{ at } z = 0 \text{ and } z = 1. \tag{10.36}$$

The trial functions are

$$W = z^2(1-z)^2 W_0,\ \Sigma = z(1-z)\Sigma_0,\ \Phi = z(1-z)\Phi_0,\ \Theta = z(1-z)\Theta_0. \tag{10.37}$$

Substituting Equation 10.37 in the system of Equations 10.31–10.34, we get the following linear system of equation in matrix form:

$$\begin{bmatrix} (1-sF)2(12+a^2)+12T_a & -9R_a a^2 & 9R_n a^2 & -9\frac{R_S}{L_e}a^2 \\ 3N_A/\varepsilon & \frac{14N_A}{L_n}(10+a^2) & \frac{14}{L_n}\left[(10+a^2)+\frac{sL_n}{\sigma}\right] & 0 \\ 3 & -14(10+a^2+s) & 0 & -14N_{TC}(10+a^2) \\ 3/\varepsilon & -14N_{CT}(10+a^2) & 0 & -\frac{14}{L_e}(10+a^2)-\frac{sL_e}{\sigma} \end{bmatrix} \begin{bmatrix} W_0 \\ \Theta_0 \\ \Phi_0 \\ \Sigma_0 \end{bmatrix} = \begin{bmatrix} 0 \\ 0 \\ 0 \\ 0 \end{bmatrix}. \tag{10.38}$$

A nontrivial solution to the linear system Equation 10.38 exists if and only if

$$R_a = \frac{1}{\left[\left(\frac{10+a^2}{L_e}+\frac{s}{\sigma}\right)\varepsilon - N_{TC}(10+a^2)\right]} \times \left\{ \begin{aligned} & \frac{-R_n}{\left(\frac{10+a^2}{L_n}+\frac{s}{\sigma}\right)} \left[\begin{aligned} & N_A(10+a^2+s)\left(\frac{10+a^2}{L_e}+\frac{s}{\sigma}\right) - N_A N_{TC} N_{CT}(10+a^2)(10+a^2) \\ & +\frac{N_A}{L_n}(10+a^2)\left(\frac{10+a^2}{L_e}+\frac{s}{\sigma}\right)\varepsilon - \frac{N_A}{L_n} N_{TC}(10+a^2)(10+a^2) \end{aligned} \right] \\ & +\frac{28\varepsilon}{27a^2}(12+a^2)(1-sF)+12T_a\left[(10+a^2+s)\left(\frac{10+a^2}{L_e}+\frac{s}{\sigma}\right) - N_{TC}N_{CT}(10+a^2)(10+a^2)\right] \\ & +\frac{R_S}{L_e}\left[\varepsilon N_{CT}(10+a^2)-(10+a^2+s)\right] \end{aligned} \right\} \tag{10.39}$$

10.3.1.1 *Stationary Convection*

Substituting $s = 0$, Equation 10.39 becomes

$$R_a = \frac{28\varepsilon}{27a^2(\varepsilon - N_{TC}L_e)}(12+a^2+12T_a)(10+a^2)(1-L_e N_{TC} N_{CT}) + \frac{R_S(\varepsilon N_{CT}-1)}{(\varepsilon - N_{TC}L_e)} - \frac{\{N_A(L_n+\varepsilon) - N_A L_e(L_n N_{CT}+1)N_{TC}\}}{(\varepsilon - N_{TC}L_e)} R_n. \tag{10.40}$$

10.3.2 For Rigid–Free Boundaries

The boundary conditions are given by [17] and [21]

$$W = D^2W = \Theta = \Phi = \Sigma = 0 \text{ at } z = 0. \tag{10.41}$$

$$W = DW = \Theta = \Phi = \Sigma = 0 \text{ at } z = 1. \tag{10.42}$$

The trial functions are

$$W = z^2(1-z)(3-2z)W_0,\ \Phi = z(1-z)\Phi_0,\ \Sigma = z(1-z)\Sigma_0,$$
$$\Theta = z(1-z)\Theta_0. \tag{10.43}$$

Substituting Equation 10.43 in the system of Equations 10.31–10.34, we get the following linear system of equation in matrix form:

$$\begin{bmatrix} 2(1-sF)(216+19a^2)+216T_a & -39a^2R_a & 39a^2R_n & -39\frac{R_S}{L_e}a^2 \\ \frac{13}{\varepsilon}N_A & 14\frac{N_A}{L_n}(10+a^2) & \frac{14(10+a^2)}{L_n}+\frac{14s}{\sigma} & 0 \\ 13 & -14(10+a^2+s) & 0 & -14N_{TC}(10+a^2) \\ \frac{13}{\varepsilon} & -14N_{CT}(10+a^2) & 0 & -\frac{14(10+a^2)}{L_e}-\frac{14s}{\sigma} \end{bmatrix} \begin{bmatrix} W_0 \\ \Theta_0 \\ \Phi_0 \\ \Sigma_0 \end{bmatrix} = \begin{bmatrix} 0 \\ 0 \\ 0 \\ 0 \end{bmatrix}. \tag{10.44}$$

A nontrivial solution to the linear system Equation 10.44 exists if and only if

$$R_a = \frac{1}{\left[\left(\frac{10+a^2}{L_e}+\frac{s}{\sigma}\right)\varepsilon - N_{TC}(10+a^2)\right]} \times \left\{ \begin{aligned} & \frac{-R_n}{\left(\frac{10+a^2}{L_n}+\frac{s}{\sigma}\right)} \left[\begin{aligned} & N_A(10+a^2+s)\left(\frac{10+a^2}{L_e}+\frac{s}{\sigma}\right) - N_A N_{TC} N_{CT}(10+a^2)(10+a^2) \\ & +\frac{N_A\varepsilon}{L_n}(10+a^2)\left(\frac{10+a^2}{L_e}+\frac{s}{\sigma}\right) - \frac{N_A}{L_n}N_{TC}(10+a^2)(10+a^2) \end{aligned} \right] \\ & +\frac{28\varepsilon}{507a^2}(216+19a^2)(1-sF)+216T_a\left[(10+a^2+s)\left(\frac{10+a^2}{L_e}+\frac{s}{\sigma}\right) - N_{TC}N_{CT}(10+a^2)(10+a^2)\right] \\ & +\frac{R_S}{L_e}\left[\varepsilon N_{CT}(10+a^2)-(10+a^2+s)\right] \end{aligned} \right\}. \tag{10.45}$$

10.3.2.1 Stationary Convection

Putting $s=0$ in Equation 10.45, it reduces to

$$R_a = \frac{28\varepsilon}{507a^2(\varepsilon - N_{TC}L_e)}(216+19a^2+216T_a)(10+a^2)(1-L_eN_{TC}N_{CT}) + \frac{R_S(N_{TC}\varepsilon - 1)}{(\varepsilon - L_eN_{TC})} - \frac{\{N_A(L_n+\varepsilon) - N_AL_e(L_nN_{CT}+1)N_{TC}\}}{(\varepsilon - N_{TC}L_e)}R_n. \tag{10.46}$$

10.4 Result and Discussion

In this research, we analyzed the impact of ε, R_s, N_{TC}, N_{CT}, L_n, R_n, N_A, L_e, and T_a on the onset of stationary convection for two boundary conditions. Figure 10.2 displays the graph of R_a in relation to a for numerous values of $\varepsilon = 0.2, 0.4, 0.6$. Setting another parameter as R_s=100, N_{TC}=5, N_{CT}=5, N_A=5, L_e=500, L_n=500, R_n=−1, T_a=100. It is obvious that ε goes on

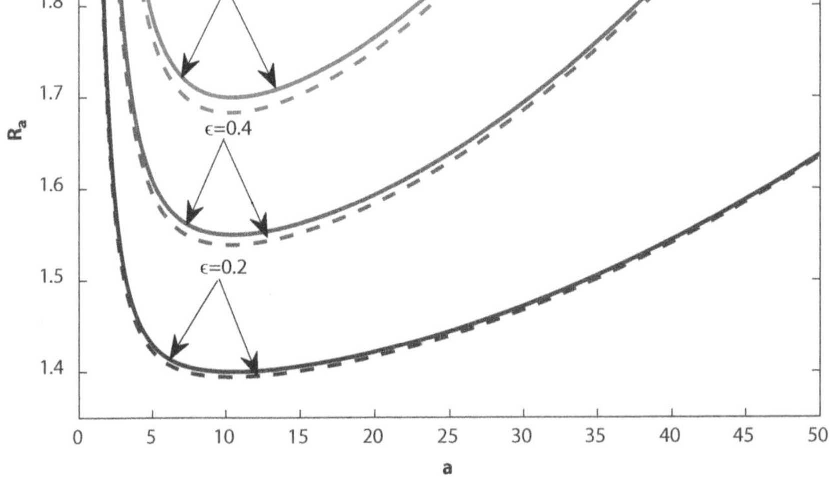

Figure 10.2 Deviation of R_a versus a for different values of ε.

increasing with the growth in the value of R_a. Thus, ε takes stabilizing effect, delays the beginning of convection. Figure 10.3 illustrates the graph of R_a w. r. t. a for various values of $R_s = 10, 50, 100$. Adjusting another parameter as $\varepsilon = 0.4, N_{TC} = 5, N_{CT} = 5, N_A = 5, L_e = 500, L_n = 500, R_n = -1, T_a = 100$. It is clear that R_s goes on decreasing with the increase in the value of R_a. Thus, R_s takes destabilizing effect, increases the beginning of convection.

Figure 10.4 demonstrates the graph of R_a in relation to a for several values of $N_{TC} = 5, 10, 15$. Adjusting another parameter as $\varepsilon = 0.4, R_s = 100, N_{CT} = 5, N_A = 5, L_e = 500, L_n = 500, R_n = -1, T_a = 100$. It is obvious that N_{TC} goes on decreasing with the increase in the value of R_a. Thus, N_{TC} takes destabilizing effect, enhances the beginning of convection.

Figure 10.5 illustrates the graph of R_a w. r. t. a for various values of $N_{CT} = 5, 10, 15$. Adjusting another parameter as $\varepsilon = 0.4, R_s = 100, N_{TC} = 5, N_A = 5, L_e = 500, L_n = 500, R_n = -1, T_a = 100$. It is clear that N_{CT} goes on increasing with the increase in the value of R_a. Thus, N_{TC} takes stabilizing effect, stays the beginning of convection. Figure 10.6 indicates the graph of R_a in relation to a for several values of $L_n = 500, 1000, 1500$.

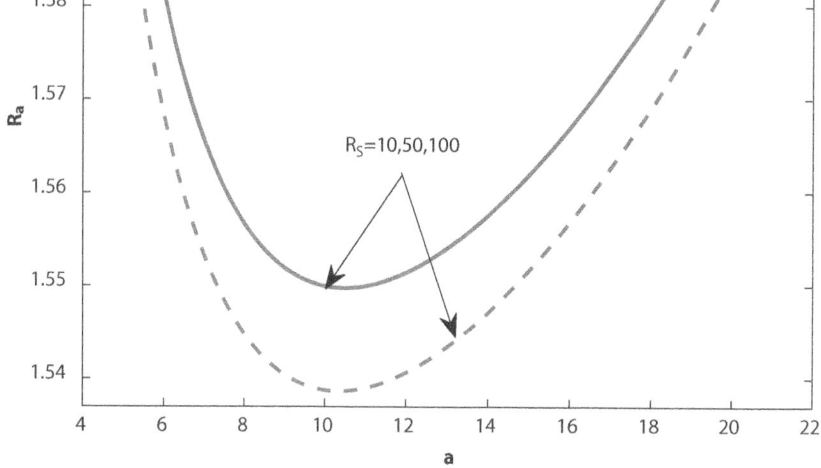

Figure 10.3 Deviation of R_a against a for numerous values of R_s.

Adjusting another parameter as $\varepsilon = 0.4, R_s = 100, N_{TC} = 5, N_{CT} = 5, N_A = 5,$ $L_e = 500, R_n = -1, T_a = 100$. It is obvious that L_n goes on increasing with the increase in the value of R_a. Thus, L_n takes stabilizing effect, delays the beginning of convection. Figure 10.7 illustrates the graph of R_a in relation to a for various values of $R_n = -1, -2, -3$. Adjusting another parameter as $\varepsilon = 0.4, R_s = 100, N_{TC} = 5, N_{CT} = 5, N_A = 5, L_e = 500, L_n = 500, T_a = 100$. It is obvious that R_n goes on decreasing with the rise in the value of R_a. Thus, R_n takes destabilizing effect, improves the beginning of convection. Figure 10.8 demonstrates the graph of R_a w. r. t. a for various values of $N_A = 1,5,10$. Adjusting another parameter as $\varepsilon = 0.4, R_s = 100, N_{TC} = 5, N_{CT} = 5, L_e = 500, L_n = 500, R_n = -1, T_a = 100$. It is clear that N_A goes on increasing with the increase in the value of R_a. Thus, N_A takes stabilizing effect, stays the beginning of convection. Figure 10.9 illustrates the graph of R_a w. r. t. a for various values of $L_e = 500, 1000, 1500$. Adjusting another parameter as $\varepsilon = 0.4, R_s = 100, N_{TC} = 5, N_{CT} = 5, N_A = 5, L_n = 500, R_n = -1, T_a = 100$. It is obvious that L_e goes on decreasing with the increase in the value of R_a. Thus, L_e takes destabilizing effect, enhancing the start of convection

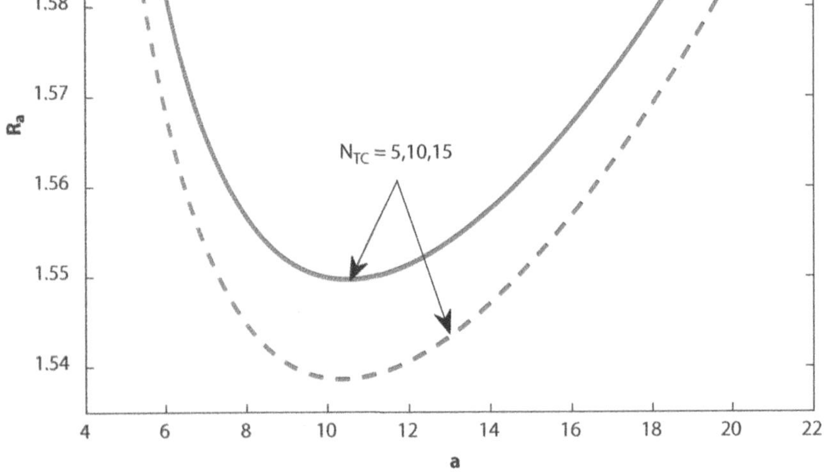

Figure 10.4 Variation of R_a against a for numerous values of N_{TC}.

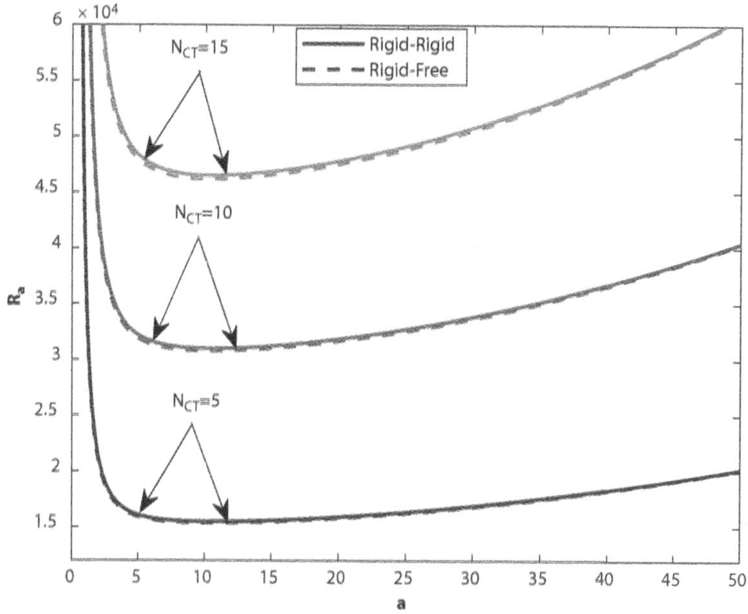

Figure 10.5 Variation of R_a against a for numerous values of N_{CT}.

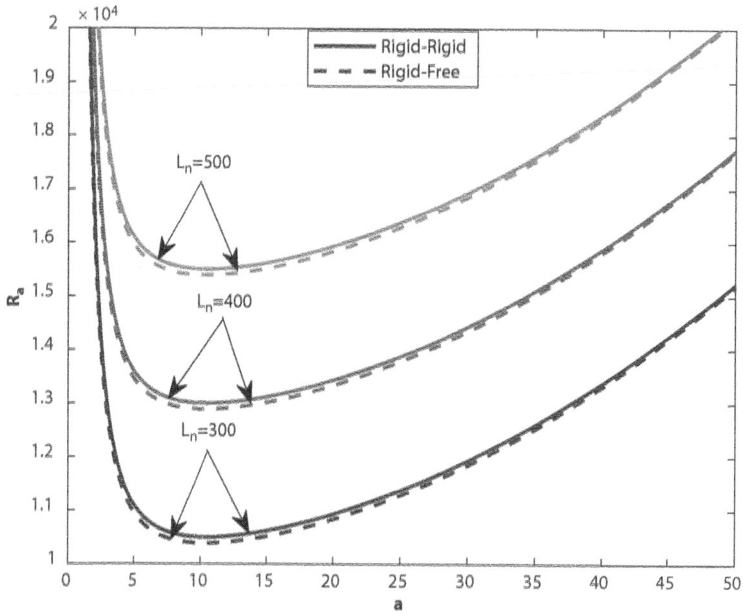

Figure 10.6 Deviation of R_a with a for different values of L_n.

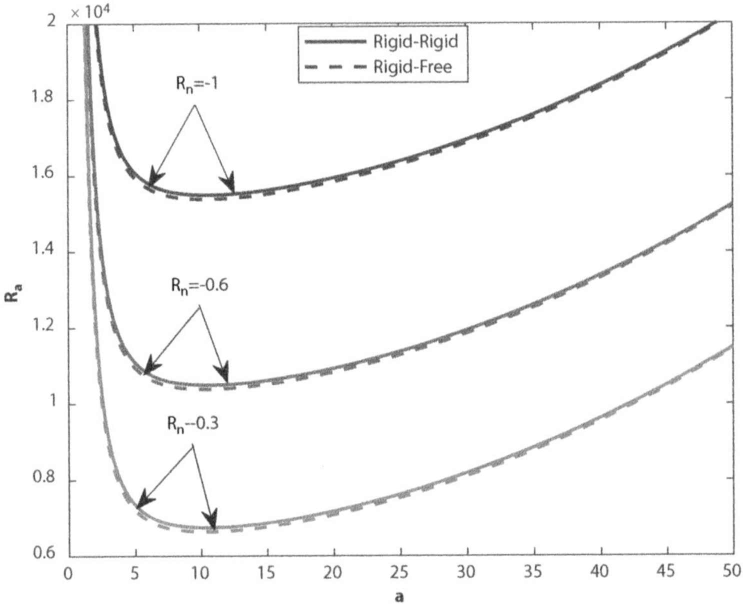

Figure 10.7 Variation of R_a against a for numerous values of R_n.

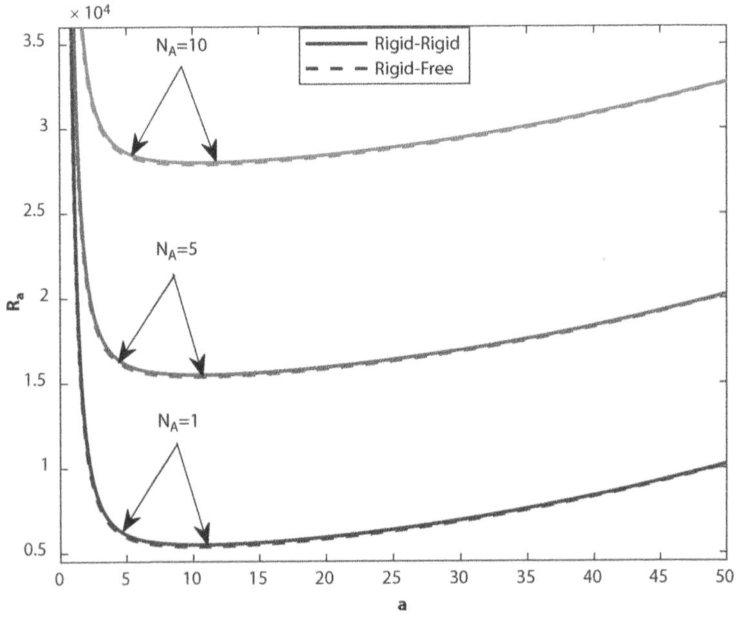

Figure 10.8 Deviation of R_a with a for numerous values of N_A.

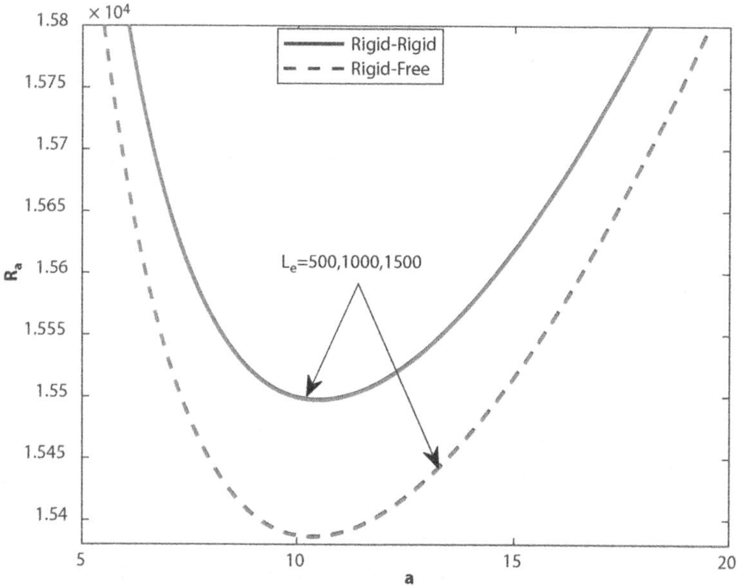

Figure 10.9 Variation of R_a against a for various values of L_e.

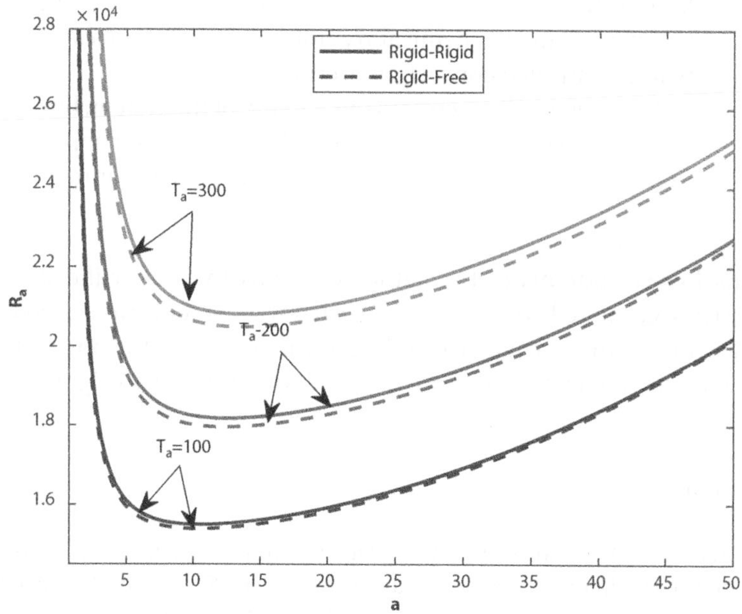

Figure 10.10 Deviation of R_a with a for different values of T_a.

of the system. Figure 10.10 displays the graph of R_a w. r. t. a for various values of $T_a = 100, 200, 300$. Taking another parameter as $\varepsilon = 0.4, R_s = 100, N_{TC} = 5, N_{CT} = 5, N_A = 5, L_e = 500, L_n = 500, R_n = -1$. It is clear that T_a goes on increasing with the increase in the value of R_a. Thus, T_a takes stabilizing effect, postpones the beginning of convection of the system.

10.5 Conclusion

Our primary conclusions are as follows:

- Porosity, Soret parameter, modified diffusivity ratio, thermo-nanoparticle Lewis number, and Taylor number have stabilizing effect on the system.
- Solutal Rayleigh number, Dufour parameter, thermosolutal Lewis number, and concentration Rayleigh number have destabilizing influence on the system.
- The findings suggest that for systems utilizing nanofluids in porous media, careful consideration of rotational effects and boundary conditions can lead to improved designs in thermal management, chemical reactors, and other applications where controlled convection is crucial.
- Rotation generally increases the stability of the system, limiting the growth of convection cells. This is critical for applications requiring controlled thermal and solutal environments.

AI disclosure statement: The author confirms that no artificial intelligence (AI)-generated content or AI-assisted writing tools were used in the preparation of this chapter (including summaries and the abstract). All text, analysis, and conclusions are the result of the author's original work.

References

1. Acharya, D.S. and Reddy, G.S., Double diffusive natural convection in power-law fluid saturated porous medium with Soret and Dufour effects. *J. Brazil. Soci. Mech. Sci. Eng.*, 34, 4, 525–530, 2012. https://doi.org/10.1590/S1678-58782012000400014.

2. Buongiorno, J., Convective Transport in Nanofluids. *J. Heat Transfer-transactions Asme.*, 128, 3, 240–250, 2006. doi.org/10.1115/1.2150834.
3. Chand, R. and Kango, S.K., Thermosolutal instability of dusty rotating Maxwell visco-elastic fluid in porous medium. *Adv. App. Sci. Res.*, 2, 6, 541–553, 2011.
4. Chand, R. and Rana, G.C., On the onset of thermal convection in rotating nanofluid layer saturating a Darcy–Brinkman porous medium. *Inter. J. Heat Mass Transf.*, 55, 21–22, 5417–24, 2012.
5. Chandrasekhar, S., *Hydrodynamic and hydromagnetic stability*, Courier Corporation, USA, 2013.
6. Kumar, A., Sharma, P.L., Deepak, Thakur, P., Soret and Dufour Effects on Thermosolutal Convection in Jeffrey Nanofluid in the Presence of Porous Medium. *Stru. Inte. Life*, 24, 1, 33–39, 2024a.
7. Kumar, A., Sharma, P., Lata, P., Bains, D., Thakur, P., Effect of magnetic field on the onset of thermal convection in a Jeffery nanofluid layer saturated by a porous medium: free-free, rigid-rigid and rigid-free boundary conditions. *J. Nig. Soci. Phys. Sci.*, 1934–1934, 2024b. https://doi.org/10.46481/jnsps.2024.1934.
8. Kuznetsov, A.V. and Nield, D.A., The Onset of Double-Diffusive Nanofluid Convection in a Layer of a Saturated Porous Medium. *Trans. Porous Media*, 85, 3, 941–951, 2010a. https://doi.org/10.1007/s11242-010-9600-1.
9. Kuznetsov, A.V. and Nield, D., The onset of double-diffusive nanofluid convection in a layer of a saturated porous medium. *Trans. Porous Media, 85*, 941–951, 2010b. DOI 10.1007/s11242-010-9600-1.
10. Kuznetsov, A.V. and Nield, D.A., Thermal instability in a porous medium saturated by a nanofluid: Brinkman model. *Transp. Porous Med.*, 81, 409–422, 2010c. DOI 10.1007/s11242-009-9413-2.
11. Malashetty, M.S., Tan, W., Swamy, M., The onset of double diffusive convection in a binary viscoelastic fluid saturated anisotropic porous layer. *Phy. Fluids*, 21, 8, 2009a. https://doi.org/10.1063/1.3194288.
12. Malashetty, M.S., Swamy, M., Heera, R., The onset of convection in a binary viscoelastic fluid saturated porous layer. *ZAMM-J. Appl. Mathe. Mech./ Zeitschrift für AngewandteMathematik und Mechanik Appl. Mathe. Mech.*, 89, 5, 356–369, 2009b. https://doi.org/10.1002/zamm.200800199.
13. Nield, D.A. and Kuznetsov, A.V., The onset of double-diffusive convection in a nanofluid layer. *Int. J. Heat Fluid Flow*, 32, 4, 771–776, 2011, doi: 10.1016/j.ijheatfluidflow.2011.03.010.
14. Nield, D.A. and Kuznetsov, A.V., The onset of double-diffusive convection in a nanofluid layer. ASME J. Heat Transf. submitted, 2010.
15. Nithyadevi, N. and Yang, R.J., Double diffusive natural convection in a partially heated enclosure with Soret and Dufour effects. *Int. J. Heat Fluid Flow*, 30, 5, 902–910, 2009. https://doi.org/10.1016/j.ijheatfluidflow.2009.04.001.
16. Rana, G.C. and Chand, R., Stability analysis of double-diffusive convection of Rivlin-Ericksen elastico-viscous nanofluid saturating a porous medium: a

revised model. *ForschungimIngenieurwesen*, 79, 1, 87–95, 2015. DOI 10.1007/s10010-015-0190-5.

17. Sharma, P.L., Kumar, A., Bains, D., Lata, P., Rana, G.C., Thermal convective instability in a Jeffrey nanofluid saturating a porous medium: rigid-rigid and rigid-free boundary conditions. *Struc. Inte. Life*, 23, 3, 351–356, 2023a.
18. Sharma, P.L., Kumar, A., Bains, D., Rana, G.C., Effect of Magnetic Field on Thermosolutal Convection in Jeffrey Nanofluid with Porous Medium. *Spec. Top. Rev. Poro. Media — Inte. J.*, 14, 3, 17–29, 2023b. 10.1615/SpecialTopicsRevPorousMedia.2023046929.
19. Sharma, P.L., Kumar, A., Kapalta, M., Bains, D., Effect of magnetic field on thermosolutal convection in a rotating non-Newtonian nanofluid with porous medium. *Inter. J. Appl. Mathe. Statis. Sci. (IJAMSS)*, 12, 1, 19–30, 2023c.
20. Sharma, P.L. and Kumar, A., Effects of rotation and magnetic field on thermosolutal convection in Jeffrey nanofluid with porous medium. *Spec. Top. Rev. Poro. Media — Inte. J.*, 15, 3, 43–60, 2024. 10.1615/SpecialTopicsRevPorousMedia.2023048400.
21. Sharma, P.L., Lata, P., Kumar, A., Effect of magnetic field on thermal instability in a porous medium layer saturated by a Jeffrey nanofluid using brinkman model: free-free, rigid-rigid, rigid-free boundary conditions. *Struc. Inte. Life*, 25, 1, 53–59, 2025.
22. Sharma, P.L., Katiyar, M., Deepak, Kumar, A., Effects of rotation and magnetic Field on thermosolutal convection in elastico-viscous Walters' (model B') nanofluid with porous medium. *Stoc. Model. App.*, 26, 21, 2022.
23. Tzou, D.Y., Instability of Nanofluids in Natural Convection. *J. Heat Transf.*, 130, 7, 072401, 2008a. https://doi.org/10.1115/1.2908427.
24. Tzou, D.Y., Thermal Instability of Nanofluids in Natural Convection. *Int. J. Heat Mass Transf.*, 51, 11-12, 2967–2979, 2008b. https://doi.org/10.1016/j.ijheatmasstransfer.2007.09.014.
25. Veronis, G., On finite amplitude instability in thermohaline convection. *J. Mar. Rese.*, 23, 1–17, 1965. https://elischolar.library.yale.edu/journal_of_marine_research/1038.
26. Wang, S. and Tan, W., Stability analysis of double-diffusive convection of Maxwell fluid in a porous medium heated from below. *Phys. Lett. A*, 372, 17, 3046–3050, 2008. https://doi.org/10.1016/j.physleta.2008.01.024.
27. Yadav, D., Lee, D., Cho, H.-H., Lee, J., The Onset of Double-Diffusive Nanofluid Convection in a Rotating Porous Medium Layer with Thermal Conductivity and Viscosity Variation: A Revised Model. *J. Poro. Media*, 19, 1, 31–46, 2016. https://doi.org/10.1615/JPorMedia.v19.i1.30.
28. Yang, Z., Wang, S., Zhao, M., Li, S., Zhang, Q., The onset of double diffusive convection in a viscoelastic fluid-saturated porous layer with non-equilibrium model. *PLoS One*, 8, 11, 2013. Article ID e79956 https://doi.org/10.1371/journal.pone.0079956.

11

Study of Two-Phase Flow Characteristics Due to Stretching Sheet

Aswin Kumar Rauta

Department of Mathematics, Binayak Acharya Degree College, Berhampur, Odisha, India

Abstract

This chapter deals with the momentum boundary layer two-phase flow and heat transfer of incompressible fluid with special attention to electrification of particles over stretching sheets because the heat transfer characteristics are important in the application of industrial processes involved with high-temperature regimes. Its good knowledge will help to design pertinent equipment. The effects of various parameters, variable viscosity parameter, heat generation/absorption parameter, and volume fraction on velocity and temperature profiles are considered. Governing partial differential equations (PDEs) of both phases are generated. A set of nonlinear ordinary differential equations are obtained using similarity transformations to reduce the PDEs. The boundary value problem is converted into an initial value problem using the shooting technique. Then the problem is solved using Runge–Kutta fourth-order method. The result obtained is compared with the other researchers output, which is in good agreement with them. Graphical representation of the result is illustrated. The Nusselt and skin friction coefficient are computed and compared with other authors'. Flows due to a continuously moving surface are encountered in various engineering applications, viz. the polymer processing unit of a chemical engineering plant, annealing of copper wires, glass fiber, and drawing of plastic films. The industrial sector plays an important role that has potential contributions toward the progress of the society. The investigation done in this paper is original and new in the area of inclined stretching sheet problems.

Email: aswinmath2003@gmail.com; ORCID: 0000-0002-8482-9478

Dhananjay Yadav, Mukesh Kumar Awasthi and Harith Mohamed Al-Azri (eds.) Non-Newtonian Fluids for Industrial Applications: Modeling and Simulations, (243–262) © 2026 Scrivener Publishing LLC

Keywords: Boundary layer, electrification of particles, heat–mass transfer, Nusselt number, shooting technique, stretching sheet

Nomenclature

x,y	Cartesian coordinates
η	Similarity variable
u	Velocity component of fluid along the x-axis
v	Velocity component of fluid along the y-axis
u_p	Velocity component of the particle along the x-axis
v_p	Velocity component of the particle along the y-axis
$U_w(x)$	Stretching sheet velocity
M	Electrification parameter
E	Electric field of force
m	Mass of particle
e	Charge of particle
T	Temperature of fluid phase
T_p	Temperature of particle phase
T_w	Wall temperature
T_∞	Temperature at large distance from the wall.
A^* and B^*	Nondimensional heat generation/absorption parameter for fluid phase
A^*_p and B^*_p	Nondimensional heat generation/absorption parameter for particle phase
θ	Nondimensional fluid phase temperature
θ_p	Nondimensional dust phase temperature
Ra	Radiation parameter
q_{rp}	Radiative heat flux for particle phase
q_r	Radiative heat flux for fluid phase
β^*	volumetric coefficient of thermal expansion
k^*	Mean absorption coefficient.
σ^*	Stefan–Boltzman constant
P_r	Prandtl number
G_r	Grashof number
F_r	Froud number
E_c	Eckert number
ω	Density ratio
φ	Volume fraction
c	Stretching rate
l	Characteristic length

c_p	Specific heat of fluid
c_s	Specific heat of particles
k_s	Thermal conductivity of particle
k	Thermal conductivity of fluid
A	Unsteady parameter
a	Constant
β	Fluid particle interaction parameter
τ_T	Thermal relaxation time, i.e., the time required by the dust particle to adjust its temperature relative to the fluid
τ_p	Velocity relaxation time, i.e., the time required by the dust particle to adjust its velocity relative to the fluid
τ	Relaxation time of particle phase
μ	Dynamic viscosity of fluid
ν	Kinematic viscosity of fluid
γ	Ratio of specific heat
ε	Diffusion parameter
ρ	Density of the fluid phase
ρ_s	Material density
ρ_p	Density of the particle phase

11.1 Introduction

The flow over stretching sheet has several applications such as cooling of nuclear reactor, aerospace component production, metal casting, food processing, lubrication, the aerodynamic extrusion of plastic sheet, glass blowing, cooling or drying of papers, drawing plastic films, extrusion of polymer melt-spinning process, rolling, manufacturing of plastic films and artificial fibers, cooling industry, dying industry, etc. The importance of such flow problems involving high-temperature regime lies in the fact that the mechanical properties of the final product are influenced by the stretching rate and the rate of cooling. The rate of stretching is also important because the rapid stretching results in sudden solidification, thereby destroying the properties expected for outcomes. In fact, the rate of heat transfer over a surface has a pivotal role in the quality of final products. Due to a wide area of applications and simple geometry, many researchers have studied boundary layer flow and heat transfer over a stretched surface and have presented the numerical and analytical solutions for various flow characteristics. The research development in this area has been reviewed as explained below.

The study of stretching sheet problem was initiated by B.C. Sakiadis [4] in his study of boundary layer flow over a continuous moving flat surface. Exact and approximate methods are adopted by him to investigate the problem. F.K. Tsou *et al.* [13] verified Sakiadis's theoretical predictions for Newtonian fluids and extended the same work to the heat transfer case. Later on, L.J. Crane [23] extended this problem to a stretching sheet whose surface velocity varies linearly with the distance x from a fixed origin. An exact similarity solution for the dimensionless differential system was obtained. Then many authors [1, 5, 8, 9, 11, 12, 21, 24, 30] studied such closed-form similarity solution for several other features like viscoelastic magnetohydrodynamics (MHD), suction, porosity and heat source/sink. etc. L.J. Grubka *et al.* [22] discussed the temperature field in the flow over a stretching surface when a uniform heat flux is exerted to the surface. M. M. Rashidi *et al.* [26] studied the free convective heat and mass transfer for MHD flow over a permeable vertical stretching sheet in the presence of radiation and buoyancy effects. K. Bhattacharyya *et al.* [19] investigated the effects of suction/blowing on steady boundary layer stagnation-point flow and heat transfer toward a shrinking sheet with thermal radiation. Many processes in engineering applications occur at high temperatures, and the radiative heat transfer becomes very important for the design of the pertinent equipment. However, the time-dependent boundary layer flows have been scarcely studied. B.J. Gireesha *et al.* [6] numerically investigated the boundary layer flow of an unsteady dusty fluid and heat transfer over a stretching surface with non-uniform heat source/sink. Mohmmod Safique Nazar *et al.* [29] discussed the unsteady boundary layer rotating flow due to a stretching surface. S. Mukhopadhyay [36] examined the effect of thermal radiation on the unsteady mixed convection flow and heat transfer bounded by a porous stretching surface embedded in a porous medium. P. R. Sharma *et al.* [31] studied unsteady MHD forced convection flow and mass transfer along a vertical stretching sheet with heat source/sink and variable fluid. There have been many theoretical models developed to describe flow properties and heat transfer toward a stretching sheet, but here, it is not possible to present all the literatures. So, some of the referred papers are cited in the reference section [14–18, 20, 25, 27–30, 32–34]. Due to entrainment of the ambient fluid, this boundary layer flow is quite different from Blasius flow past a flat plate. S.L. Soo [35] studied the effect of electrification on the dynamics of a particulate system. At low temperature, electrification of solid particles occurs because of impact with the wall. Even a very slight charge on the solid particles will have a pronounced effect on concentration distribution in the flow of a gas–solid system. Although electric charge on the solid particles can be excluded by definition in theoretical analysis or when dealing truly with a boundless system, electrification of the solid particles always occurs when contact and separation are made between the solid particles and a wall of different materials or similar materials but different surface condition. The electric charges on the solid particles cause deposition of the solid particles on a wall in a more significant manner than the gravity effect and are expected to affect the motion of a metalized propellant and its product of reaction through a rocket nozzle and the jet at the exit of the nozzle.

Recently, a new idea is added to the study of boundary layer fluid flow and heat transfer. Indeed, in such studies, the effects of thermal radiation and temperature-dependent viscosity are included. The equation of the particle phase, the momentum equation for particulate phase in normal direction, heat due to conduction, and viscous dissipation in the energy equation of the particle phase have been considered for better understanding of the boundary layer characteristics. The effects of volume fraction, fluid interaction parameter, radiation, electrification, and unsteady parameter on skin friction and heat transfer have been studied. The problem of two-phase suspension flow is solved in the framework of a model of a two-way coupling model or a two-fluid approach. In this chapter, radiation and electrification of particles and source/sink parameters are addressed with reasonable simplification. Runge–Kutta method integrated with shooting technique is used to solve the system of ordinary differential equations that are obtained from governing partial differential equations.

11.2 Modeling of the Problem

Let us consider an unsteady two-dimensional laminar boundary layer flow of a viscous, incompressible dusty fluid over a stretching sheet. The flow is caused by an impermeable stretching sheet. A Cartesian coordinate system is used. The sheet is considered along the x-axis and y-axis being normal to it. Two equal and opposite forces are applied along the stretching sheet. The sheet is being stretched with the velocity $U_w(x)$ along the x-axis, keeping the origin fixed in the fluid of ambient temperature T_∞. Both the fluid and the dust particle clouds are supposed to be static at the beginning. The dust particles are assumed to be spherical in shape and uniform in size.

11.3 Flow Analysis and Coordinate System

The governing equations of unsteady two-dimensional boundary layer incompressible flows of dusty fluids are given by

$$\frac{\partial u}{\partial x}+\frac{\partial v}{\partial y}=0 \tag{11.1}$$

$$\frac{\partial}{\partial t}\rho_p+\frac{\partial}{\partial x}\left(\rho_p u_p\right)+\frac{\partial}{\partial y}\left(\rho_p v_p\right)=0 \tag{11.2}$$

$$\begin{aligned}(1-\varphi)\rho\left[\frac{\partial u}{\partial t}+u\frac{\partial u}{\partial x}+v\frac{\partial u}{\partial y}\right]&=(1-\varphi)\mu\frac{\partial^2 u}{\partial y^2}-\frac{1}{\tau_p}\varphi\rho_s\left(u-u_p\right)\\&\quad-(1-\varphi)\rho g\beta^*\left(T-T_\infty\right)+\varphi\rho_s\left(\frac{e}{m}\right)E\end{aligned} \tag{11.3}$$

$$\varphi\rho_s\left(\frac{\partial u_p}{\partial t}+u_p\frac{\partial u_p}{\partial x}+v_p\frac{\partial u_p}{\partial y}\right)=\frac{\partial}{\partial y}\left(\varphi\mu_s\frac{\partial u_p}{\partial y}\right)+\frac{1}{\tau_p}\varphi\rho_s\left(u-u_p\right)+\varphi\left(\rho_s-\rho\right)g+\varphi\rho_s\left(\frac{e}{m}\right)E \tag{11.4}$$

$$\varphi\rho_s\left(\frac{\partial v_p}{\partial t}+u_p\frac{\partial v_p}{\partial x}+v_p\frac{\partial v_p}{\partial y}\right)=\frac{\partial}{\partial y}\left(\varphi\mu_s\frac{\partial v_p}{\partial y}\right)+\frac{1}{\tau_p}\varphi\rho_s\left(v-v_p\right) \tag{11.5}$$

$$\begin{aligned}(1-\varphi)\rho c_p\left[\frac{\partial T}{\partial t}+u\frac{\partial T}{\partial x}+v\frac{\partial T}{\partial y}\right]&=(1-\varphi)k\frac{\partial^2 T}{\partial y^2}+\frac{1}{\tau_T}\varphi\rho_s c_s\left(T_p-T\right)+\frac{1}{\tau_p}\varphi\rho_s\left(u-u_p\right)^2\\&\quad+(1-\varphi)\mu\left(\frac{\partial u}{\partial y}\right)^2+\varphi\rho_s\left(\frac{e}{m}\right)Eu_p\end{aligned} \tag{11.6}$$

$$\varphi\rho_s c_s\left[\frac{\partial T_p}{\partial t}+u_p\frac{\partial T_p}{\partial x}+v_p\frac{\partial T_p}{\partial y}\right]=\frac{\partial}{\partial y}\left(\varphi k_s\frac{\partial T_p}{\partial y}\right)-\frac{1}{\tau_p}\varphi\rho_s c_s\left(T_p-T\right)-\frac{1}{\tau_p}\varphi\rho_s\left(u-u_p\right)^2$$
$$+\varphi\mu_s\left[u_p\frac{\partial^2 u_p}{\partial y^2}+\left(\frac{\partial u_p}{\partial y}\right)^2\right]+\varphi\rho_s\left(\frac{e}{m}\right)Eu_p$$

(11.7)

with boundary conditions $u=U_w(x)=\dfrac{cx}{1-at},\ v=0\quad as\quad y\to 0$

$$\rho_p=\omega\rho, u=0, u_p=0, v_p\to v\quad as\quad y\to\infty$$

where ω is the density ratio in the main stream.

The following nondimensional temperature boundary conditions are used in order to solve the temperature Equations 11.6 and 11.7.

$$T=T_w=T_\infty+T_0\frac{cx^2}{\nu(1-at)^2}\quad at\quad y=0$$

$$T\to T_\infty,\quad T_p\to T_\infty\quad as\quad y\to\infty$$

For most of the gases, $\tau_p\approx\tau_T$, $k_s=k\dfrac{c_s}{c_p}\dfrac{\mu_s}{\mu}$ if $\dfrac{c_s}{c_p}=\dfrac{2}{3P_r}$.

Introducing the following nondimensional variables in Equations 11.1–11.7:

$$u=\frac{cx}{1-at}f'(\eta),\quad v=-\sqrt{\frac{cv}{1-at}}f(\eta),\quad \frac{\varphi\rho_s}{\rho}=\frac{\rho_p}{\rho}=\rho_r=H(\eta)$$
$$u_p=\frac{cx}{1-at}F(\eta),\quad v_p=\sqrt{\frac{cv}{1-at}}G(\eta),\quad \eta=\sqrt{\frac{c}{\nu(1-at)}}y$$
$$P_r=\frac{\mu c_p}{k},\quad \beta=\frac{1-at}{c\tau_p},\quad \epsilon=\frac{v_s}{v},\quad \varphi=\frac{\rho_p}{\rho_s},\quad A=\frac{a}{c},\quad E_c=\frac{cv}{C_pT_0},\quad M=\frac{(1-at)^2}{c^2x}\left(\frac{e}{m}\right)E,$$
$$F_r=\frac{c^2x}{g(1-at)^2},\quad G_r=\frac{g\beta^*(T_w-T_\infty)(1-at)^2}{c^2x},\quad \gamma=\frac{\rho_s}{\rho},\quad v=\frac{\mu}{\rho},$$
$$\theta(\eta)=\frac{T-T_\infty}{T_w-T_\infty},\quad \theta_p(\eta)=\frac{T_p-T_\infty}{T_w-T_\infty},\quad \text{Where}\quad T-T_\infty=T_0\frac{cx^2}{\nu}\frac{1}{(1-at)^2}\theta,\quad T_p-T_\infty=T_0\frac{cx^2}{\nu}\frac{1}{(1-at)^2}\theta_p$$

Then Equations 11.1 and 11.7 become

$$H'(\eta) = -\left(H(\eta)F(\eta) + H(\eta)G'(\eta)\right) / \left(A\frac{\eta}{2} + G(\eta)\right) \quad (11.8)$$

$$f''' = -ff'' + (f')^2 + A\left(f' + \frac{\eta}{2}f''\right) - \frac{\beta H}{(1-\varphi)}(F - f') - \frac{HM}{(1-\varphi)} - G_r\theta(\eta) \quad (11.9)$$

$$F'' = \left[A\left\{\frac{\eta}{2}F' + F\right\} + (F)^2 + GF' + \beta(F - f') - \left(1 - \frac{1}{\gamma}\right)\frac{1}{F_r} - M\right] / \epsilon \quad (11.10)$$

$$G'' = \left[\frac{A}{2}(\eta G' + G) + GG' + \beta(f + G)\right] / \epsilon \quad (11.11)$$

$$\theta'' = \begin{array}{l} Pr(2f'\theta - f\theta') - \frac{2}{3}\frac{\beta H}{1-\varphi}(\theta_p - \theta) - \frac{PrE_c\beta H}{1-\varphi}(F - f')^2 - PrE_c(f'')^2 \\ + \frac{A}{2}Pr(\eta\theta'(\eta) + 4\theta(\eta)) - \frac{P_rE_cHMF(\eta)}{(1-\varphi)} - (A^* f'(\eta) + B^*\theta(\eta)) \end{array} \quad (11.12)$$

$$\theta_p'' = \frac{\left[\begin{array}{l} \frac{A}{2}(\eta\theta_p' + 4\theta_p) + 2F\theta_p + G\theta_p' + \beta(\theta_p - \theta) + \frac{3}{2}E_cP_r\beta(f' - F)^2 \\ -\frac{3}{2}\epsilon E_cP_r(FF'' + (F')^2) - \frac{3}{2}E_cP_rMF - \frac{3}{2}\frac{1}{\gamma}(A_p^*F(\eta) + B_p^*\theta_p(\eta)) \end{array}\right]}{\left(\frac{\varepsilon}{P_r}\right)} \quad (11.13)$$

$$G' = 0, f = 0, f' = 1, F' = 0, \theta = 1, \theta_p' = 0 \ as\eta \to 0 \ \& \ f' = 0, F = 0, G = -f, H = \omega, \theta \to 0, \theta_p \to 0 \ as\eta \to \infty$$

The physical quantities of interest are the skin friction coefficient C_f and the local Nusselt number Nu_x, which are defined as

$C_f = \frac{\tau_w}{\rho U_w^2}$, $Nu_x = \frac{x q_w}{k(T_w - T_\infty)}$, where the surface shear stress τ_w and surface heat flux q_w are given by $\tau_w = \mu \left(\frac{\partial u}{\partial y}\right)_{y=0}$, $q_w = -k\left(\frac{\partial T}{\partial y}\right)_{y=0}$.

11.4 Solution Method

The set of nonlinear differential Equations 11.8 to 11.12 subject to the boundary conditions 11.13 were solved numerically using the Runge–Kutta method with shooting technique. The advantage of the shooting technique is to solve a boundary value problem by converting it into an initial value problem. Here, in this problem, the values of $f''(0), F(0), G(0), H(0), \theta'(0), \theta_p(0)$ are not known, but $f'(\infty)=0, F(\infty)=0, G(\infty)=-f(\infty), H(\infty)=\omega, \theta(\infty)=0, \theta_p(\infty)=0$ are given. We use the shooting method to determine the values of $f''(0), F(0), G(0), H(0), \theta'(0), \theta_p(0)$. In this problem, the missing values of $f''(0), \theta'(0), F(0), G(0), H(0)$ and $\theta_p(0)$ for different set of values of the parameter are chosen on a hit-and-trial basis such that the boundary condition at the other end, i.e., the boundary condition at infinity (η_∞), is satisfied. The most important step in this method is to choose an appropriate finite value of $\eta \to \infty$. In order to determine $\eta \to \infty$ for the boundary value problem described by Equations 11.8–11.12, we start with initial guess values for a particular set of physical parameters to obtain $F(0), G(0), H(0), \theta_p(0), \theta'(0)$, and $f''(0)$. The solution procedure is repeated with another large value of $\eta \to \infty$ until two successive values of $F(0), G(0), H(0), \theta_p(0), \theta'(0)$, and $f''(0)$ differ only by a specified significant digit. We have supplied $f''(0) = \propto_0$ and $f''(0) = \propto_1$. The improved value of $f''(0) = \propto_2$ is determined by utilizing linear interpolation formula. Then, the value of $f'(\propto_2, \infty)$ is determined by using the Runge–Kutta method. If $f'(\propto_2, \infty)$ is equal to $f'(\infty)$ up to a certain decimal accuracy, then $\propto_2$ i.e $f''(0)$ is determined; otherwise, the above procedure is continued with $\propto_0 = \propto_1$ and $\propto_1 = \propto_2$ until a correct $\propto_2$ is obtained. The same procedure described above is adopted to determine the correct values of $F(0), G(0), H(0), \theta'(0), \theta_p(0)$. If they agreed to about six significant digits, the last value of η_∞ is used; otherwise, the procedure is repeated until further change in η_∞ did not lead to any more change in the value of $F(0), G(0), H(0), \theta_p(0), \theta'(0)$, and $f''(0)$. Depending upon the initial guess and number of steps N=81.0, the solution of the present problem is obtained by numerical computation after finding the infinite value for η. So, we ran our bulk of computations for $\eta_\infty = 5.0$ with step size $\Delta\eta = 0.125$.

11.5 Discussion

The following table shows an excellent agreement between our numerical results and the previously reported results. The accuracy of results obtained by our investigation is compared with the results obtained by C.H. Chen [10], L.J. Grubka *et al.* [22], A. Subhas *et al.* [3], S. Mukhopadhyay *et al.* [36], A. Ishak *et al.* [2], and B.J. Gireesha *et al.* [6] for the local Nusselt number in the limiting condition $\beta, A^*, B^*, A_p{}^*, B_p{}^*, R_a, G_r, E_c, A, M, \phi = 0.0$. It is observed from the table that our result is in good agreement with their results. The computations were done by the computer language FORTRAN-77. The shear stress (skin friction coefficient), which is proportional to $f''(0)$, and the rate of heat transfer (Nusselt number), which is proportional to $\theta'(0)$, are tabulated in Table 11.1 for different values of

Table 11.1 Showing initial values of wall velocity gradient $-f''(0)$ and temperature gradient $-\theta'(0)$.

φ	β	A	A*	B*	A_p^*	B_p^*	$-f''(0)$	$-\theta'(0)$
0	0	0	0	0	0	0	1.001396	1.082315
0.01	0.01	0.1	0.1	0.1	0.1	0.1	1.011790	0.553447
0.1							1.010030	0.553648
0.3							1.004963	0.556359
0.5							0.995560	0.563478
0.01	0.01	0.1	0.1	0.1	0.1	0.1	1.011790	0.553447
	0.02						1.010760	0.554952
	0.03						1.012920	0.549767
0.01	0.01	0.1	0.1	0.1	0.1	0.1	1.011790	0.553447
		0.2					1.048937	0.597137
		0.3					1.084461	0.640866

(Continued)

Table 11.1 Showing initial values of wall velocity gradient $-f''(0)$ and temperature gradient $-\theta'(0)$. (*Continued*)

φ	β	A	A*	B*	A_p^*	B_p^*	$-f''(0)$	$-\theta'(0)$
			-0.5				1.012777	0.755196
			-0.1				1.012232	0.619318
			0.0				1.012135	0.585140
0.01	0.01	0.1	0.1	0.1	0.1	0.1	1.011790	0.553447
			0.5				1.011339	0.414973
				-0.5			1.013463	0.783452
				-0.1			1.012409	0.641113
				0.0			1.012396	0.595400
0.01	0.01	0.1	0.1	0.1	0.1	0.1	1.011790	0.553447
				0.5			1.010519	0.308891
					-0.5		1.011781	0.553444
					-0.1		1.011785	0.553445
					0.0		1.011789	0.553446
0.01	0.01	0.1	0.1	0.1	0.1	0.1	1.011790	0.553447
					0.5		1.011791	0.553448
						-0.5	1.011785	0.553411
						-0.1	1.011788	0.553438
						0.0	1.011789	0.553443
0.01	0.01	0.1	0.1	0.1	0.1	0.1	1.011790	0.553447
						0.5	1.011792	0.553461

parameters used. It is observed from the table that shear stress and rate of heat transfer decrease with the increase in 'Ec'. The shear stress and rate of heat transfer increase for increasing values of Pr. The Nusselt number increases with the increasing values of the unsteady parameter 'A' and electrification parameter 'M', but the Nusselt number decreases when the radiation parameter 'Ra' increases. The skin friction coefficient decreases for increasing values of 'M'. It is also observed from the table that the increasing value of diffusion parameter ϵ decreases both the skin friction and the Nusselt number. A comparative study is given in Table 11.2 below.

Figure 11.1 explains that the temperature of fluid phase decreases with the increase in the unsteady parameter 'A'. Physically, when the unsteadiness increases, the sheet loses more heat, which causes the decrease in temperature. We also observe that as the distance from the stretching sheet within the dynamic region increases, the temperature field decreases as unsteadiness increases. But Figure 11.2 describes that the temperature profile of the particle phase increases with the increase in the unsteady parameter because getting more time of collision of particles produces more frictional heating.

Figure 11.3 demonstrates the effect of β, which infers that the increase in β increases the particle phase velocity. Figure 11.4 illustrates that the increase in β increases the temperature of both fluid as well as particle

Table 11.2 Comparison results for the wall temperature gradient $-\theta'(0)$ in the case of $\beta, A^*, B^*, A_p^*, B_p^*, R_a, G_r, E_c, A, M, \phi = 0.0$.

P_r	C.H. Chen [9]	Grubka, L.J. *et al.* [22]	Mukhopadhyay, S. *et al.* [36]	B.J. Gireesha *et al.* [7]	Present study $-\theta'(0)$
0.72	1.0885	1.0885	1.0885	1.0885	1.088635
1.0	1.3333	1.3333	1.3333	1.3333	1.333927
3.0	2.5097	-	2.5097	2.5097	2.509564

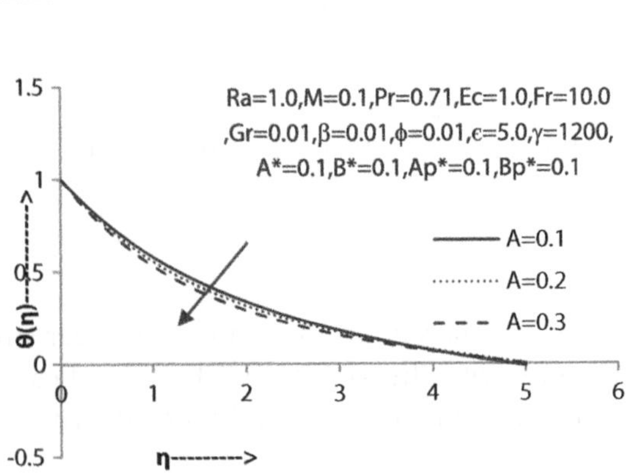

Figure 11.1 Nondimensional temperature profile of fluid phase w.r.t. 'A'.

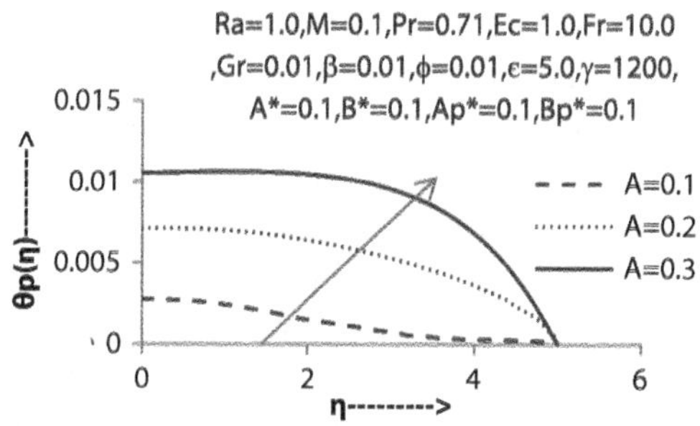

Figure 11.2 Nondimensional temperature profile of particle phase w.r.t. 'A'.

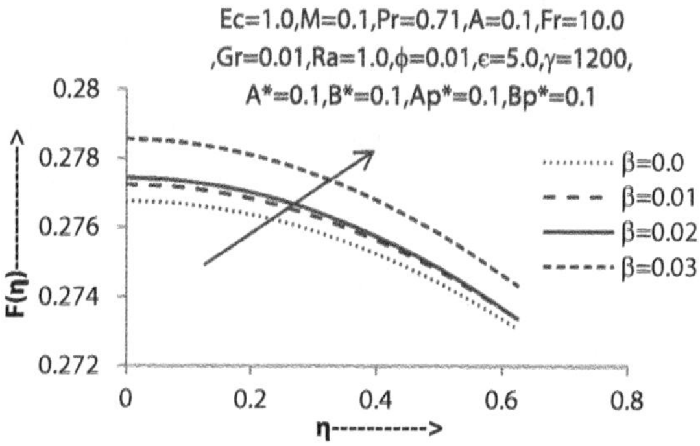

Figure 11.3 Nondimensional velocity profile of particle phase w.r.t. 'β'.

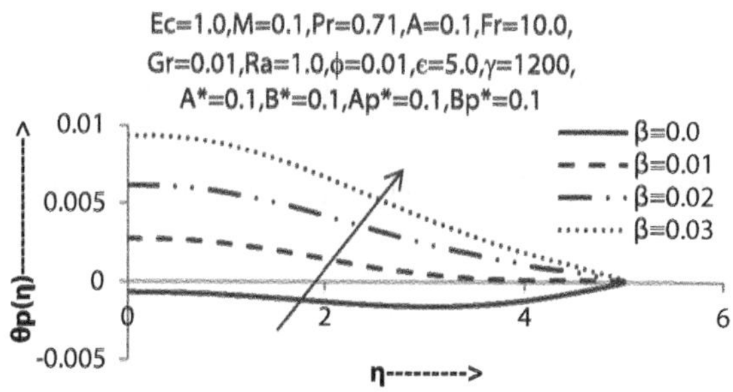

Figure 11.4 Nondimensional temperature profile of particle phase w.r.t. 'β'.

phase because delaying the accumulation of high-volume fractions of particles allows greater access of the fluid to the particles, increasing the convective heat and mass transfer.

Figure 11.5 indicates the increase in the velocity profile particle phase with the increase in Pr. Figure 11.6 depicts the effect of Pr on the temperature profile of the fluid phase and particle phase. From the figure, we observe that, when Pr increases, the temperature of both phases decreases. This is due to the fact that for smaller values of *Pr* equivalent to larger values of thermal conductivities, this phenomenon leads to the decrease in energy ability that reduces the thermal boundary layer, and therefore, heat is able to diffuse away from the stretching sheet.

Figures 11.7 and 11.8 demonstrate that the temperature of both phases increases with the increasing value of A*. The presence of the heat source generates energy in the thermal boundary layer, and as a consequence, the temperature rises. In the case of heat absorption A*<0, the temperature

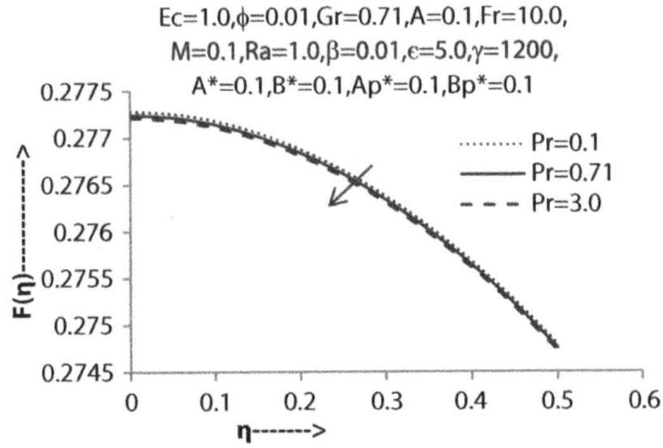

Figure 11.5 Nondimensional velocity profile of particle phase w.r.t. 'Pr'.

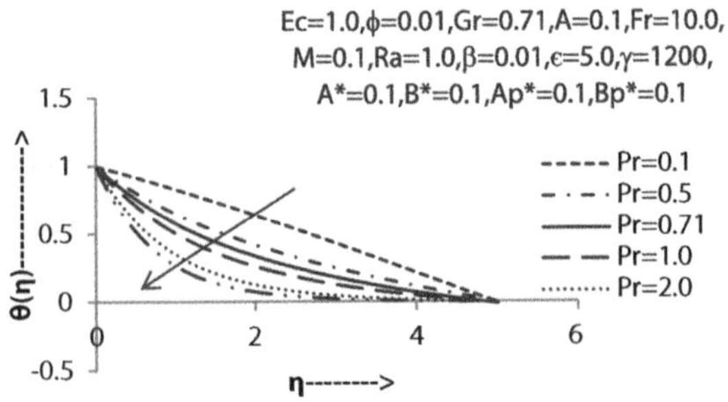

Figure 11.6 Nondimensional temperature profile of fluid phase w.r.t. 'Pr'.

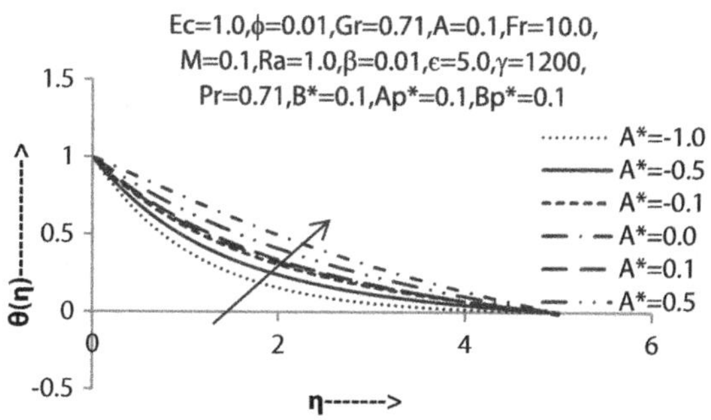

Figure 11.7 Nondimensional temperature profile of fluid phase w.r.t. 'A*'.

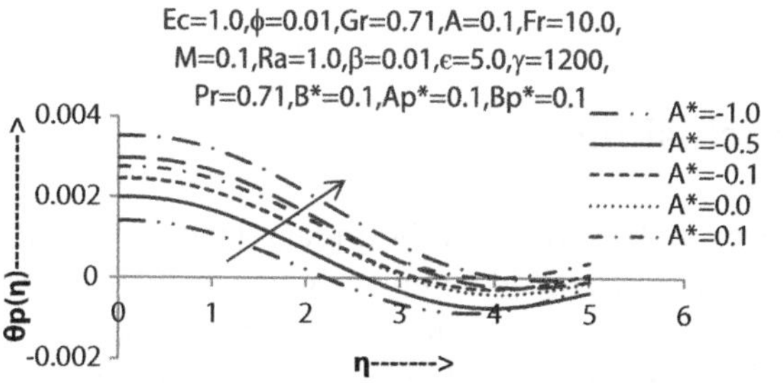

Figure 11.8 Nondimensional temperature profile of particle phase w.r.t. 'A*'.

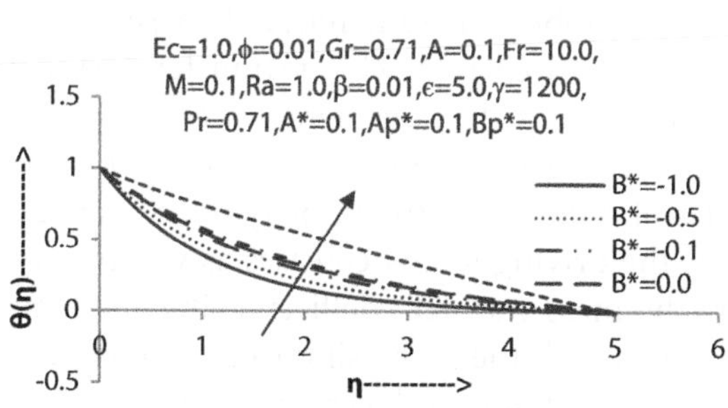

Figure 11.9 Nondimensional temperature profile of fluid phase w.r.t. 'B*'.

falls with decreasing values of A*<0 owing to the absorption of energy in the boundary layer.

Figures 11.9 and 11.10 explain the behavior of temperature of both phases w.r.t. B*, which indicates that the increase in the value of B* increases the temperature of both phases. As in the case of space-dependent heat source, the temperature increases due to the release of thermal energy for B*>0,

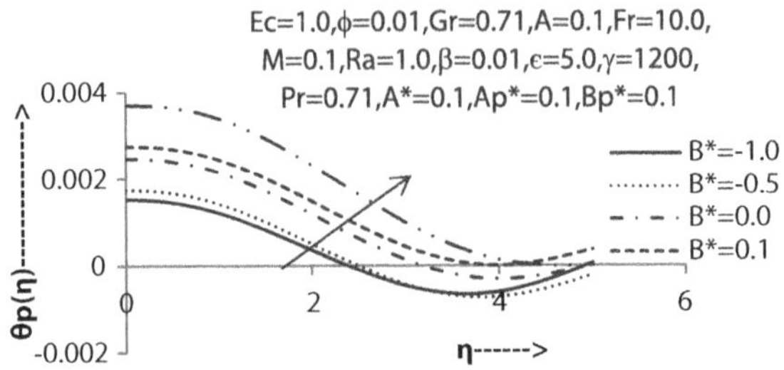

Figure 11.10 Nondimensional temperature profile of particle phase w.r.t. 'B*'.

while the temperature drops for decreasing values of $B^*<0$ owing to the absorption of energy.

11.6 Conclusions

We conclude by noting that the boundary layer flow and heat transfer characteristics were discussed extensively by considering various physical parameters like unsteady parameter 'A', Prandtl number 'Pr', Eckert number 'Ec', electrification parameter 'M', radiation parameter 'Ra', Grashof number 'Gr', heat source/sink parameters A^* and B^*, volume fraction 'φ' and particle interaction parameter 'β', ratio of specific heat γ, diffusion parameter 'ε', and Froud number 'Fr'. Our model is based on the motion and collisions of particles, as well as the charge transfer during these collisions. The novelty of the study is the consideration of the diffusion parameter, which is scarcely investigated by other researchers. It is observed that the increasing values of ϵ decreases the velocity of the particle phase. As the length increases, the driving force decreases as we approach equilibrium, and this lowers the mass transfer coefficient. But the temperature of the fluid phase increases with the increasing value of ϵ because the diffusion of particles through the fluid produces frictional heating due to collision of particles. Also, the temperature of the particle phase decreases with the increase in the value of ϵ due to heat transfer from the particle phase to the fluid phase. Another important investigation of the present study is the inclusion of the radiation term and heat source/sink term in the particle phase. The influence of these terms is the same as that of the fluid phase, which was interpreted in detail. In our study, the claim is that the velocity profile increases with the increase in the electrification parameter, which arises due to electrification of particles, whereas the analysis made by authors taking electrically conducting fluid in the presence of external magnetic field shows a decrease in fluid velocity due to the presence of opposing Lorentz force. The increasing value of the unsteady parameter A decreases the temperature profiles of the fluid phase and increases the temperature of the dust phase. Also, the increase in the value of A decreases the velocity of both phases. The rate of cooling is much faster for higher values of the unsteady parameter, but it takes a long time for cooling during the steady flow.

Finally, we observed the following results:

i. The increasing value of the unsteady parameter A decreases the temperature profiles of the fluid phase and increases the temperature of the dust phase. Also, the increase in the value of A decreases the velocity of both phases. The rate of cooling is much faster for higher values of the unsteady parameter, but it takes a long time for cooling during the steady flow.
ii. The thermal boundary layer thickness decreases on the effect of Pr. The temperature decreases at a faster rate for higher values of Pr, which implies that the rate of cooling is faster in case of a higher Prandtl number.
iii. Increasing β increases the velocity and temperature of the particle phase.
iv. The increasing value of φ decreases the temperature profile of both phases but increases the momentum boundary layer thickness of the particle phase.
v. We investigated the problem assuming the values γ=1200.0, T= 5.0, and F_r = 10.0.

AI disclosure statement: The author acknowledges the use of AI-based tools for minor language editing or grammatical correction. However, all intellectual contributions, analyses, and conclusions presented in this chapter are entirely the author's own.

References

1. Adhikari, A. and Sanyal, D.C., Heat transfer on MHD viscous flow over a stretching sheet with prescribed heat flux. *Bull. Int. Math. Virtual Inst.*, 3, 35–47, 2013. ISSN 1840-4367.
2. Ishak, A., Nazar, R., Pop, I., Hydromagnetic flow and heat transfer adjacent to a stretching vertical sheet. *Heat Mass Transfer*, 44, 921–927, 2008.

3. Subhas, A.M. and Mahesh, Heat transfer in MHD visco-elastic fluid flow over a stretching sheet with variable thermal conductivity, non-uniform heat source and radiation. *Appl. Math. Model.*, 32, 1965–83, 2008.
4. Sakiadis, B.C., Boundary Layer behavior on continuous solid surface; boundary layer equation for two dimensional and axisymmetric flow. *A.I.Ch.E. J.*, 7, 26–28, 1961.
5. Gireesha, B.J., Chamakha, A.J., Manjunatha, S., Bagewadi, C.S., Mixed convective flow of a dusty fluid over a vertical stretching sheet with non-uniform heat source/sink and radiation. *Int. J. Numer. Methods Heat Fluid Flow*, 23, 4, 598–612, 2013.
6. Gireesha, B.J., Roopa, G.S., Bagewadi, C.S., Boundary Layer flow of an unsteady Dusty fluid and Heat Transfer over a stretching surface with non-uniform heat source/sink. *Engineering*, 3, 726–735, 2011. http://www.SciRP.org/Journal/am, Scientific Research.
7. Gireesha, B.J., Manjunatha, S., Bagewadi, C.S., Effect of radiation on boundary layer flow and heat transfer over a stretching sheet in the presence of a free stream velocity. *J. Appl. Fluid Mech.*, 7, 1, 15–24, 2014.
8. Lavanya, B. and Leela Ratnam, A., Dufour and Soret effects on steady MHD free convective flow past a vertical porous plate embedded in a porous medium with chemical reaction, radiation heat generation and viscous dissipation, Pelagia Research Library. *Adv. Appl. Sci. Res.*, 5, 1, 127–142, 2014.
9. Prasanna Kumara, C., Ramesh, G.K., Chamkha, A.J., Gireesha, B.J., Stagnation-point flow of a viscous fluid towards a stretching surface with variable thickness and thermal radiation. *Int. J. Ind. Math.* (ISSN 2008-5621), 7, 1, 9, 2015. Article ID IJIM-00582.
10. Chen, C.H., Laminar Mixed convection Adjacent to vertical continuity stretching sheet. *Heat Mass Transfer*, 33, 5–6, 471–476, 1998.
11. Datta, and Mishra, S.K., Boundary layer flow of a dusty fluid over a semi-infinite flat plate. *Acta Mech.*, 42, 1–2, 71–83, 1982. doi:10.1oo7/BF01176514.
12. Dimian, M.F. and Megahed, A.M., Effects of variable fluid properties on unsteady heat transfer over a stretching surface in the presence of thermal radiation. *Ukr. J. Phys.*, 58, 345, 2013.
13. Tsou, F.K., Sparrow, E.M., Glodstein, R.J., Flow and Heat Transfer in the boundary layer on a continous moving surface. *Int. J. Heat Masstransfer*, 10, 219–235, 1967.
14. Sarojamma, G., Mahaboobjan, S., Nagendramma, V., Influence of hall currents on cross diffusive convection in a MHD boundary layer flow on stretching sheet in porous medium with heat generation. *Int. J. Math. Arch.*, 6, 3, 227–248, 2015. www.ijma.info ISSN 2229–5046.
15. Shit, G.C. and Haldar, R., combined effects of thermal radiation and hall current on MHD free-convective flow and mass transfer over a stretching sheet with variable viscosity. *J. Appl. Fluid Mech.*, 5, 2, 113–121, 2012. www.jafmonline.net, ISSN 1735-3572, EISSN 1735-3645.
16. Schlichting, H., *Boundary Layer Theory*, McGraw-Hill, New York, 1968.

17. Dessie, H. and Kishan, N., Scaling group analysis on MHD Free convective heat and mass transfer over a stretching surface with suction/injection, heat source /sink considering viscous dissipation and chemical reaction effects. *Appl. Appl. Mathematics*, 9, 553–572, 2014. Hunegnawd@yahoo.com, kishan_n@rediffmail.com.
18. Konch, J. and Hazarika, G.C., Effects of Variable Viscosity and Thermal Conductivity on MHD Free Convection Flow of Dusty Fluid along a Vertical Stretching Sheet with Heat Generation. *Int. Res. J. Eng. Technol. (IRJET)*, 03, 02, 2016. e-ISSN: 2395-0056 Feb-2016 www.irjet.net.
19. Bhattacharyya, K. and Layek, G.C., Effects of suction/blowing on steady boundary layer stagnation-point flow and heat transfer towards a shrinking sheet with thermal radiation. *Int. J. Heat Mass Transf.*, 54, 302–307, 2011.
20. Prasad, K.V., Vajravelu, K., Datti, P.S., The effects of variable fluid properties on the hydromagnetic flow and heat transfer over a non-linearly stretching sheet. *Int. J. Ther. Sci.*, 49, 603–610, 2010.
21. Vajravelu, K. and Nayfeh, J., Hydromagnetic flow of a dusty fluid over a stretching sheet. *Int. J. Nonlinear Mech.*, 27, 6, 937–945, 1992.
22. Grubka, L.J. and Bobba, K.M., Heat Transfer characteristics of a continuous stretching surface with variable temperature. *Int. J. Heat Mass Transfer*, 107, 248–250, 1985.
23. Crane, L.J., Flow past a stretching plate. *Z. fur AngewandteMathematik und physic ZAMP*, 2, 4, 645–647, 1970.
24. Das, M., Mahatha, B.K., Nandkeolyar, R., Mandal, B.K., Saurabh, K., Unsteady Hydromagnetic Flow of a Heat Absorbing Dusty Fluid Past a Permeable Vertical Plate with Ramped Temperature. *J. Appl. Fluid Mech.*, 7, 3, 485–492, 2014.
25. Khader, M.M. and Megahed, A.M., Numerical solution for boundary layer flow due to a nonlinearly stretching sheet with variable thickness and slip velocity. *Eur. Phys. J.*, 2013. Plus128.
26. Rashidi, M.M., Rostami, B., Freidoonimehr, N., Abbasbandy, S., Free convective heat and mass transfer for MHD flow over a permeable vertical stretching sheet in the presence of the radiation and buoyancy effects. *Ain Shams Eng. J.*, 5, 901–912, 2014.
27. Rashidi, M.M. and Mohimanian Pour, S.A., Analytic approximate solutions for unsteady boundary-layer flow and heat transfer due to a stretching sheet by Homotopy analysis method. *Nonlinear Anal.-Model. Control*, 15, 1, 83–95, 2010.
28. M., Aziz, W., Ahmed, M., Abbasi, A., Taj, M., MHD flow of a viscous fluid over an exponentially stretching sheet in a porous medium. *J. Appl. Math.*, 8, 2014. Article ID 256761.
29. Shafique, M., Nazir, Abbas, F., Numerical solution of two-dimensional stagnation flows of micropolar fluids towards a shrinking sheet by using SOR iterative procedure. *RESEARCH ORGANISATION J. Progressive Res. Math. (JPRM)*, 3, 1, 2015. SCITECH ISSN: 2395-0218.

30. Mukherjee, B. and Prasad, N., Effect of radiation and porosity parameter on hydromagnetic flow due to exponentially stretching sheet in a porous media. *Inter. J. Eng. Sci. Tech.*, 6, 58, 2014.
31. Sharma, P.R., Sharma, M., Yadav, R.S., Unsteady MHD forced convection flow and mass transfer along a vertical stretching sheet with heat source /sink and variable fluid properties. *Int. Res. J. Eng. Technol. (irjet),* 02, 03, 2015. E-ISSN: 2395-0056, www.irjet.net P-ISSN: 2395-0072.
32. Barik, R.N., Dash, G.C., Rath, P.K., Heat and mass transfer on MHD flow through a porous medium over a stretching surface with heat source. *Math. Theory Model,* 2, 7, 2012. ISSN 2224-5804.
33. Sharidan, S., Mahmood., J., Pop, I., Similarity solutions for the unsteady boundary layer flow and Heat Transfer due to a stretching sheet. *Inc. Appl. Mech. Engeenering,* 11, 3, 647–654, 2008.
34. Ahmed, S. and Chamkha, A.J., Effects of chemical reaction, heat and mass transfer and radiation on mhd flow along a vertical porous wall in the present of induced magnetic field. *Int. J. Ind. Math.*, 2, 245–261, 2010.
35. Soo, S.L., Effect of electrification on the dynamics of a particulate system. *I. EC. Fund.*, 3, 75–80, 1964.
36. Mukhopadhyay, S. and Anderssson, H.I., Effect of slip and heat transfer analysis of flow over an unsteady stretching surface. *Heat Mass Transfer*, 45, 11, 1447–1452, 2009. doi:10.1007/s11012-009-0516-7.

12

Thermophoresis and Brownian Movement Impact on Maxwell Fluid Flow Over Permeable Stretching Sheet with Variable Magnetic Field

S.M. Sachhin[1], G. M. Sachin[1], K. R. Harshitha[1], U.S. Mahabaleshwar[1]* and M. K. Awasthi[2]

[1]*Department of Studies in Mathematics, Davangere University, Shivagangothri, Davangere, India*
[2]*Department of Mathematics, Babasaheb Bhimrao Ambedkar University, Lucknow, India*

Abstract
This study explores the effects of thermophoresis and Brownian movement on the movement of Maxwell fluid across a permeable surface under a variable magnetic field, enhancing efficiency in various engineering systems. The present work analyses the magnetic field and porous media impact in momentum and thermophoresis and Brownian motion impact in mass transfer analysis; the considered PDEs are transformed to ODEs utilising similarity terms; numerical BVP4c shooting method is utilised to solve ODEs; findings of the present analysis reveals that enhancing the MHD and porous media decays the velocity, increasing the solar radiation and heat absorption enhances the energy, thermophoresis enhances the temperature and concentration, and enhancing Brownian motion decreases the concentration and enhances the temperature. The present study has many useful applications in in industrial cooling, biomedical engineering, and nanofluid dynamics. The findings contribute to optimizing heat and mass transfer processes.

**Corresponding author*: u.s.m@davangereuniversity.ac.in; ORCID: 0000-0003-1380-6057
S.M. Sachhin: ORCID: 0009-0006-4101-1074
M. K. Awasthi: ORCID: 0000-0002-6706-5226

Dhananjay Yadav, Mukesh Kumar Awasthi and Harith Mohamed Al-Azri (eds.) Non-Newtonian Fluids for Industrial Applications: Modeling and Simulations, (263–278) © 2026 Scrivener Publishing LLC

Keywords: Thermophoresis, brownian motion, maxwell fluid, magnetic field, porous media

12.1 Introduction

Maxwell fluid exhibits both viscosity and elasticity, making it a fundamental model in rheology. It is described by the Maxwell constitutive equation, which accounts for stress relaxation over time. Unlike Newtonian fluids, which have a constant viscosity, Maxwell fluids respond to deformation with a time-dependent stress. This behaviour is particularly important in polymer processing, biological fluids, and industrial applications involving viscoelastic materials. Understanding Maxwell fluids helps in designing materials with controlled flow properties, improving efficiency in engineering and medical applications. Their study also contributes to advancements in complex fluid mechanics and material science. Samuel [1] examined the chemical reaction impact on Maxwell radiative fluid movement across autocatalytic expanding surface. Kumar [2] studied the radiation and heat source impact on electrically conducting Maxwell fluid flow across expanded surface. Rubab [3] examined the Cattaneo–Christov model on Maxwell fluid movement across expanding sheet using the HAM method. L1 scheme was utilised by Liu [4] to explore the heat transfer influence on Maxwell fluid movement across expanding sheet.

A magnetic field in fluid flow plays a crucial role in magnetohydrodynamics (MHD), where electrically conducting fluids interact with magnetic forces. This phenomenon is significant in various fields, including astrophysics, geophysics, and industrial processes like plasma confinement and cooling systems. Magnetic fields can control turbulence, enhance heat transfer, and improve flow regulation in engineering applications. Understanding their effects helps in designing efficient systems in power generation, aerospace, and biomedical technologies. Tzirtzilakis [5] examined the magnetic field influence on laminar fluid movement across expanding surface by using the numerical method. Pourmehran [6] examined the buoyancy force influence on convective fluid flow across vertical expanding surface. Hayat [7] explored the solar radiation influence on mixed convective movement of fluid across point of stagnation. Misra [8] explored the impact of heat generation and magnetic field on fluid movement across expanding sheet.

Porous media in fluid flow refer to materials containing interconnected voids through which a fluid can move. It is defined as a solid matrix with a network of pores that influence fluid transport, affecting permeability and flow behaviour. This concept is essential in various fields, including petroleum engineering, groundwater hydrology, and chemical processing. The study of fluid flow in porous media helps in optimizing oil recovery, designing efficient filtration systems, and understanding natural water movement in soils and rocks. Its applications extend to biomedical engineering, environmental science, and industrial processes, making it a critical area of research in fluid mechanics. Subhas [9] explored the heat generation impact on viscoelastic fluid flow across a porous medium. Mohanty [10] examined incompressible micropolar fluid movement across porous media with the influence of heat transfer. Megahed [11] studied the Eckert number influence on Maxwell fluid flow across permeable sheet. Jena [12] explored the heat absorption impact on Jeffrey fluid movement across porous media using the Darcy method.

Thermophoresis effect plays a crucial role in aerosol dynamics, semiconductor manufacturing, and environmental pollution control. Understanding thermophoresis helps in designing efficient filtration systems, improving nanoparticle deposition techniques, and enhancing thermal management in industrial processes. It is widely applied in biomedical research, atmospheric sciences, and advanced engineering systems, making it an important subject in fluid mechanics and heat transfer studies. Tawade [13] explored the Lewis number with Brownian motion impact on Casson fluid flow across expanding surface. Mehta [14] explored the Brownian movement impact on viscoelastic fluid movement on expandable surface. Nayak [15] examined the velocity slip impact on Williamson fluid flow across permeable sheet with thermophoresis. References [16–23]

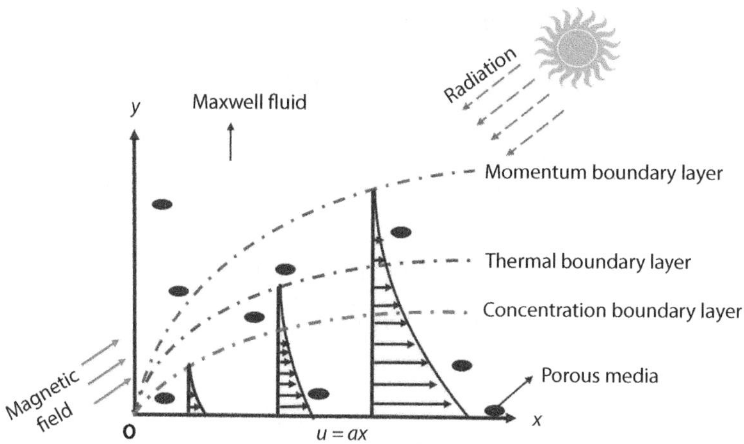

Figure 12.1 Schematic representation of the fluid flow.

explored the heat transfer impact on magnetic liquid flow across various boundaries.

Inspired by the above-mentioned studies, the authors noticed that there is a lack of study on Maxwell fluid flow across permeable sheet with the impact of thermophoresis, magnetic field, porous media, and Brownian motion; the authors utilized the research gap and studied the magnetic field and porous media impact in momentum and thermophoresis and Brownian motion impact in mass transfer analysis; governing PDEs are transformed to ODEs *via* similarity equations, and numerical BVP4c shooting method is utilised to solve ODEs. The present study has many useful applications in in industrial cooling, biomedical engineering, and nanofluid dynamics. The findings contribute to optimizing mass transfer processes. Figure 12.1 represents the schimatic diagram.

12.2 Mathematical Analysis

Governing equations for the proposed problems are [1, 5, 9, 13] as follows:

$$\frac{\partial u}{\partial x}+\frac{\partial v}{\partial y}=0, \tag{12.1}$$

$$\begin{aligned} u\frac{\partial u}{\partial x}+v\frac{\partial u}{\partial y} &= \nu\frac{\partial^2 u}{\partial y^2}-k_0\left(u^2\frac{\partial^2 u}{\partial x^2}+v^2\frac{\partial^2 u}{\partial y^2}+2uv\frac{\partial^2 u}{\partial x\partial y}\right) \\ &\quad -\frac{\sigma B_0^2 Sin^2(\pi)}{\rho}u-\frac{\mu}{\rho K}u. \end{aligned} \tag{12.2}$$

$$\begin{aligned} u\frac{\partial T}{\partial x}+v\frac{\partial T}{\partial y} &= \alpha\left(\frac{\partial^2 T}{\partial y^2}\right)-\frac{1}{\rho_f Cp}\frac{\partial qr}{\partial y}+\frac{Q_0}{\rho Cp}(T-T_\infty) \\ &\quad +\tau\left\{D_B\left(\frac{\partial C}{\partial y}\frac{\partial T}{\partial y}\right)+\left(\frac{D_T}{T_\infty}\right)\left[\left(\frac{\partial T}{\partial y}\right)^2\right]\right\}, \end{aligned} \tag{12.3}$$

$$u\frac{\partial C}{\partial x}+v\frac{\partial C}{\partial y}=D_B\frac{\partial^2 C}{\partial y^2}+\left(\frac{D_T}{T_\infty}\right)\left(\frac{\partial^2 T}{\partial y^2}\right), \tag{12.4}$$

The associated B. Cs for the present problem are [1, 5, 9, 13]

$$u = U_w(x), \quad v = 0, \quad -k\frac{\partial T}{\partial y} = h_f(T_f - T), \quad C = C_w \quad \text{at} \quad y = 0,$$
$$u \to 0, \quad v \to 0, \quad T \to T_\infty, \quad C \to C_\infty \quad \text{as} \quad y \to 0. \tag{12.5}$$

The similarity transformations for the present problem are [1, 5, 9, 13]

$$\eta = \left(\frac{U_w}{\nu x}\right)^{0.5} y, \quad u = axf_\eta(\eta), v = -\sqrt{a\nu}\, f(\eta),$$
$$\theta(\eta) = \frac{T - T_\infty}{T_w - T_\infty}, \quad \varphi(\eta) = \frac{C - C_\infty}{C_w - C_\infty}. \tag{12.6}$$

Where η is the similarity variable, and f, θ, and φ are dimensionless. The velocity components u and v in Equation 12.6 automatically satisfy the continuity Equation 12.1. Other governing equations are transformed as follows:

$$f_\eta^2(\eta) - f(\eta) f_{\eta\eta}(\eta) = f_{\eta\eta\eta}(\eta) - \beta\begin{bmatrix} f^2(\eta) f_{\eta\eta\eta}(\eta) \\ -2f(\eta) f_\eta(\eta) f_{\eta\eta}(\eta) \end{bmatrix} - \left(M \operatorname{Sin}^2(\pi) + L\right) f_\eta(\eta), \tag{12.7}$$

$$\frac{1}{\Pr}(1 + Nr)\theta_{\eta\eta}(\eta) + f(\eta)\theta_\eta(\eta) + Ni\theta(\eta) + Nb\theta_\eta(\eta)\varphi_\eta(\eta) + Nt\theta_\eta^2(\eta) = 0, \tag{12.8}$$

$$\varphi_{\eta\eta}(\eta) + Le\Pr f(\eta)\varphi_\eta(\eta) + \frac{Nt}{Nb}\theta_{\eta\eta}(\eta) = 0, \tag{12.9}$$

Modified B. Cs are

$$f(\eta)=0, \quad f'(\eta)=1, \quad \theta'(\eta)=-Bi(1-\theta(0)), \quad \varphi(\eta)=1, \quad \text{at} \quad \eta=0,$$
$$f'(\eta)\to 0, \quad \theta(\eta)\to 0, \quad \varphi(\eta)\to 0, \quad \text{as} \quad \eta\to\infty, \tag{12.10}$$

The engineering coefficients are calculated as [1, 5, 9, 13] follows:

$$\frac{\mathrm{Re}_x^{1/2}}{2}C_f=(1+\beta)f_{\eta\eta}(0),\mathrm{Re}_x^{-1/2}Nu_x=-\theta_\eta(0),\mathrm{Re}_x^{-1/2}Sh_x=-\varphi_\eta(0).$$

where $\mathrm{Re}_x=\frac{xU_w(x)}{\nu}$ is the local Reynolds number.

12.3 Numerical Method and Solution

Additionally, to simplify the nonlinear equations, new terms are introduced, facilitating their transformation into linear differential equations and reducing higher-order complexity.

$$y_1=f, y_2=f', y_3=f'', y_4=\theta, y_5=\theta', y_6=\varphi, y_7=\varphi'. \tag{12.11}$$

Equations 12.9–12.11 are calculated to first-order ODEs as follows:

$$y_2^1=y_3, y_3^1=-\left(\left(\frac{1}{1+y_1^2\beta}\right)\left(y_1y_3-y_2^2-M\,\mathrm{Sin}^2(\pi)y_2-Ly_2-2\beta y_1y_2y_3\right)\right),$$
$$y_4^1=y_5, y_5^1=-\left(\left(\frac{1}{(1+Nr)}\right)\left(\mathrm{Pr}\,y_1y_5+Ni\,\mathrm{Pr}\,y_4\right)+\mathrm{Pr}\left(Nby_5y_7+Nty_5y_5\right)\right),$$
$$y_6^1=y_7, y_7'=-\left(Le\,\mathrm{Pr}\,y_1y_7-\frac{Nt}{Nb}y_5^1\right), \tag{12.12}$$

Table 12.1 The dimensionless parameters with expressions.

$\beta = k_0 a$	Maxwell parameter
$Bi = (\nu / a)^{1/2} h_f / k$	Biot number
$Ni = \dfrac{Q_0}{a\rho Cp}$	Heat source parameter
$Nb = \dfrac{\tau D_B (C_w - C_\infty)}{\nu}$	Brownian motion
$Nt = \dfrac{\tau D_T (T_w - T_\infty)}{\nu T_\infty}$	Thermophoretic parameter
$Le = \dfrac{\alpha}{D_B}$	Lewis number
$M = \dfrac{\sigma B_0^2}{\rho a}$	Magnetic field
$L = \dfrac{\mu}{\rho K a}$,	Porosity parameter
$Nr = \dfrac{16\sigma^* T_\infty^3}{3k^* \kappa}$	Thermal radiation

Table 12.2 Comparison values of $-\theta'(0)$ and $-\varphi'(0)$ with $Le = 10, \Pr = 10, Bi = \infty$, $M = 0, L = 0$.

Nb	*Nt*	$-\theta'(0)$ **Khan *et al.*, [25]**	$-\theta'(0)$ **Makinde [24]**	$-\theta'(0)$ **Present results**	$-\varphi'(0)$ **Khan *et al.*, [25]**	$-\varphi'(0)$ **Makinde [24]**	$-\varphi'(0)$ **Present results**
0.1	0.1	0.9524	0.9524	0.952231	2.1294	2.1294	2.129081
0.2	0.1	0.5056	0.5056	0.505599	2.3819	2.3819	2.381899
0.3	0.1	0.2522	0.2522	0.252218	2.4100	2.4100	2.410000
0.4	0.1	0.1194	0.1194	0.119412	2.3997	2.3997	2.399732
0.5	0.1	0.0543	0.0543	0.054224	2.3836	2.3836	2.383629
0.1	0.2	0.6932	0.6932	0.693111	2.2740	2.2740	2.274199
0.1	0.3	0.5201	0.5201	0.520008	2.5286	2.5286	2.528607
0.1	0.4	0.4026	0.4026	0.402245	2.7952	2.7952	2.795221
0.1	0.5	0.3211	0.3211	0.321212	3.0351	3.0351	3.035009

12.4 Results and Discussion

The present work analyses the magnetic field and porous media impact in momentum and thermophoresis and Brownian motion impact in mass transfer analysis; the considered PDEs are transformed to ODEs *via* similarity equations, and numerical BVP4c shooting method is utilised to solve ODEs; several physical parameters are explained through graphs in this section. Table 12.1 represents the dimensionless parameters with expressions, Table 12.2 represents the comparison of present results with previous publications.

Figure 12.2 represents the L impact on momentum of the fluid movement; enhancing the porosity term decays the velocity, and an increase in the porosity parameter generally leads to a higher permeability of the medium, allowing fluid to flow more easily. This results in an increase in velocity due to reduced resistance. Conversely, lower porosity restricts flow, decreasing velocity as the medium becomes more compact. The relationship between porosity and momentum is crucial in applications such as filtration, groundwater movement, and industrial fluid transport.

Figure 12.3 represents the magnetic parameter impact on momentum; enhancing the magnetic term decays the momentum, and the magnetic parameter influences fluid velocity by introducing a resistive force known as the Lorentz force. As the magnetic term enhances, this force restricts the motion of the conducting fluid, reducing its velocity. In electrically conductive flows, such as magnetohydrodynamics (MHD), stronger magnetic fields

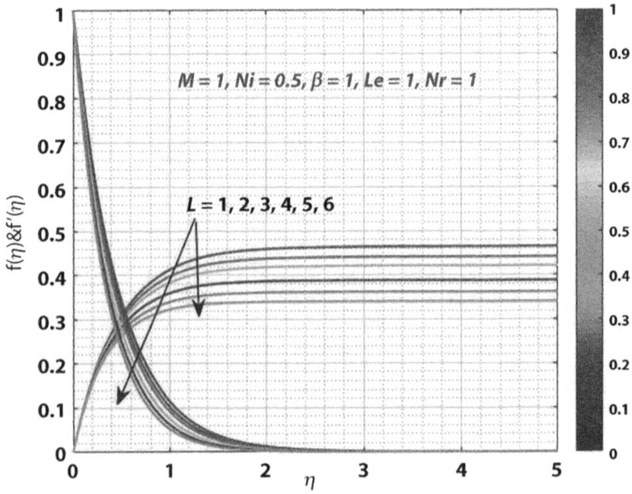

Figure 12.2 Porosity parameter impact on velocity profiles.

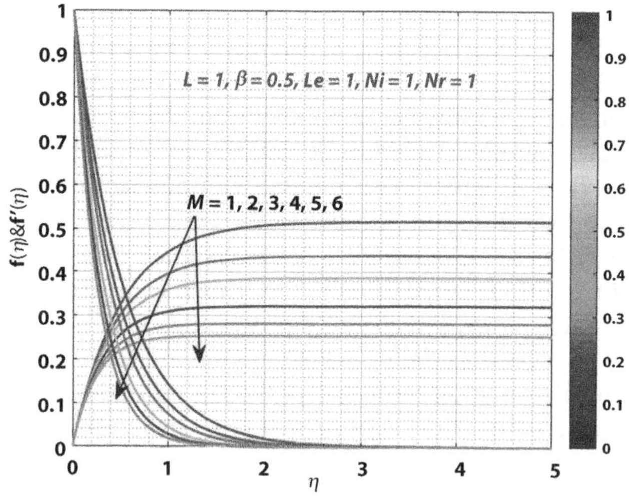

Figure 12.3 Magnetic parameter impact on velocity profiles.

can significantly slow down the fluid. This effect is widely studied in applications like plasma physics, cooling systems, and electromagnetic pumps.

Figure 12.4 represents the Maxwell parameter impact on momentum; enhancing the Maxwell term decays the momentum, and the Maxwell parameter, related to viscoelastic fluids, affects velocity by influencing the fluid's ability to store and exhibits elastic energy. Higher Maxwell parameter values indicate stronger elastic effects, which can resist deformation

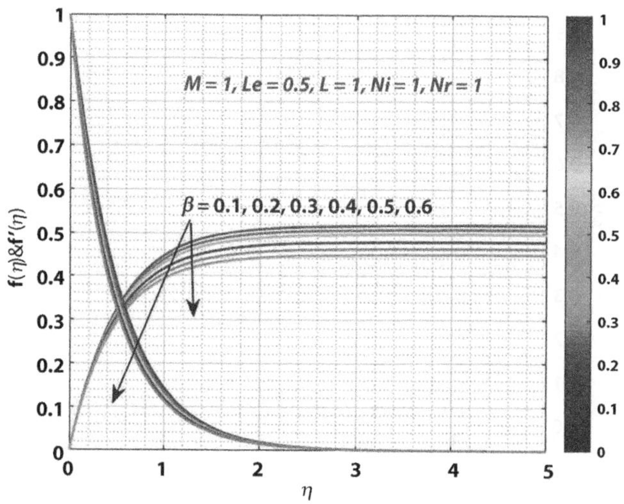

Figure 12.4 Maxwell parameter impact on velocity profiles.

and reduce the fluid's velocity. In contrast, lower values allow the fluid to behave more like a Newtonian fluid, leading to smoother and faster flow.

Figure 12.5 represents the Lewis number impact on concentration of the fluid movement; rising the Lewis number term decays the concentration, and the Lewis number significantly influences concentration distribution in fluid flow. A higher Lewis number indicates slower mass diffusion compared to heat diffusion, leading to steeper concentration gradients. Conversely, a lower Lewis number enhances mass transfer, resulting in more uniform concentration distribution. This effect is crucial in combustion, atmospheric flows, and chemical processing.

Figure 12.6 represents the Brownian motion impact on $\varphi(\eta)$; enhancing the Brownian motion term decays the concentration, and Brownian motion enhances the diffusion of particles in a fluid, leading to increased mixing and a more uniform concentration distribution. As particle movement intensifies, concentration gradients become less steep, promoting mass transfer. This effect is particularly significant in nanofluids, where smaller particles experience stronger Brownian motion, improving thermal and mass diffusion. It plays a crucial role in applications like drug delivery, chemical mixing, and heat transfer enhancement.

Figure 12.7 represents the thermophoresis impact on concentration; enhancing the thermophoresis term decays the concentration, and thermophoresis influences fluid concentration by causing particles to migrate from hotter to cooler regions due to temperature gradients. This movement affects

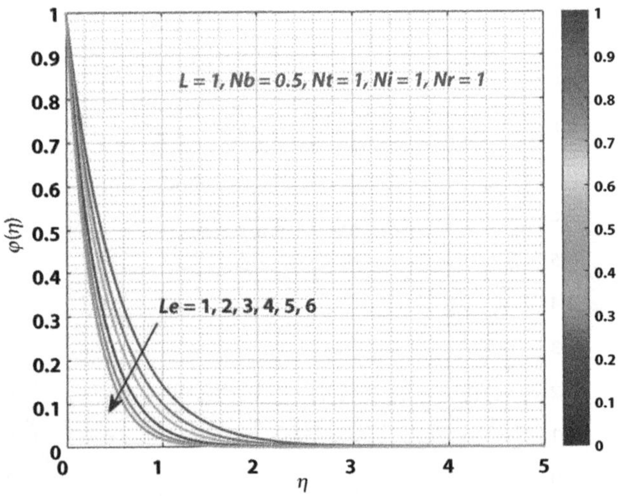

Figure 12.5 Lewis number impact on concentration profiles.

concentration distribution, often creating a non-uniform profile. Stronger thermophoretic forces enhance particle separation, reducing concentration near heated surfaces. This phenomenon is significant in aerosol dynamics, semiconductor manufacturing, and environmental pollution control.

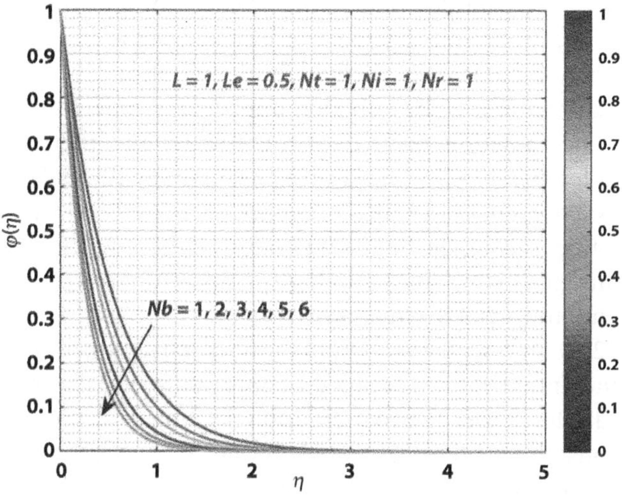

Figure 12.6 Brownian motion impact on concentration profiles.

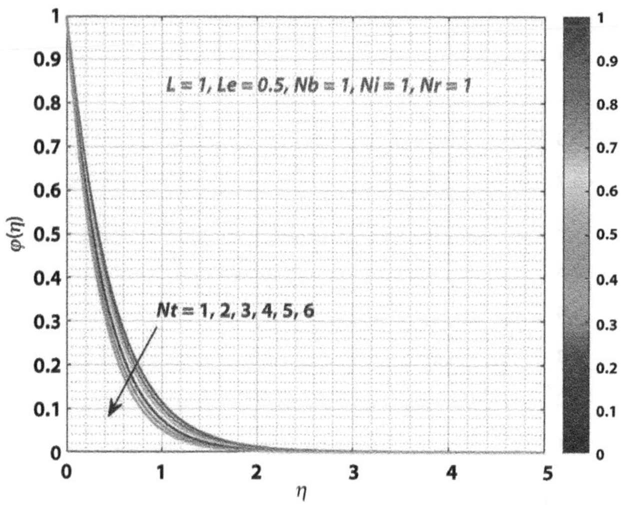

Figure 12.7 Thermophoresis impact on concentration profiles.

Figure 12.8 represents the heat source impact on temperature; rising the heat source term rises the temperature, and a heat source raises the energy of the fluid by introducing thermal energy, which enhances heat transfer and affects flow behaviour. As heat input increases, temperature gradients become more pronounced, leading to stronger convective currents. This process plays a crucial role in industrial heating, energy systems, and thermal regulation. Proper management of heat sources is essential for maintaining efficiency and stability in fluid flow applications.

Figure 12.9 represents the radiation impact on temperature of the flow; enhancing the radiation term rises the energy, and thermal radiation affects fluid temperature by transferring heat energy through electromagnetic waves, independent of direct contact. Increased radiation enhances heat absorption, raising the fluid temperature and influencing flow characteristics. This effect is significant in high-temperature systems, astrophysical flows, and industrial thermal processes.

Figure 12.10 represents the Prandtl number impact on temperature; rising the Prandtl number term decays the temperature, and the greater Prandtl number indicates slower thermal diffusion, leading to steeper temperature gradients and a thicker thermal boundary layer. Conversely, a lower Prandtl number enhances heat diffusion, resulting in a more uniform temperature profile. This parameter is crucial in heat transfer analysis for applications like cooling systems, aerodynamics, and industrial fluid flows.

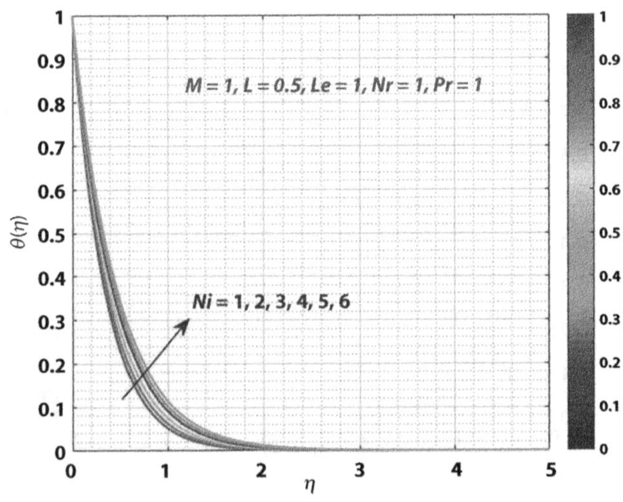

Figure 12.8 Heat source impact on temperature profiles.

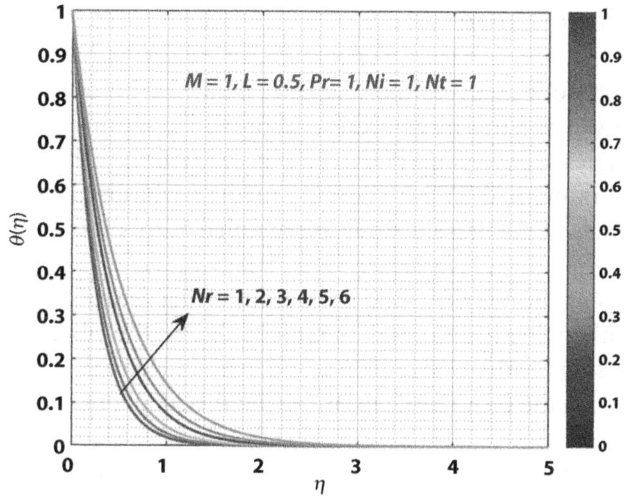

Figure 12.9 Radiation impact on temperature profiles.

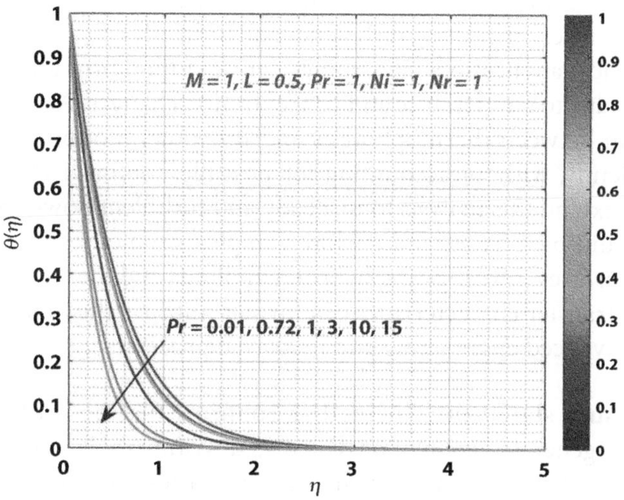

Figure 12.10 Prandtl number impact on temperature profiles.

12.5 Conclusion

The present work analyses the magnetic field and porous media impact in momentum and thermophoresis and Brownian motion impact in mass transfer analysis; governing PDEs are transformed to ODEs *via* similarity terms, and numerical BVP4c shooting method is utilised to solve ODEs. The outcomes of the present study are revealed as follows:

- Rising the magnetic field and porous media decays the velocity.
- Enhancing the Maxwell parameter decays the velocity.
- Rising the thermal radiation and heat source rises the energy.
- Rising the Brownian movement and thermophoresis enhances the energy.
- Enhancing the Prandtl number decays the energy profile.
- Rising the Brownian movement and thermophoresis decays the concentration.

This study is limited to steady, laminar Maxwell fluid flow over a porous extended surface, neglecting turbulence and unsteady effects that may influence practical applications. It assumes constant fluid properties, ignoring variations due to temperature and concentration changes. Additionally, the interaction between thermophoresis and Brownian motion is considered independently, without accounting for possible nonlinear effects. Future work can extend the analysis to unsteady and turbulent flows, incorporate variable fluid properties, and explore complex geometries like curved or rotating surfaces. Investigating hybrid nanofluids and their enhanced heat and mass transfer characteristics can further improve the model's applicability in advanced engineering and industrial processes.

AI disclosure statement: The author confirms that no artificial intelligence (AI)-generated content or AI-assisted writing tools were used in the preparation of this chapter (including summaries and the abstract). All text, analysis, and conclusions are the result of the author's original work.

References

1. Samuel, D.J., Numerical Investigations of Thermal Radiation and Activation Energy Imparts on Chemically Reactive Maxwell Fluid Flow Over an Exothermal Stretching Sheet in a Porous Medium. *Int. J. Appl. Comput. Math.*, 8, 3, 148, May 2022. doi: 10.1007/s40819-022-01356-8.

2. Kumar, M.A. and Reddy, Y.D., Computational modelling of radiative Maxwell fluid flow over a stretching sheet containing nanoparticles with chemical reaction. *J. Indian Chem. Soc.*, 100, 1, 100877, Jan. 2023. doi: 10.1016/j.jics.2022.100877.
3. Rubab, K. and Mustafa, M., Cattaneo-Christov Heat Flux Model for MHD Three-Dimensional Flow of Maxwell Fluid over a Stretching Sheet. *PLoS One*, 11, 4, e0153481, Apr. 2016. doi: 10.1371/journal.pone.0153481.
4. Liu, L. and Liu, F., Boundary layer flow of fractional Maxwell fluid over a stretching sheet with variable thickness. *Appl. Math. Lett.*, 79, 92–99, May 2018. doi: 10.1016/j.aml.2017.10.008.
5. Tzirtzilakis, E.E. and Kafoussias, N.G., Three-Dimensional Magnetic Fluid Boundary Layer Flow Over a Linearly Stretching Sheet. *J. Heat Transf.*, 132, 011702, Oct. 2009. doi: 10.1115/1.3194765.
6. Pourmehran, O., Rahimi-Gorji, M., Ganji, D.D., Heat transfer and flow analysis of nanofluid flow induced by a stretching sheet in the presence of an external magnetic field. *J. Taiwan Inst. Chem. Eng.*, 65, 162–171, Aug. 2016. doi: 10.1016/j.jtice.2016.04.035.
7. Hayat, T., Abbas, Z., Pop, I., Asghar, S., Effects of radiation and magnetic field on the mixed convection stagnation-point flow over a vertical stretching sheet in a porous medium. *Int. J. Heat Mass Transf.*, 53, 1, 466–474, Jan. 2010. doi: 10.1016/j.ijheatmasstransfer.2009.09.010.
8. Misra, S. and Kamatam, G., Effect of magnetic field, heat generation and absorption on nanofluid flow over a nonlinear stretching sheet. *Beilstein J. Nanotechnol.*, 11, 1, 976–990, Jul. 2020. doi: 10.3762/bjnano.11.82.
9. Subhas, A. and Veena, P., Visco-elastic fluid flow and heat transfer in a porous medium over a stretching sheet. *Int. J. Non-Linear Mech.*, 33, 3, 531–540, May 1998. doi: 10.1016/S0020-7462(97)00025-5.
10. Mohanty, B., Mishra, S.R., Pattanayak, H.B., Numerical investigation on heat and mass transfer effect of micropolar fluid over a stretching sheet through porous media. *Alex. Eng. J.*, 54, 2, 223–232, Jun. 2015. doi: 10.1016/j.aej.2015.03.010.
11. Megahed, A.M., Improvement of heat transfer mechanism through a Maxwell fluid flow over a stretching sheet embedded in a porous medium and convectively heated. *Math. Comput. Simul*, 187, 97–109, Sep. 2021. doi: 10.1016/j.matcom.2021.02.018.
12. Jena, S., Mishra, S.R., Dash, G.C., Chemical Reaction Effect on MHD Jeffery Fluid Flow over a Stretching Sheet Through Porous Media with Heat Generation/Absorption. *Int. J. Appl. Comput. Math.*, 3, 2, 1225–1238, Jun. 2017. doi: 10.1007/s40819-016-0173-8.
13. Tawade, J.V., Guled, C.N., Noeiaghdam, S., Fernandez-Gamiz, U., Govindan, V., Balamuralitharan, S., Effects of thermophoresis and Brownian motion for thermal and chemically reacting Casson nanofluid flow over a linearly stretching sheet. *Results Eng.*, 15, 100448, Sep. 2022. doi: 10.1016/j.rineng.2022.100448.

14. Mehta, R. and Kataria, H.R., Brownian motion and thermophoresis effects on MHD flow of viscoelastic fluid over stretching/shrinking sheet in the presence of thermal radiation and chemical reaction. *Heat Transf.*, 51, 1, 274–295, 2022. doi: 10.1002/htj.22307.
15. Nayak, B., Panda, M., Mishra, S.R., Analysis of Brownian motion and thermophoresis effects with chemical reaction on Williamson nanofluid flow over permeable stretching sheet. *Partial Differ. Equ. Appl. Math.*, 11, 100908, Sep. 2024. doi: 10.1016/j.padiff.2024.100908.
16. Babu, M.J. and Sandeep, N., 3D MHD slip flow of a nanofluid over a slendering stretching sheet with thermophoresis and Brownian motion effects. *J. Mol. Liq.*, 222, 1003–1009, Oct. 2016. doi: 10.1016/j.molliq.2016.08.005.
17. Bognár, G., On similarity solutions of boundary layer problems with upstream moving wall in non-Newtonian power-law fluids. *IMA J. Appl. Math.*, 77, 4, 546–562, Jun. 2012. doi: 10.1093/imamat/hxr033.
18. Lone, S.A., Ali, F., Saeed, A., Bognár, G., Irreversibility analysis with hybrid cross nanofluid of stagnation point and radiative flow (TiO2+CuO) based on engine oil past a stretchable sheet. *Heliyon*, 9, 4, e15056, Apr. 2023. doi: 10.1016/j.heliyon.2023.e15056.
19. Barna, I.F., Bognár, G., Guedda, M., Mátyás, L., Hriczó, K., Analytic Self-Similar Solutions Of The Kardar-Parisi-Zhang Interface Growing Equation with Various Noise Terms. *Math. Model. Anal.*, 25, 2, 241–256, Mar. 2020. doi: 10.3846/mma.2020.10459.
20. Klazly, M. and Bognár, G., A novel empirical equation for the effective viscosity of nanofluids based on theoretical and empirical results. *Int. Commun. Heat Mass Transf.*, 135, 106054, Jun. 2022. doi: 10.1016/j.icheatmasstransfer.2022.106054.
21. Bognar, G., Ferrofluid Flow in Magnetic Field Above Stretching Sheet with Suction and Injection, in: *Math. Model. Anal*, Jul. 2020, Accessed: Feb. 20, 2025. [Online]. Available: https://www.academia.edu/121605678/Ferrofluid_Flow_in_Magnetic_Field_Above_Stretching_Sheet_with_Suction_and_Injection.
22. Sachhin, S.M., Mahabaleshwar, U.S., Chan, A., Effect of slip and thermal gradient on micropolar nano suspension flow across a moving hydrogen fuel-cell membrane. *Int. J. Hydrog. Energy*, 63, 59–81, Apr. 2024. doi: 10.1016/j.ijhydene.2024.02.332.
23. Sachhin, S.M., Mahabaleshwar, U.S., Nayakar, S.N.R., Patil, H.S.R., Souayeh, B., Darcy–Brinkman model for Boussinesq–Stokes suspension tetra dusty nanofluid flow over stretching/shrinking sheet with suction/injection. *Numer. Heat Transf. Part Appl.*, 0, 0, 1–23, 2024. doi: 10.1080/10407782.2024.2359047.
24. Makinde, O.D. and Aziz, A., Boundary layer flow of a nanofluid past a stretching sheet with a convective boundary condition. *Int. J. Therm. Sci.*, 50, 7, 1326–1332, 2011.
25. Khan, W.A. and Pop, I., Boundary-layer flow of a nanofluid past a stretching sheet. *Int. J. Heat Mass Transf.*, 53, 11, 2477–2483, May 2010. doi: 10.1016/j.ijheatmasstransfer.2010.01.032.

13

Arrhenius Activation Energy and Viscosity Ratio Impact on Casson Fluid Flow Across Porous Stretching Surface with Variable Magnetic Field

S.M. Sachhin[1], G. M. Sachin[1], U.S. Mahabaleshwar[1]* and M. K. Awasthi[2]

[1]*Department of Studies in Mathematics, Davangere University, Shivagangothri, Davangere, India*
[2]*Department of Mathematics, Babasaheb Bhimrao Ambedkar University, Lucknow, India*

Abstract

Casson fluid is a non-Newtonian fluid that exhibits yield stress, meaning it behaves as a solid at low stress levels and flows as a liquid once the applied stress exceeds a certain threshold. This unique behavior makes Casson fluid models useful in a variety of applications across industries, particularly in biomedical, engineering, and industrial processes. Inspired by the above applications, the authors examined the Arrhenius activation energy and viscosity ratio impact on Casson fluid movement across porous stretching surface with variable magnetic field, thermophoresis and Brownian movement was examined using mass transfer analysis and radiation and heat source effects in energy equation, and considered PDEs were calculated to ODEs vis similarity variables later solved by using numerical method; the results of the present study show that enhancing the magnetic field and porous media reduces the velocity. The study of Casson fluid flow influenced by Arrhenius activation energy, viscosity variations, and a changing magnetic field has significant applications in several fields, including industrial processes, biomedical engineering, and energy systems. This research is particularly useful in optimizing heat and mass transfer mechanisms, improving material processing, and enhancing technological advancements in magnetohydrodynamics.

**Corresponding author*: u.s.m@davangereuniversity.ac.in; ORCID: 0000-0003-1380-6057
S.M. Sachhin: ORCID: 0009-0006-4101-1074
M. K. Awasthi: ORCID: 0000-0002-6706-5226

Dhananjay Yadav, Mukesh Kumar Awasthi and Harith Mohamed Al-Azri (eds.) Non-Newtonian Fluids for Industrial Applications: Modeling and Simulations, (279–296) © 2026 Scrivener Publishing LLC

Keywords: Casson fluid, inclined magnetic field, activation energy, thermophoresis, Brownian motion

13.1 Introduction

Casson fluid flow originates from Casson's model, which exhibits yield stress, meaning it behaves as a solid at low stress but flows like a liquid once the yield stress is exceeded. This concept is important in industries dealing with biological fluids, food processing, and polymer solutions, as it helps predict flow characteristics accurately. Applications include biomedical engineering (blood flow analysis), food industries (chocolate and ketchup processing), and industrial coatings. Understanding Casson fluid dynamics aids in optimizing designs in these fields, ensuring efficient material handling and processing. Bhattacharyya [1] examined the impact of radiation and viscosity ratio on two-dimensional Casson fluid flow across the stagnation point of expanding sheet. Shawky [2] explored the mass exchange impact on the Casson fluid movement across porous expanding sheet. Devi [3] examined the MHD impact on Casson liquid boundary layer movement across stretching surface with the help of shooting technique. Spectral relaxation technique was utilized by Mahanta [4] to examine the convective Casson fluid movement across expanding surface with porous media.

The study of magnetic fields on fluid flow originates from magnetohydrodynamics (MHD), which combines principles of electromagnetism and fluid dynamics. Magnetic fields play a key role in astrophysics, nuclear fusion research, and cooling systems in high-temperature reactors. Applications include MHD power generation, magnetic drug targeting in medicine, and metallurgical processes. Understanding this phenomenon helps in developing efficient energy systems and industrial applications. Das [5] examined the Joule heating impact on couple stress fluid movement across slip boundary with viscous dissipation. Patel [6] explored the heat generation impact on micropolar fluid movement across permeable surface using the HAM method. Liu [7] examined the magnetic field impact on laminar second-grade fluid over power law surface. Misra [8] explored the magnetic dipole impact on Walters B fluid movement across expanding surface by using the finite-difference method.

Porous media refer to materials containing interconnected voids or pores that allow fluid to pass through, such as soil, rocks, and biological tissues. This concept is crucial in various fields, including petroleum engineering, hydrogeology, and biomedical sciences, as it governs fluid transport in natural and industrial systems. Applications include oil and gas recovery, groundwater filtration, carbon sequestration, and drug delivery in medical treatments. Understanding fluid behavior in porous structures helps optimize resource extraction, environmental protection, and engineering designs. Subhas [9] examined the heat transfer impact on viscoelastic fluid movement across porous media with wall heat flux conditions. Makinde [10] explored the magnetic field influence on incompressible fluid movement across slippery expanding surface. Akshatha [11] examined the magnetic non-Newtonian fluid movement inside porous media. Mandal [12] explored the suction parameter impact on the boundary layer movement across exponentially expanding surface. Turkyilmazoglu [13] examined the couple stress impact on micropolar fluid flow across expanding sheet.

Thermal radiation refers to energy emitted by a surface due to its temperature, influencing heat transfer in fluid motion. This phenomenon is crucial in engineering and environmental sciences, as it affects energy transport in high-temperature systems. It plays a significant role in astrophysics, combustion processes, and industrial cooling. Applications include spacecraft thermal control, fire safety analysis, solar energy systems, and high-temperature manufacturing. Understanding thermal radiation in fluids helps optimize energy efficiency and system performance in various industries. Ishak [14] explained the thermal radiation impact on incompressible micropolar fluid movement across constant sheet. Mukhopadhyay [15] examined the velocity slip impact on magnetic fluid movement exponential expanding surface. Liu [16] explored the variable viscosity impact on laminar thin liquid film across expanding surface. Cortell [17] studied heat flux condition and thermal radiation impact on expanded surface by using numerical analysis. [18–24] examined the heat transfer impact on fluid flow across various boundaries.

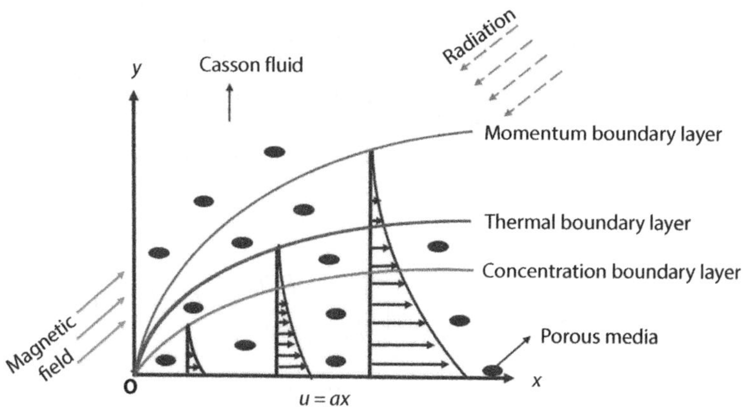

Figure 13.1 Schematic diagram.

13.2 Mathematical Analysis

The rheology of Casson fluid is given as follows:

$$\tau_{ij} = \begin{cases} 2(\mu_B + p_y / \sqrt{2\pi})e_{ij}, & \pi > \pi_c, \\ 2(\mu_B + p_y / \sqrt{2\pi_c})e_{ij}, & \pi < \pi_c, \end{cases}$$

The dimensionless Casson parameter is given below:

$$\lambda = \frac{\mu_B \sqrt{2\pi_c}}{p_y}.$$

The governing equations for the proposed problems are

$$\frac{\partial u}{\partial x} + \frac{\partial v}{\partial y} = 0, \tag{13.1}$$

$$u\frac{\partial u}{\partial x} + v\frac{\partial u}{\partial y} = \frac{\mu_{eff}}{\rho}(1+\lambda^{-1})\frac{\partial^2 u}{\partial y^2} - \frac{\sigma B_0^2 \operatorname{Sin}^2(\pi)}{\rho}u - \frac{\mu}{\rho K}u. \tag{13.2}$$

$$\begin{aligned} u\frac{\partial T}{\partial x} + v\frac{\partial T}{\partial y} &= \alpha\left(\frac{\partial^2 T}{\partial y^2}\right) - \frac{1}{\rho_f Cp}\frac{\partial q_r}{\partial y} + \frac{Q_0}{\rho Cp}(T - T_\infty) \\ &+ \tau\left\{D_B\left(\frac{\partial C}{\partial y}\frac{\partial T}{\partial y}\right) + \left(\frac{D_T}{T_\infty}\right)\left[\left(\frac{\partial T}{\partial y}\right)^2\right]\right\}, \end{aligned} \tag{13.3}$$

$$\begin{aligned} u\frac{\partial C}{\partial x} + v\frac{\partial C}{\partial y} &= D_B\frac{\partial^2 C}{\partial y^2} + \left(\frac{D_T}{T_\infty}\right)\left(\frac{\partial^2 T}{\partial y^2}\right) - Kr^2(C - C_\infty) \\ &\left(\frac{T}{T_\infty}\right)^n \exp\left(\frac{-E_a}{kT}\right), \end{aligned} \tag{13.4}$$

The associated B. Cs for the present problem are

$$u=U_w(x),\quad v=0,\quad -k\frac{\partial T}{\partial y}=h_f\left(T_f-T\right),\quad C=C_w\quad \text{at}\quad y=0,$$
$$u\to 0,\quad v\to 0,\quad T\to T_\infty,\quad C\to C_\infty\quad \text{as}\quad y\to 0. \tag{13.5}$$

The similarity transformations for the present problem are given by [1, 5, 9, 13]

$$\eta=\left(\frac{U_w}{vx}\right)^{0.5}y,\quad u=axf_\eta(\eta), v=-\sqrt{av}\,f(\eta),$$
$$\theta(\eta)=\frac{T-T_\infty}{T_w-T_\infty},\qquad \varphi(\eta)=\frac{C-C_\infty}{C_w-C_\infty}. \tag{13.6}$$

Radiative heat flux is solved by Rosseland's estimation as follows:

$$q_r=-\frac{4\sigma^*}{3k^*}\frac{\partial T^4}{\partial y},$$

on expanding T^4 by using Taylor's series to obtain

$$T^4\cong 4T_\infty{}^3T-3T_\infty{}^4,$$

ODEs of current analysis are calculated as follows:

$$f_\eta^2(\eta)-f(\eta)f_{\eta\eta}(\eta)=\Lambda\left(1+\lambda^{-1}\right)f_{\eta\eta\eta}(\eta)-\left(M\,\mathrm{Sin}^2(\pi)+L\right)f_\eta(\eta), \tag{13.7}$$

$$\frac{1}{\Pr}(1+Nr)\theta_{\eta\eta}(\eta)+f(\eta)\theta_\eta(\eta)+Ni\theta(\eta)+Nb\theta_\eta(\eta)\varphi_\eta(\eta)$$
$$+Nt\theta_\eta^2(\eta)=0, \tag{13.8}$$

$$\begin{aligned}&\varphi_{\eta\eta}(\eta)+Le\Pr f(\eta)\varphi_\eta(\eta)+\frac{Nt}{Nb}\theta_{\eta\eta}(\eta)\\&-Le\Pr Cr\left(1+\gamma_1\theta(\eta)\right)^n\exp\left(\frac{-E}{1+\gamma_1\theta(\eta)}\right)\varphi(\eta)=0,\end{aligned}\tag{13.9}$$

Modified B. Cs are

$$\begin{aligned}&f(\eta)=0,\quad f'(\eta)=1,\quad \theta'(\eta)=-Bi(1-\theta(0)),\quad \varphi(\eta)=1,\quad at\quad \eta=0,\\&f'(\eta)\to0,\quad \theta(\eta)\to0,\quad \varphi(\eta)\to0,\quad as\quad \eta\to\infty,\end{aligned}\tag{13.10}$$

The engineering coefficients are calculated as follows:

$$\frac{\mathrm{Re}_x^{1/2}}{2}C_f=\Lambda f_{\eta\eta}(0),\mathrm{Re}_x^{-1/2}Nu_x=-\theta_\eta(0),\mathrm{Re}_x^{-1/2}Sh_x=-\varphi_\eta(0).$$

where $\mathrm{Re}_x=\dfrac{xU_w(x)}{\nu}$ is the local Reynolds number.

13.3 Numerical Method and Solution

Additionally, to simplify the nonlinear equations, new terms are introduced, facilitating their transformation into linear differential equations and reducing higher-order complexity.

$$y_1=f,y_2=f',y_3=f'',y_4=\theta,y_5=\theta',y_6=\varphi,y_7=\varphi'.\tag{13.11}$$

Equations 13.9–13.11 are calculated to first-order ODEs as follows:

Table 13.1 The dimensionless parameters with expressions.

$E = \frac{E_a}{kT_\infty}$	Activation energy,
$Bi = (\nu / a)^{1/2} h_f / k$	Biot number
$Ni = \frac{Q_0}{a\rho Cp}$	Heat source parameter
$Nb = \frac{\tau D_B (C_w - C_\infty)}{\nu}$	Brownian motion
$Nt = \frac{\tau D_T (T_w - T_\infty)}{\nu T_\infty}$	Thermophoretic parameter
$Le = \frac{\alpha}{D_B}$	Lewis number
$M = \frac{\sigma B_0^2}{\rho a}$	Magnetic field
$L = \frac{\mu}{\rho Ka}$,	Porosity parameter
$Nr = \frac{16\sigma^* T_\infty^3}{3k^* \kappa}$	Thermal radiation

Table 13.2 Comparison $-\theta'(0)$ for different values of Pr, while taking $Nt = 0, Nb = 0, Ni = 0, Nr = 0$.

Pr	Waini [25]	Sreedevi [26]	Present results
2.0	0.911353	0.911341	0.911113
6.13	1.759682	1.759676	1.759610
7.0	1.895400	1.895397	1.895458
20.0	3.353902	3.353915	3.353991

$$y_2^1 = y_3,\quad y_3^1 = -\left(\left(\frac{1}{\Lambda\left(1+\lambda^{-1}\right)}\right)\left(y_1y_3 - y_2^2 - M\,\mathrm{Sin}^2(\pi)\,y_2 - Ly_2\right)\right),$$

$$y_4^1 = y_5,$$

$$y_5^1 = -\left(\left(\frac{1}{(1+Nr)}\right)\left(\Pr y_1y_5 + Ni\Pr y_4\right) + \Pr\left(Nby_5y_7 + Nty_5y_5\right)\right),$$

$$y_6^1 = y_7,$$

$$y_7' = -\left(Le\Pr y_1y_7 - \frac{Nt}{Nb}y_5^1 - Le\Pr Cr\left(1+\gamma_1y_4\right)^n \exp\left(\frac{-E}{1+\gamma_1y_4}\right)y_5\right),$$

(13.12)

13.4 Results and Discussion

The present analysis examined the Arrhenius activation energy and viscosity ratio impact on Casson fluid movement across porous stretching surface with variable MHD; thermophoresis and Brownian movement was examined using mass transfer analysis and solar radiation and heat source effects were considered. PDEs are calculated to ODEs vis similarity variables later solved by using the numerical method, Figure 13.1 represents the Schematic diagram of the fluid flow, Table 13.1 represents the non-dimensional parameters, and Table 13.2 represents the comparison values of Nusselt number and several physical parameters were examined through graphs in this section.

Figure 13.2 represents the viscosity ratio effect on velocity of the fluid flow; rising the viscosity ratio term reduces the velocity of the flow. Viscosity ratio significantly affects the velocity distribution in various fluid dynamics applications. The viscosity ratio is defined as the ratio of dynamic viscosities between different fluid layers or interacting fluids in a system. Changes in this ratio can alter the resistance to flow, impacting velocity profiles and overall fluid motion.

Figure 13.3 represents the Casson term effect on momentum of the fluid; rising the Casson term reduces the velocity. An increase in the Casson term generally leads to greater resistance to flow, reducing the momentum of the fluid. Since Casson fluids require a certain stress threshold to start flowing, a higher Casson parameter means a stronger internal structure, which restricts movement and lowers velocity. Conversely, a lower Casson parameter reduces resistance, allowing the fluid to flow more freely at higher velocities.

Figure 13.4 represents the porosity term effect on momentum; rising the porosity term reduces the momentum. The porosity parameter plays a crucial role in fluid flow through porous media, affecting velocity distribution and resistance to movement. It represents the extent to which a medium allows fluid to pass through it, with higher values indicating greater permeability and lower values suggesting more obstruction.

Figure 13.5 represents the magnetic term effect on momentum; rising the magnetic term reduces the velocity of the flow. An increase in the magnetic parameter generally leads to greater resistance to fluid movement due to the formation of Lorentz forces, which act as a damping mechanism. This results in a reduction in velocity, as the applied magnetic field opposes

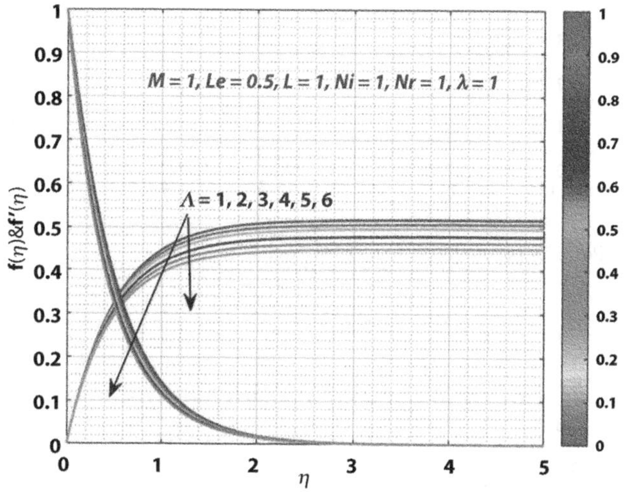

Figure 13.2 Viscosity ratio impact on velocity.

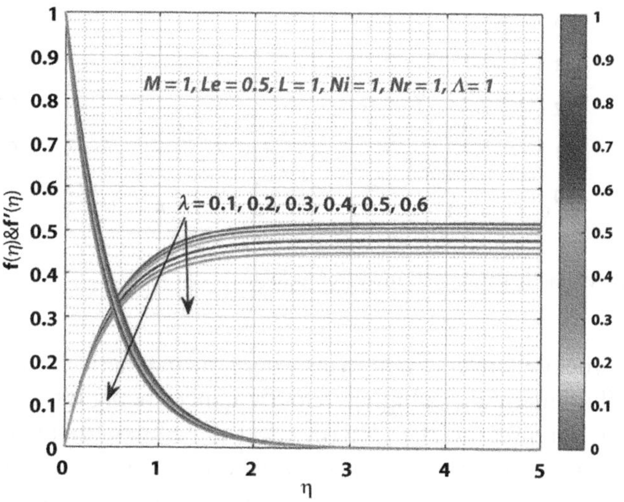

Figure 13.3 Casson parameter influence on velocity.

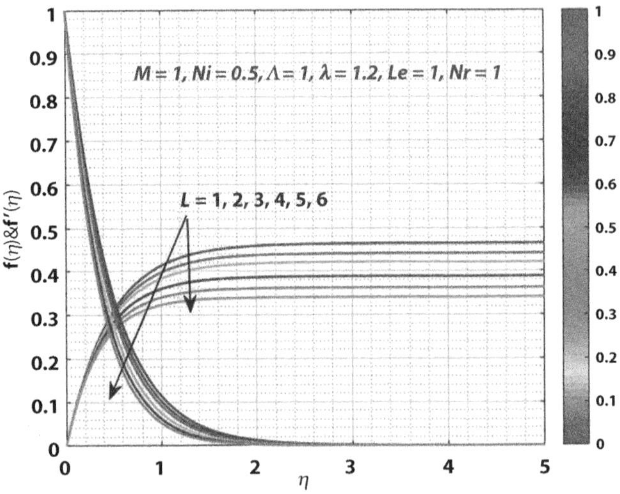

Figure 13.4 Porosity parameter impact on velocity.

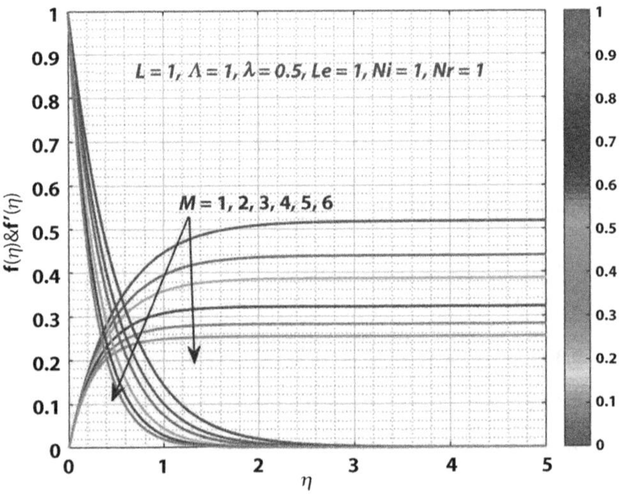

Figure 13.5 Magnetic parameter impact on velocity.

fluid movement. Conversely, in cases where the magnetic influence is minimal, the fluid maintains a higher velocity, experiencing less resistance.

Figure 13.6 represents the *Ni* term effect on temperature; rising the heat source term rises the temperature of the movement. An enhancement in the heat source parameter tends to a rise in energy within the fluid due to the continuous addition of thermal energy. This results in enhanced thermal gradients, which affect convection and heat dissipation rates. The higher the heat source intensity, the greater the temperature elevation in the system.

Figure 13.7 represents the *Nr* parameter effect on temperature; rising the thermal radiation term rises the temperature of the flow. The presence of radiation influences the thickness and characteristics of the thermal boundary layer. A higher radiation parameter results in an expanded thermal boundary layer due to increased heat absorption, leading to higher surface temperatures

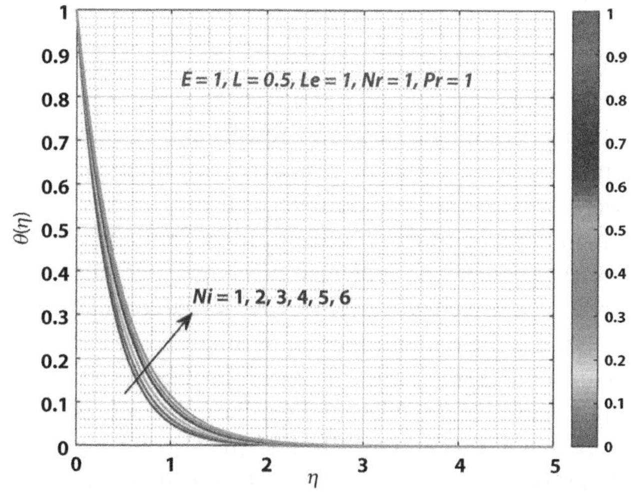

Figure 13.6 Heat source parameter impact on temperature.

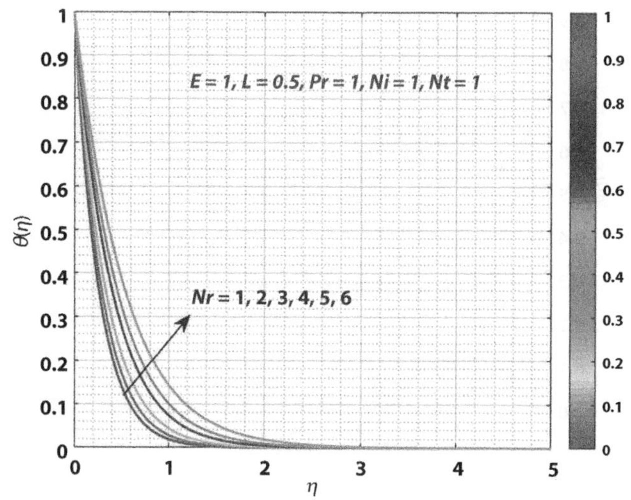

Figure 13.7 Thermal radiation parameter impact on temperature.

and smoother temperature gradients. Conversely, a lower radiation parameter limits radiative heat transfer, maintaining a thinner boundary layer.

Figure 13.8 represents the Prandtl number parameter effect on temperature; rising the Prandtl number term decays the temperature of the flow. A greater Prandtl number indicates reduced thermal diffusivity, meaning heat conduction is less effective compared to momentum transfer. As a result, temperature gradients tend to be steeper, leading to a thicker thermal boundary layer. Fluids with high Pr values, such as oils and polymers, retain heat longer, making them suitable for insulation applications.

Figure 13.9 represents the thermophoresis parameter effect on concentration; rising the thermophoresis term decays the concentration of the flow. The thermophoresis parameter affects the thickness of the concentration. A higher thermophoretic force results in a more dispersed particle distribution, leading to a thicker concentration boundary layer. In contrast, a lower thermophoresis parameter causes particles to remain more localized, maintaining a sharper concentration gradient.

Figure 13.10 represents the Brownian movement parameter effect on concentration; rising the Brownian movement decays the concentration of the flow. Brownian motion is the random motion of microscopic particles mixed in a fluid due to surrounding molecule collisions. The Brownian motion parameter quantifies the intensity of this effect and significantly impacts particle concentration distribution in various scientific and engineering applications, such as nanofluid technology, aerosol dynamics, and biomedical transport processes.

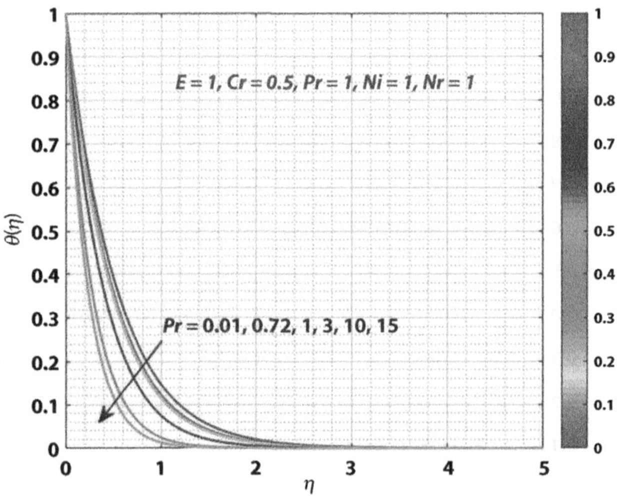

Figure 13.8 Prandtl number impact on temperature.

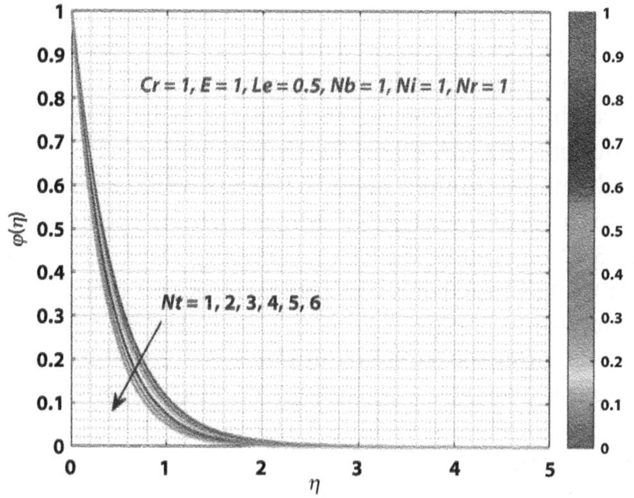

Figure 13.9 Thermophoresis impact on concentration.

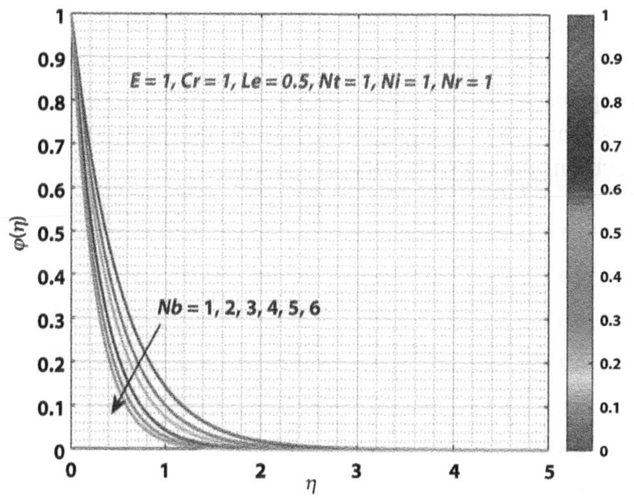

Figure 13.10 Brownian movement impact on concentration.

Figure 13.11 represents the Lewis number parameter effect on concentration; rising the Lewis number term decays the concentration of the flow. The Lewis number significantly impacts concentration distribution by regulating mass diffusivity, altering boundary layer behavior, and affecting mass transfer efficiency. A higher Lewis number leads to steeper concentration gradients and thinner boundary layers, while a lower Lewis number enhances mass diffusion, promoting uniform concentration.

Figure 13.12 represents the activation energy parameter effect on concentration; rising the activation energy term decays the concentration of the flow. Activation energy is a crucial parameter in chemical reaction kinetics, representing the required minimum energy for reactant molecules to undergo a transformation. The activation energy parameter significantly influences concentration distribution in reacting flows, combustion processes, and catalytic systems. Its impact is particularly important in chemical engineering and biomedical applications.

Figure 13.13 represents the *Cr* parameter effect on concentration; rising the *Cr* parameter decays the concentration. A higher chemical reaction parameter accelerates the reaction rate, leading to a faster depletion of reactant concentration and a corresponding increase in product concentration. Conversely, a lower reaction parameter results in a slower reaction rate, causing reactants to persist longer while delaying the accumulation of products. This effect is significant in catalytic reactions, where reaction speed directly impacts production efficiency.

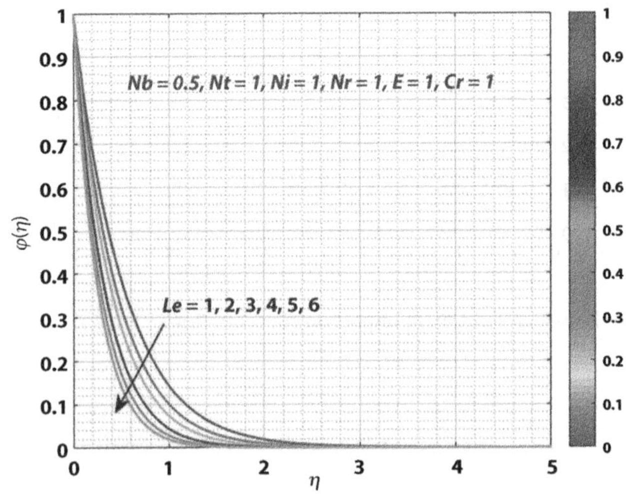

Figure 13.11 Lewis number impact on concentration.

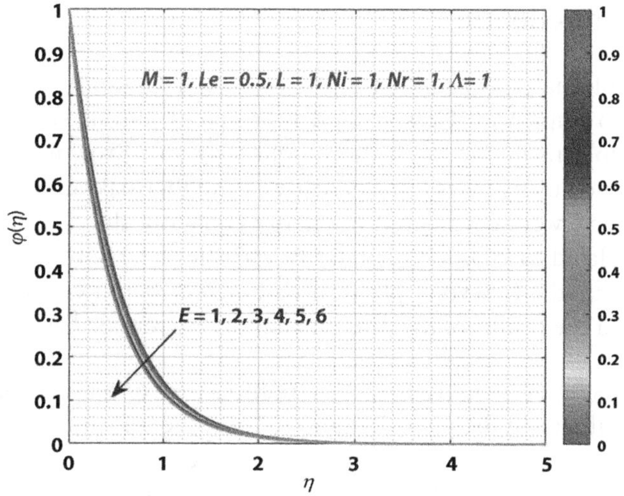

Figure 13.12 Activation energy impact on concentration.

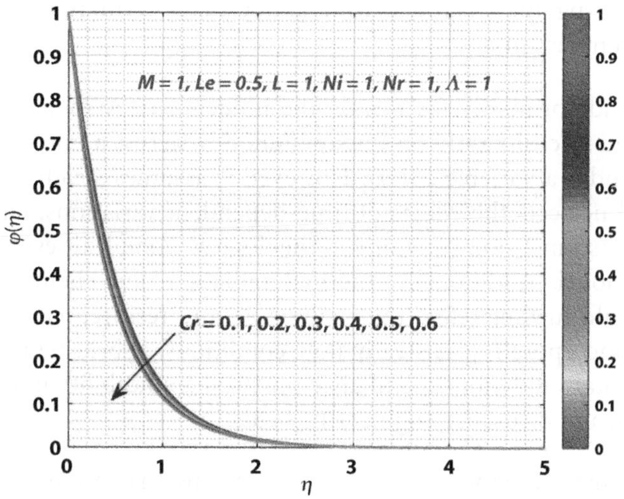

Figure 13.13 *Cr* impact on concentration.

13.5 Conclusion

The present analysis examined the Arrhenius activation energy and viscosity ratio impact on Casson fluid flow across porous stretching sheet with variable magnetic field; thermophoresis and Brownian movement was examined using mass transfer analysis and thermal radiation and heat source effects were considered in energy equation; the considered PDEs were calculated to ODEs vis similarity variables later solved by using the numerical method. Several physical parameters were examined through graphs, and the results of the present analysis are revealed as follows:

- Rising the chemical reaction reduces the concentration.
- Enhancing the viscosity ratio reduces the velocity.
- Rising the thermal radiation rises the energy.
- Rising the Prandtl number reduces the temperature.
- Rising the thermophoresis reduces the concentration.
- Enhancing the heat source parameter rises the temperature.
- Rising the magnetic field and Casson term reduces the temperature.

This study primarily focuses on the influence of Arrhenius activation energy and viscosity ratio on Casson fluid flow across a porous stretching surface under a variable MHD. However, it assumes steady-state conditions and neglects the effects of time-dependent variations, nanoparticle suspensions, and non-Newtonian rheological complexities beyond the Casson model. Additionally, thermal radiation and chemical reaction effects could further refine the analysis. Future research can extend this work by incorporating unsteady flow conditions, hybrid nanofluids, and experimental validations to enhance practical applicability in biomedical, industrial, and aerospace engineering fields.

AI disclosure statement: The author confirms that no artificial intelligence (AI)-generated content or AI-assisted writing tools were used in the preparation of this chapter (including summaries and the abstract). All text, analysis, and conclusions are the result of the author's original work.

References

1. Bhattacharyya, K., MHD Stagnation-Point Flow of Casson Fluid and Heat Transfer over a Stretching Sheet with Thermal Radiation. *J. Thermodyn.*, 2013, 1, 169674, 2013. doi: 10.1155/2013/169674.

2. Shawky, H.M., Magnetohydrodynamic Casson Fluid Flow With Heat And Mass Transfer Through A Porous Medium Over A Stretching Sheet. *J. Porous Media*, 15, 4, 2012. doi: 10.1615/JPorMedia.v15.i4.70.
3. Devi, R., Poply, V., Mala, M., Effect of aligned magnetic field and inclined outer velocity in Casson fluid flow over a stretching sheet with heat source. *J. Therm. Eng.*, 7, 4, Art. no. 4, May 2021. doi: 10.18186/thermal.930347.
4. Mahanta, G. and Shaw, S., 3D Casson fluid flow past a porous linearly stretching sheet with convective boundary condition. *Alex. Eng. J.*, 54, 3, 653–659, Sep. 2015. doi: 10.1016/j.aej.2015.04.014.
5. Das, S., Patra, R.R., Jana, R.N., The layout of Boussinesq couple-stress fluid flow over an exponentially stretching sheet with slip in porous space subject to a variable magnetic field. *Multidiscip. Model. Mater. Struct.*, 16, 5, 1131–1154, May 2020. doi: 10.1108/MMMS-09-2019-0168.
6. Patel, H.R., Patel, S.D., Darji, R., Mathematical Study of unsteady micropolar fluid flow due to non-linear stretched sheet in the presence of magnetic field. *Int. J. Thermofluids*, 16, 100232, Nov. 2022. doi: 10.1016/j.ijft.2022.100232.
7. Liu, I.-C., Flow and heat transfer of an electrically conducting fluid of second grade over a stretching sheet subject to a transverse magnetic field. *Int. J. Heat Mass Transf.*, 47, 19, 4427–4437, Sep. 2004. doi: 10.1016/j.ijheatmasstransfer.2004.03.029.
8. Misra, J.C. and Shit, G.C., Biomagnetic viscoelastic fluid flow over a stretching sheet. *Appl. Math. Comput.*, 210, 2, 350–361, Apr2009. doi: 10.1016/j.amc.2008.12.088.
9. Subhas, A. and Veena, P., Visco-elastic fluid flow and heat transfer in a porous medium over a stretching sheet. *Int. J. Non-Linear Mech.*, 33, 3, 531–540, May 1998. doi: 10.1016/S0020-7462(97)00025-5.
10. Makinde, O.D., Khan, Z.H., Ahmad, R., Khan, W.A., Numerical study of unsteady hydromagnetic radiating fluid flow past a slippery stretching sheet embedded in a porous medium. *Phys. Fluids*, 30, 8, 083601, Aug. 2018. doi: 10.1063/1.5046331.
11. Akshatha, H.D., Sachhin, S.M., Mahabaleshwar, U.S., Lodhi, R.K., Ramesh, K., Dynamics of sodium alginate-based ternary nanofluid flow over a stretching sheet with Al2O3, SiO2, and TiO2 nanoparticles. *Multiscale and Multidiscip. Model. Exp. and Des.*, 8, 1, 32, Nov. 2024. doi: 10.1007/s41939-024-00591-4.
12. Mandal, I.C. and Mukhopadhyay, S., Heat transfer analysis for fluid flow over an exponentially stretching porous sheet with surface heat flux in porous medium. *Ain Shams Eng. J.*, 4, 1, 103–110, Mar. 2013. doi: 10.1016/j.asej.2012.06.004.
13. Turkyilmazoglu, M., Flow of a micropolar fluid due to a porous stretching sheet and heat transfer. *Int. J. Non-Linear Mech.*, 83, 59–64, Jul. 2016. doi: 10.1016/j.ijnonlinmec.2016.04.004.
14. Ishak, A., Thermal boundary layer flow over a stretching sheet in a micropolar fluid with radiation effect. *Meccanica*, 45, 3, 367–373, Jun. 2010. doi: 10.1007/s11012-009-9257-4.

15. Mukhopadhyay, S., Slip effects on MHD boundary layer flow over an exponentially stretching sheet with suction/blowing and thermal radiation. *Ain Shams Eng. J.*, 4, 3, 485–491, Sep. 2013. doi: 10.1016/j.asej.2012.10.007.
16. Liu, I.-C. and Megahed, A.M., Numerical Study for the Flow and Heat Transfer in a Thin Liquid Film Over an Unsteady Stretching Sheet with Variable Fluid Properties in the Presence of Thermal Radiation. *J. Mech.*, 28, 2, 291–297, Jun. 2012. doi: 10.1017/jmech.2012.32.
17. Cortell, R., Effects of viscous dissipation and radiation on the thermal boundary layer over a nonlinearly stretching sheet. *Phys. Lett. A*, 372, 5, 631–636, Jan. 2008. doi: 10.1016/j.physleta.2007.08.005.
18. Bognár, G., On similarity solutions of boundary layer problems with upstream moving wall in non-Newtonian power-law fluids. *IMA J. Appl. Math.*, 77, 4, 546–562, Jun. 2012. doi: 10.1093/imamat/hxr033.
19. Lone, S.A., Ali, F., Saeed, A., Bognár, G., Irreversibility analysis with hybrid cross nanofluid of stagnation point and radiative flow (TiO2+CuO) based on engine oil past a stretchable sheet. *Heliyon*, 9, 4, e15056, Apr. 2023. doi: 10.1016/j.heliyon.2023.e15056.
20. Barna, I.F., Bognár, G., Guedda, M., Mátyás, L., Hriczó, K., Analytic Self-Similar Solutions Of The Kardar-Parisi-Zhang Interface Growing Equation With Various Noise Terms. *Math. Model. Anal.*, 25, 2, 241–256, Mar. 2020. doi: 10.3846/mma.2020.10459.
21. Klazly, M. and Bognár, G., A novel empirical equation for the effective viscosity of nanofluids based on theoretical and empirical results. *Int. Commun. Heat Mass Transf.*, 135, 106054, Jun. 2022. doi: 10.1016/j.icheatmasstransfer.2022.106054.
22. Bognar, G., Ferrofluid Flow in Magnetic Field Above Stretching Sheet with Suction and Injection, in: *Math. Model. Anal*, Jul. 2020, Accessed: Feb. 20, 2025. [Online]. Available: https://www.academia.edu/121605678/Ferrofluid_Flow_in_Magnetic_Field_Above_Stretching_Sheet_with_Suction_and_Injection.
23. Sachhin, S.M., Mahabaleshwar, U.S., Chan, A., Effect of slip and thermal gradient on micropolar nano suspension flow across a moving hydrogen fuel-cell membrane. *Int. J. Hydrog. Energy*, 63, 59–81, Apr. 2024. doi: 10.1016/j.ijhydene.2024.02.332.
24. Sachhin, S.M., Mahabaleshwar, U.S., Nayakar, S.N.R., Patil, H.S.R., Souayeh, B., Darcy–Brinkman model for Boussinesq–Stokes suspension tetra dusty nanofluid flow over stretching/shrinking sheet with suction/injection. *Numer. Heat Transf. Part Appl.*, 0, 0, 1–23, 2024. doi: 10.1080/10407782.2024.2359047.
25. Waini, I., Ishak, A., Pop, I., Unsteady flow and heat transfer past a stretching/shrinking sheet in a hybrid nanofluid. *Int. J. Heat Mass Transf.*, 136, 288–297, Jun. 2019. doi: 10.1016/j.ijheatmasstransfer.2019.02.101.
26. Sreedevi, P., Sudarsana Reddy, P., Chamkha, A., Heat and mass transfer analysis of unsteady hybrid nanofluid flow over a stretching sheet with thermal radiation. *SN Appl. Sci.*, 2, 7, 1222, Jun. 2020. doi: 10.1007/s42452-020-3011-x.

14

Computational Fluid Dynamics Examination of Non-Newtonian Fluid Flows over an Exponentially Extending Surface with Thermal Source/Sink

Priyanka Chandra[1] and Raja Das[2*]

[1]Department of Mathematics, SR University, Warangal, Telangana, India
[2]Department of Mathematics, School of Advanced Sciences, Vellore Institute of Technology (VIT), Vellore, India

Abstract

This investigation delves into the magnetohydrodynamic laminar boundary layer transport of a non-Newtonian fluid across an exponentially extending surface. In this research, we integrate the influences of heat sink/source, heat radiation, MHD, and exponentially stretching surface. Through the application of similarity transformations, the system of nonlinear ordinary differential equations is generated from the set of governing equations, which were originally expressed as nonlinear partial differential equations. The extensively evaluated and very efficient variational finite-element method, a computational fluid dynamics tool, is employed to solve the reduced ordinary differential equations. Critical parameters, such as the magnetic field, radiation, Casson fluid, Prandtl number, and heat sink/source, are examined to evaluate their influence on velocity field, temperature distributions, thermal transmission rates, and skin friction values. The findings indicate that elevated Casson fluid values decrease the velocity gradient, hence lowering skin friction at the surface. An increase in the magnetic field diminishes velocity owing to Lorentz forces and elevates the skin friction coefficient. The results complement prior scientific research, affirming the geometry's validity. This study is crucial for magnetohydrodynamic non-Newtonian fluids flows, applicable in technologies for cooling, metallic molding, and polymer extrusion. The incorporation of heat

**Corresponding author*: rajadasrkl@gmail.com; ORCID: 0000-0001-9945-5969
Priyanka Chandra: ORCID: 0009-0005-1706-5841

Dhananjay Yadav, Mukesh Kumar Awasthi and Harith Mohamed Al-Azri (eds.) Non-Newtonian Fluids for Industrial Applications: Modeling and Simulations, (297–318) © 2026 Scrivener Publishing LLC

radiation, magnetohydrodynamic, and heat sink/source parameters enhances its applicability to systems that require a lot of energy, including thermal insulation and nuclear reactors.

Keywords: Computational fluid dynamics, finite-element method, non-Newtonian fluid, casson fluid, exponential stretching surface

14.1 Introduction

The examination of non-Newtonian fluids has numerous applications in biomedical, industrial, and engineering sciences, particularly in the mining of crude oil made with compounds of petroleum. A non-Newtonian fluid that shows yield stress is called a Casson fluid. The blood of humans can be regarded as a Casson fluid because of the chain structure of blood cells and the presence of components such as proteins, rouleaux, and fibrinogen. Consequently, the Casson fluid has important relevance in both engineering and scientific domains. The evolution of boundary layer dynamics on a continuously stretched surface commenced with Sikiadis [1]. Gnaneswara Reddy [2] examined an unstable 2D flow of a non-Newtonian fluid across an extending surface, considering the effects of heat radiations and variable thermal conductivity. Pramanik [3] examined the thermal transfer properties of Casson fluid dynamics over an exponentially stretched surface within a porous material and under heat radiation, concluding that a rise in the Casson parameter diminishes the velocity profile. The authors in Ref. [4] examined the unsteady free convection flow of Casson fluid adjacent to an oscillating vertical plate maintained at a constant wall temperature.

Magnetohydrodynamics (MHD) [5] pertains to the examination of electrically conductive fluids' behavior under magnetic fields. The importance of magnetohydrodynamic parameter originates from its tremendous applications in disciplines like biomedical sector, geophysics, astronomy, and numerous industrial process, comprising metallic cooling, atomic power plants, and advanced coolant systems employing metallic liquids [6]. Chandna and Garg [7] researched viscous magnetohydrodynamic fluids. Immanuel *et al.* [8] examined the viscous continuous flow of Casson fluid exposed to a homogeneous magnetic field with a steady velocity. Non-Newtonian fluids, which deviate from the conventional line flow characteristics of Newtonian fluids, are frequently observed in several engineering and industrial activities, including blood circulation, polymer extrusion, and processing of food [9]. Casson fluids demonstrate yield stress behavior, rendering them especially pertinent in industries where fluids experience shear thickening or thinning phenomena. Comprehending the flow and thermal transmission properties of Casson fluids under magnetic fields is crucial for enhancing processes in relevant applications.

In recent decades, there has been an increased focus on the examination associated with non-Newtonian fluid dynamics over elongated surface, owing to its numerous scientific and economic implications. Mukhopadhyay [10] examined heat transfer and the flow of Casson fluid *via* nonlinearly expanding surface. Manideep *et al.* [11] analyzed the 2D magnetohydrodynamic free convection Casson fluid flow traversing an oscillating vertical plate with a thermally constant boundary. Salah and Almohammadi [12] examined magnetohydrodynamic boundary layer fluid flow resulting from an exponentially stretched surface along with heat radiation effects. Usman *et al.* [13] employed collocation to investigate the effects of temperature and velocity slip on the flow of Casson nanofluid across an inclined permeable stretched cylinder. Rahman *et al.* [14] investigated the flow of a hybrid nanofluid toward a sheet that is growing and contracting exponentially, influenced by radiation and the presence of a heat source or sink.

Chamkha *et al.* [15] examined the heat radiation effects on magnetohydrodynamic forced convection flow adjacent to a non-isothermal wedge, considering the influence of thermal sink/source. Sumalatha and Bandari [16] examined the impact of heat sink/source and radiation on the flow of Casson fluid *via* nonlinear extending surface. Prakash *et al.* [17] examined heat and mass transport in the flow of Casson fluid across an exponentially stretched surface with heat sink/source and hydromagnetic effects. Seth *et al.* [18] examined MHD stagnation-point flow along an exponentially stretched surface accompanied by an exponentially flowing free stream, incorporating a non-uniform heat sink/source, thermal radiation, and viscous dissipation. Recently, Aghalya and Tamizharasi [19] evaluated the mixed impacts of thermal radiation and heat source/sink on the flow of magnetohydrodynamic Casson fluid *via* exponentially extending vertically inclined surface.

Thermal radiation is a crucial element influencing heat transmission in fluid flow process, especially at elevated temperature distributions. It is essential in applications including nuclear plants, spaceship re-entry, and extreme temperatures systems for engineering. The inclusion of heat sink/source, alongside thermal radiation, adds complexity to the heat transmission action. The heat source elevates heat in the fluid, improving temperature parameter, whereas a heat sink dissipates it, resulting in temperature decline. Comprehending these impacts within the framework of magnetohydrodynamic (MHD) fluid flow across a nonlinear exponentially stretched surface provides significant insights about heat control for applications in industry.

The governing equations for flows of fluid across an exponentially stretched surface, incorporating thermal radiation and magnetic field effects, are extremely nonlinear, coupled, and complex to solve. Computational fluid dynamic (CFD) tools are utilized to surmount these challenges. A variety of fluid flow issues have been addressed with several CFD tools, including the finite-element method [9, 20], finite-difference method [21, 22], and finite-volume method [23]. Among these three CFD techniques, the finite-element method (FEM) has garnered greater attention because to its proficiency in modeling intricate geometries and materials, providing flexibility and adaptation for many engineering and scientific challenges, which other approaches, like finite-difference methods, find difficult to address.

Thus, in this study, we examined the magnetohydrodynamic boundary layer transport of a non-Newtonian fluid, across an exponentially extending surface, along with magnetohydrodynamic, heat sink and sources, and heat radiation effects. The Casson fluid is used as a non-Newtonian fluid in this study. This study utilizes a validated and highly efficient variational finite-element method to analyze the flow geometry. Graphical depictions of the effect of various flow parameters on fluid velocity (f') and temperature (θ) are provided. Tables present numerical values for the reduced Nusselt number and skin friction coefficient.

14.2 Mathematical Formulation

Examine a stable, incompressible laminar, 2D boundary layer transport of a dissipative non-Newtonian fluid (Casson fluid) across an exponentially stretched surface situated at the wall $y = 0$. The flow of Casson fluid is restricted to the region where $y > 0$. A changing magnetic field was applied, and the absence of applied voltage indicates a lack of electric field. The induced magnetic field is assumed to be insignificant relative to the exterior magnetic field. This indicates the presence of a low magnetic Reynolds number in this investigation. Additionally, we evaluated the heat source and radiation parameters affecting the flow. The sheet expands exponentially at velocity $U_w(x)$ while maintaining its original configuration. The equations presented below are intended to address fundamental issues of fluid dynamics. They encompass the conservation equations for mass and momentum. Collectively, these equations provide a comprehensive framework for the analysis and interpretation of fluid behavior in various contexts:

$$\frac{\partial u}{\partial x}+\frac{\partial u}{\partial y}=0 \tag{14.1}$$

$$\left(u\frac{\partial u}{\partial x}+v\frac{\partial u}{\partial y}\right)+\frac{B_0^2\sigma u}{\rho_f}=\left(v+\frac{v}{\beta}\right)\frac{\partial^2 u}{\partial y^2} \tag{14.2}$$

$$\rho_f C_p\left(u\frac{\partial T}{\partial x}+v\frac{\partial T}{\partial y}\right)+\frac{\partial q_r}{\partial y}=Q\left(T-T_\infty\right)+K_T\rho_f C_p\frac{\partial^2 T}{\partial y^2} \tag{14.3}$$

In this context, u and v represent the velocity components along the x and y axes, respectively; ρ_f denotes the fluid density; Q is the dimensional heat source; σ signifies electrical conductivity; β is the Casson parameter; v refers to kinematic viscosity; B_0 represents the magnetic field; C_p indicates specific heat; K_T represents thermal diffusivity; q_r denotes radiative heat flux; and T represents temperature. Employing the Rosseland approximation, we obtain

$$q_r=-\frac{\partial T^4}{\partial y}\frac{4\sigma}{3K^*}\approx 4T_\infty^3-3T_\infty^4 \tag{14.4}$$

$$q_r=-\frac{4\sigma\partial\left(4T_\infty^3T-3T_\infty^4\right)T^4}{3K^*\partial y}=-\frac{16T_\infty^3\sigma}{3K^*}\frac{\partial T}{\partial y} \tag{14.5}$$

Therefore, Equation 14.3 becomes

$$\rho_f C_p\left(u\frac{\partial T}{\partial x}+v\frac{\partial T}{\partial y}\right)-\frac{16T_\infty^3\sigma}{3K^*}\frac{\partial^2 T}{\partial y^2}=Q_0\left(T-T_\infty\right)+K_T\rho_f C_p\frac{\partial^2 T}{\partial y^2} \tag{14.6}$$

The boundary conditions are given by

$$\begin{aligned} &at \quad y=0: u=U_w(x), \quad v=-V(x), \quad T=T_w \\ &as \quad y\rightarrow\infty: u\rightarrow 0, \quad T\rightarrow T_\infty \end{aligned} \tag{14.7}$$

The wall and the outermost edge of the boundary layer exhibit distinct circumstances, represented by the symbol's "w" for wall and "∞" for outermost edge. The boundary requirements in Equation 14.7 guarantee that the temperature and velocity of both fields exhibit physical coherence with the stretched sheet, and they consider the fluid's behavior at both the surface and around infinity. Defining the similarity variables as

$$\begin{aligned} &\eta=\sqrt{\frac{u_0}{2\nu l}}ye^{\frac{x}{2l}}, \quad v=-\sqrt{\frac{u_0}{2\nu l}}e^{\frac{x}{2l}}\{f(\eta)+\eta f'(\eta)\}, \quad u=u_0e^{\frac{x}{2l}}f'(\eta), \\ &\theta(\eta)=\frac{T-T_\infty}{e^{(x/2l)}T_0}, \quad Q_0=e^{(x/2l)}Q \end{aligned} \tag{14.8}$$

Here T_∞ denotes the ambient fluid temperature. u_0 represents the fluid velocity and l is the characteristic length. Inserting Equation 14.8 into Equations 14.2 and 14.6 yields

$$\left(1+\frac{1}{\beta}\right)f'''+ff''-2(f')^2-Mf'=0 \tag{14.9}$$

$$\left(1+\frac{4R_d}{3}\right)\theta''+\Pr\left(f\theta'+Q\theta-f'\theta\right)=0 \tag{14.10}$$

along with boundary conditions

$$\begin{aligned} &f(0)=S_b, \quad f'(0)=1, \quad \theta(0)=1; \\ &as \quad \eta\rightarrow\infty; \quad f'\rightarrow 0, \quad \theta\rightarrow 0 \end{aligned} \tag{14.11}$$

where Pr is the Prandtl number, M is the magnetic field parameter, S_b is the suction/blowing parameter, R_d is the radiation parameter, and Q is the sink/source parameter. These parameters are defined as follows:

$$\Pr = \frac{\nu}{K_T}, \quad M = \frac{2lB_0^2\sigma}{\rho_f u_0}, \quad S_b = v_0\sqrt{\frac{2l}{u_0 \nu}}, \quad R_d = \frac{4\sigma T_\infty^3}{K^* K_T}, \quad Q = \frac{2lQ_0}{u_0 \rho^2 C_p} \tag{14.12}$$

The Nusselt number, used to quantify the degree of convective heat transport, is ascertained by evaluating the skin friction, which offers significant conclusions by approximating the total drag friction applied to a fluid. Furthermore, the skin friction coefficient and the Nusselt number can be stated as

$$Cf_x = -\left(1 + \frac{1}{\beta}\right) f''(0) \quad and \quad Nu_x = -\left(1 + \frac{4R_d}{3}\right)\theta'(0) \tag{14.13}$$

14.3 Computational Fluid Dynamic Tools: FEM

The finite-element method (FEM) has developed into an essential technique for modeling and simulating many technical systems across industries such as residential, transportation, and networking in recent decades. It is a numerical approach that aims to provide a nearly perfect solution for the dissemination of field variables in a region that is analytically challenging to understand. The core concept of this method, namely the discretization of the field into a collection of simple subfields, called finite elements, develops after structural analysis. The FEM process generally comprises four stages:

i. **Meshing:** Mesh is employed to decompose geometry into small components called elements, utilizing a set of nodes, allowing for the approximate solution of an element by fundamental polynomial functions. Upon identifying the system's components, one may utilize fundamental physical reasoning to formulate the element equations in relation to pertinent variables.
ii. **Formulation of element equations:** A fundamental element is pulled out of the meshes, and the variational expression of the specified issue is formulated over this basic element. By employing a linear arrangement of node values and an estimated function, we will derive an

approximation of the solution for every single component, which will subsequently be substituted into the system to formulate the element equations.

iii. **Development of the global stiffness matrix:** The system's comprehensive equations are formulated by integrating each individual element equation and subsequently solving the system of equations for the unknown nodal variable. The assembly should be predicated on the interconnections of the elements.

iv. **Establishing boundaries and revealing solutions to the constructed equations:** The gathered equations have the requisite natural and essential boundary conditions imposed, and they were subsequently solved using numerical techniques such as Gauss elimination, row reduction, or the Lower–Upper decomposition method (LU).

To solve a set of differential Equations 14.9 and 14.10, along with boundary conditions 14.11, we will assume that $f' = g$ [24]. Then, Equations 14.9–14.11 are converted into

$$\left(1+\frac{1}{\beta}\right)g'' + fg' - 2(g)^2 - Mg = 0 \tag{14.14}$$

$$\left(1+\frac{4R_d}{3}\right)\theta'' + \Pr\left(f\theta' + Q\theta - g\theta\right) = 0 \tag{14.15}$$

$$\begin{aligned} &f(0) = S_b, \quad g(0) = 1, \quad \theta(0) = 1; \\ &as \quad \eta \to \infty; \quad g \to 0, \quad \theta \to 0 \end{aligned} \tag{14.16}$$

14.3.1 Variational Formulation

The variational version of Equations 14.14 and 14.15 over a standard linear element (η_i, η_{i+1}) is given by the following equations:

$$\int_{\eta_i}^{\eta_{i+1}} w_1 \left(f' - g\right)\partial\eta = 0 \tag{14.17}$$

$$\int_{\eta_i}^{\eta_{i+1}} w_2\left(\left(1+\frac{1}{\beta}\right)g'' + fg' - 2(g)^2 - Mg\right)\partial\eta = 0 \quad (14.18)$$

$$\int_{\eta_i}^{\eta_{i+1}} w_3\left(\left(1+\frac{4R_d}{3}\right)\theta'' + \Pr\left(f\theta' + Q\theta - g\theta\right)\right)\partial\eta = 0 \quad (14.19)$$

Here w_1, w_2 and w_4 are stochastic test functions that denote variations in f, g, and θ, respectively.

14.3.2 Finite-Element Formulation

Given that the domain is divided into two-noded line elements, a finite-element framework can be constructed by substituting estimated finite-element solutions into the previously described equations of the specified form [25].

$$f = \sum_{j=1}^{2} f_j N_j, \quad g = \sum_{j=1}^{2} g_j N_j, \quad \theta = \sum_{j=1}^{2} \theta_j N_j \qquad (14.20)$$
$$w_1 = w_2 = w_3 = N_i \qquad (i = 1,2)$$

where N_i represents the trial functions of a standard linear element (η_i, η_{i+1}), defined as follows:

$$N_1^i = \frac{(\eta_{i+1} - \eta)}{(\eta_{i+1} - \eta_i)}, \quad N_2^i = \frac{(\eta - \eta_i)}{(\eta_{i+1} - \eta_i)}, \quad \eta_i \le \eta \le \eta_{i+1} \quad (14.21)$$

Consequently, the overall stiffness matrix of the finite-element model for Equations 14.17–14.19 pertaining to a standard line element is expressed as

$$\begin{bmatrix} [K^{11}] & [K^{12}] & [K^{13}] \\ [K^{21}] & [K^{22}] & [K^{23}] \\ [K^{31}] & [K^{32}] & [K^{33}] \end{bmatrix} \begin{bmatrix} f \\ g \\ \theta \end{bmatrix} = \begin{bmatrix} [b^1] \\ [b^2] \\ [b^3] \end{bmatrix}$$

side this element matrix, each $[K^{mn}]$ and $[b^m](m,n=1,2,3)$ are of order 2×2 and 2×1, respectively.

In this study, each element matrix has a dimension of 10×10. For intermediate levels of $\eta(>6)$, no significant effect on the outcomes has been observed. Consequently, infinity has been designated as 6 for computational purposes. The domain is divided into 1000 linear elements, each measuring 0.003 in length, resulting in the formation of a 3003×3003 global matrix after the assembly of the element matrix. The resulting system of equations is very nonlinear; hence, a dependable iterative method is required for its solution. The system is linearized by incorporating the assumed known functions. The solution is executed with a dependable Gauss elimination method, maintaining a precision of 0.00001. The algorithm's software was executed on a PC utilizing MATLAB R2022b. This analysis revealed remarkable convergence.

14.4 Results Analysis

In this section, we present the results obtained from the computational fluid dynamic analysis of the magnetohydrodynamic boundary layer transport of non-Newtonian Casson fluid across an exponentially extending surface by integrating the influences of thermal source/sink and heat radiation. The governing ordinary differential equations, which are extremely nonlinear and coupled, were resolved utilizing the highly validated variational finite-element method, a computational fluid dynamic tool, which yielded precise and effective results. The outcomes are analyzed concerning essential flow parameters, namely the Casson fluid parameter, Prandtl number, magnetic field parameter, heat source/sink parameter, and thermal radiation parameter.

To corroborate the findings, a comparison was conducted with prior studies on analogous topics. In this regard, Pramanik [3] investigated fluid dynamics on stretching sheets in simple configurations. The findings of the current study correspond with their results under conditions of minimal heat radiation and consistent thermal sink/source, therefore verifying the precision of the finite-element approach. Moreover, this study improves comprehension of the issue by integrating exponential extending alongside the mixed influences of thermal sink/source and heat radiation. The precision of the employed finite-element approach was first corroborated by juxtaposing the thermal transmission rates with the findings of Pramanik [3], as illustrated in Table 14.1, across different Prandtl number values.

The finite-element approach produced solutions that were comparably precise while exhibiting greater effectiveness in managing nonlinear boundary circumstances. Such comparisons underscore the effectiveness of the current technique and its possible application to more intricate fluid circulation and transmission of heat scenarios. The present article fills a vacuum in the scientific research by delivering a thorough examination of the flow of MHD Casson fluid, integrating heat radiation and the effects of heat sources or sinks, while highlighting the collective impact of these variables on the dynamics of flow and heat transfer. The measured values of the Nusselt number and skin friction coefficient various magnetic fields, with default values $Q = \beta = R_d = S_b = 0.1$, are presented in Table 14.2. It is evident from Table 14.2 that a rise in the magnetic field intensifies the Lorentz force, thereby restricting fluid motion and amplifying the velocity gradient nearer the surface, leading to increased skin friction. Convective heat transmission is also decreased by the decreased fluid velocity; consequently, the thermal boundary layer becomes thicker and the Nusselt number gets lowered.

Unless specified differently, the simulations in this context assume the following parameter default values: $M = 1$, $R_d = 0.1$, $S_b = 0.1$, $Q = 0.1$, $\Pr = 0.7$, and $\beta = 1$. Figures 14.1 and 14.2 show the effect of the magnetic field parameter (M) on velocity $f'(\eta)$ and temperature $\theta(\eta)$ profiles, respectively. From the figures, we can see that when the magnetic field increases, the velocity profile decreases; however, the temperature profile increases. This fluctuation arises from a phenomenon called

Table 14.1 Comparison of heat transfer rate with benchmark studies [3] for varying values of Prandtl numbers by taking default parameter values as $Q = \beta = R_d = S_b = 0.1$.

	$-\theta'(0)$	
Prandtl number	**FEM**	**Pramanik [3]**
1	0.9911	0.9547
2	1.4729	1.4714
3	1.8701	1.8691
5	2.5021	2.5001
10	3.6591	3.6603

Table 14.2 Skinfriction coefficient and Nusselt number values with varying magnetic field parameters by taking default values as $Q = \beta = R_d = S_b = 0.1$.

M	Nu_x	Cf_x
0.1	27.8923	0.343381
0.3	29.9001	0.338992
0.5	30.5311	0.335991
0.7	31.6450	0.332752
1	31.9012	0.329923

magnetohydrodynamic drag, largely induced by the Lorentz force, which exacerbates the drop in velocity profiles. The Lorentz force results from an association among the magnetic field and the movement of ions in the liquid. This force opposes fluid mobility by working perpendicular towards both the magnetic flux and the stream's direction, thereby reducing the velocity. Additionally, the Lorentz force induces circulating currents within the fluid, which create characteristic flow patterns.

Figure 14.1 The diagram depicts the magnetic field effects on the velocity profile. It examines four specific magnetic field values, namely M = 1, 2, 3, 4. The X-axis denotes the (η) values, whereas the vertical Y-axis signifies the associated velocity values.

Figure 14.2 The diagram illustrates the magnetic field effects on the temperature profile. It examines four specific magnetic field values, namely M = 1, 2, 3, 4. The X-axis denotes the (η) values, whereas the Y-axis signifies the associated temperature values.

The impact of the Casson fluid (β) on the velocity $f'(\eta)$ and temperature $\theta(\eta)$ fields are illustrated in Figures 14.3 and 14.4, respectively. The figures show that the velocity profile lowers and the temperature profile increases as the Casson fluid parameter is increased. The increased plastic viscosity linked to the Casson fluid is responsible for this behavior, which enhances resistance to deformation. Within the boundary layer, the fluid's velocity decreases due to the higher resistance, which hinders its smooth flow. Since a higher Casson fluid parameter makes the fluid work harder to overcome its own resistance, a higher value for this parameter has a more noticeable effect on the suppression of fluid motion. There is a direct

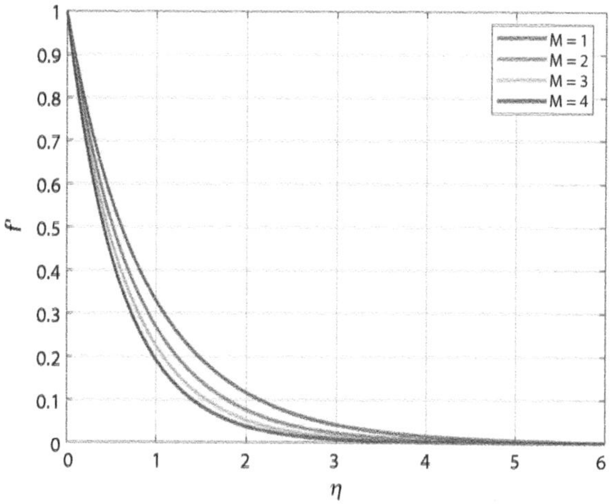

Figure 14.1 Effect of M on $f'(\eta)$.

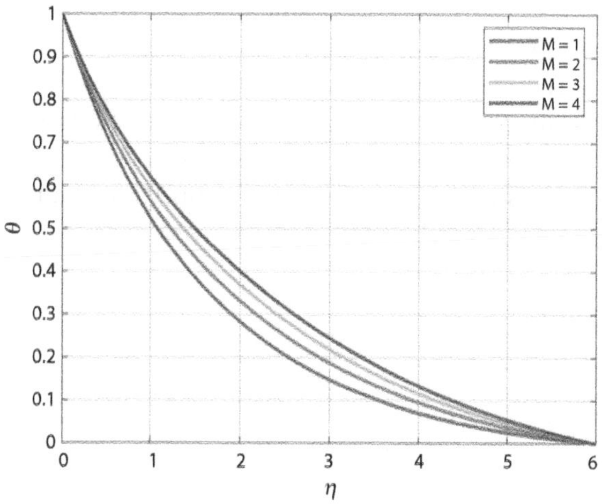

Figure 14.2 Effect of M on $\theta(\eta)$.

correlation between the increased friction and the amount of heat dissipated as a result of interactions between fluid particles and the increased effort needed to start and maintain fluid flow in the presence of higher resistance. The temperature of the fluid is increased due to the dissipated heat, which in turn increases the overall temperature profile.

Figure 14.3 The diagram depicts the effects of the Casson fluid parameter on the velocity profile. It examines four specific Casson fluid parameter values,

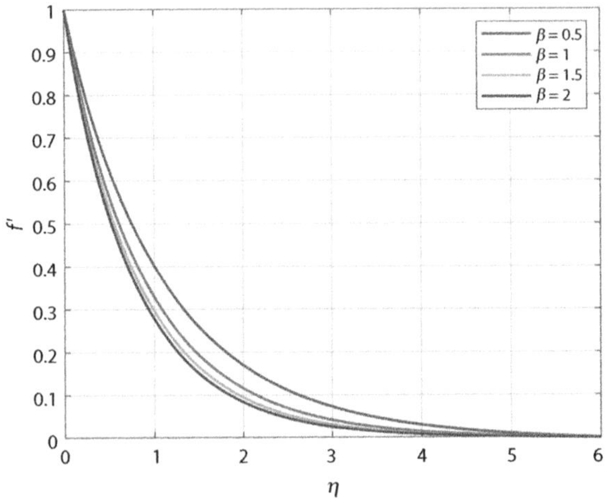

Figure 14.3 Effect of β on $f'(\eta)$.

namely $\beta = 0.5, 1, 1.5, 2$. The X-axis denotes the (η) values, whereas the vertical Y-axis signifies the associated velocity values.

Figure 14.4 The diagram illustrates the Casson fluid parameter impact on the temperature profile. It examines four specific Casson fluid parameter values, namely $\beta = 0.5, 1, 1.5, 2$. The X-axis denotes the (η) values, whereas the Y-axis signifies the associated temperature values.

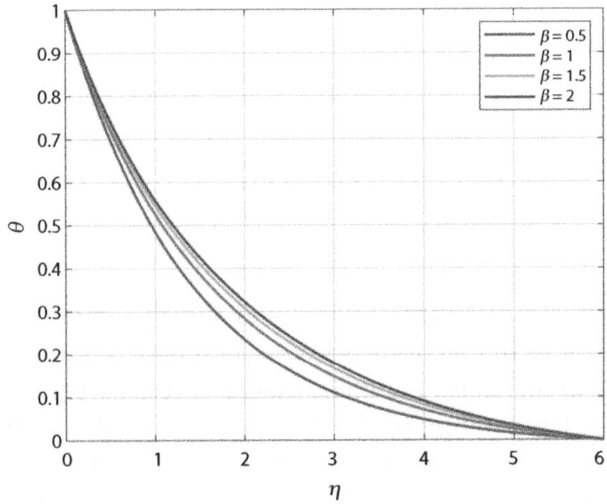

Figure 14.4 Effect of β on $\theta(\eta)$.

Figures 14.5 and 14.6 show the effect of the Prandtl number (Pr) on velocity $f'(\eta)$ and temperature $\theta(\eta)$ profiles, respectively. From the figures, we can see that as the Prandtl number escalates, the temperature distribution decreases. However, no significant changes have been observed in velocity profile. This proves that the fluid's momentum transport capabilities outperform its heat transport capabilities. As a result, the temperature profiles drop because convective heat transfer is no longer the primary mechanism. The impact of the radiation parameter (R_d) on the velocity $f'(\eta)$ and temperature $\theta(\eta)$ fields is depicted in Figures 14.7 and 14.8. The figures show that the temperature profile grows in conjunction with the radiation parameter. Nevertheless, the velocity profile has not shown any notable change. Without the need for direct physical contact, heat can be transferred from warmer to colder areas through the process of radiative heat transfer, which makes use of electromagnetic waves to transmit energy. Radiative effects raise the temperature profile by increasing the energy exchange in the fluid.

Figure 14.5 The diagram depicts the influence of the Prandtl number on the velocity profile. It examines four specific values of the Prandtl number, namely $Pr = 0.5$, 0.7, 1, 1.8. The X-axis denotes the (η) values, whereas the vertical Y-axis signifies the associated velocity values.

Figure 14.6 The diagram depicts the influence of the Prandtl number on the temperature profile. It examines four specific values of the Prandtl

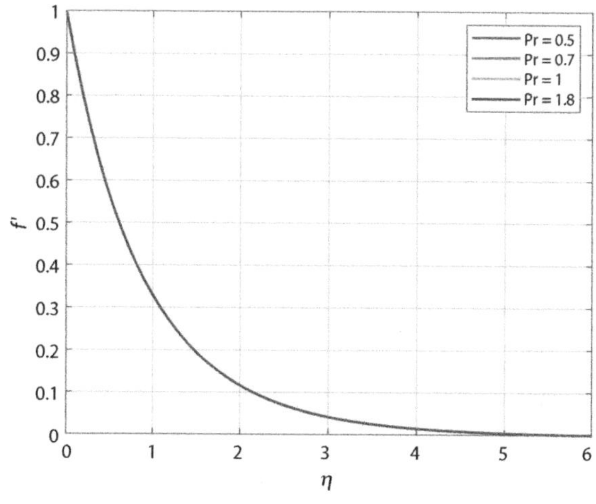

Figure 14.5 Effect of Pr on $f'(\eta)$.

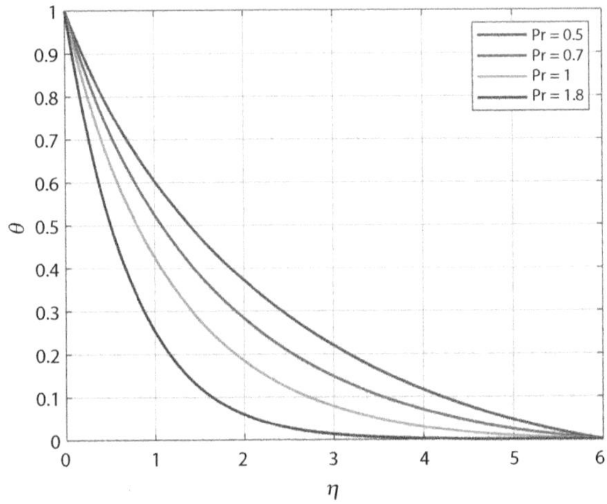

Figure 14.6 Effect of Pr on $\theta(\eta)$.

number, namely $Pr = 0.5$, 0.7, 1, 1.8. The X-axis denotes the (η) values, whereas the Y-axis signifies the associated temperature values.

Figure 14.7 The diagram depicts the effects of the radiation parameter on the velocity profile. It examines four specific values of the radiation parameter, namely $R_d = 0.1$, 0.4, 0.7, 1. The X-axis denotes the (η) values, whereas the vertical Y-axis signifies the associated velocity values.

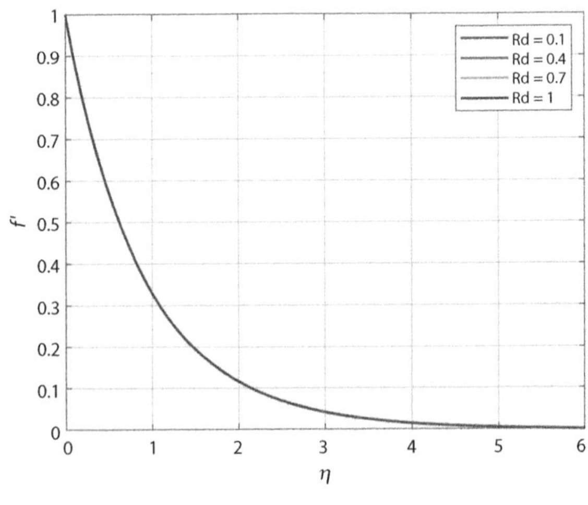

Figure 14.7 Effect of R_d on $f'(\eta)$.

Figure 14.8 The diagram depicts the radiation parameter impacts on the temperature profile. It examines four specific values of the radiation parameter, namely $R_d = 0.1, 0.4, 0.7, 1$. The X-axis denotes the (η) values, whereas the Y-axis signifies the associated temperature values

Figures 14.9 and 14.10 show the impacts of the heat source/sink parameter (Q) on velocity $f'(\eta)$ and temperature $\theta(\eta)$ plots, respectively. The temperature profile rises as the heat source/sink parameter increases, as seen in the figures. However, there has been no discernible shift in the velocity profile. One of the parameters that changes the internal energy of the fluid is the thermal source/sink parameter. A heat source increases the temperature of the fluid by supplying thermal energy to the system, and a heat sink decreases the temperature by removing thermal energy from the system. This impact, which significantly affects the temperature distribution, is proportional to the size of the heat source or sink.

Figure 14.9 The diagram depicts the impact of the heat sink/source parameter on the velocity plots. It examines four specific values of the heat sink/source parameter, namely $Q = 0.1, 0.2, 0.3, 0.5$. The X-axis denotes the (η) values, whereas the vertical Y-axis signifies the associated velocity values.

Figure 14.10 The diagram depicts the influence of the heat sink/source parameter on the temperature profile. It examines four specific values of the heat sink/source parameter, namely $Q = 0.1, 0.2, 0.3, 0.5$. The X-axis denotes the (η) values, whereas the Y-axis signifies the associated temperature values

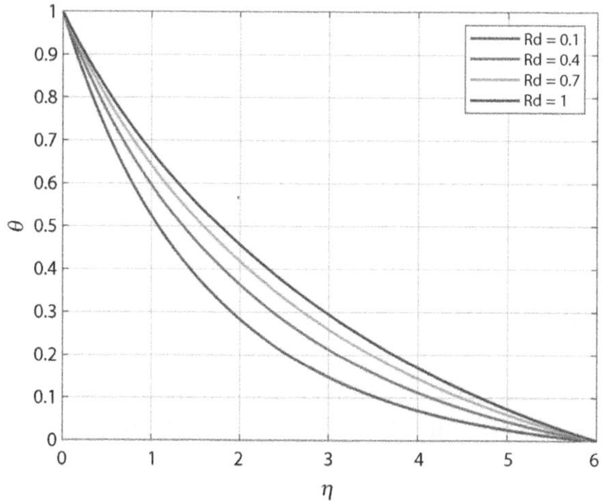

Figure 14.8 Effect of R_d on $\theta(\eta)$.

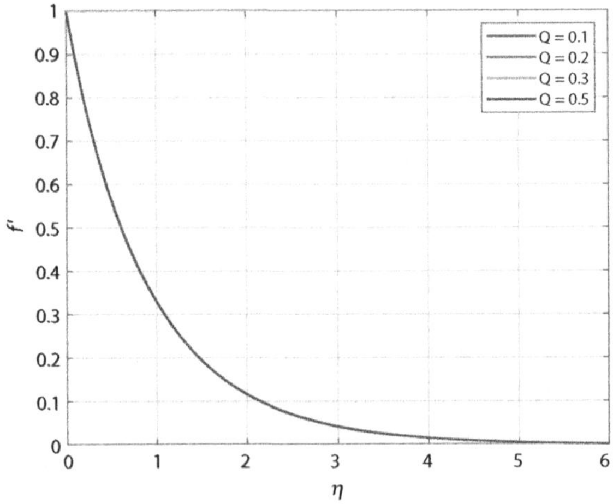

Figure 14.9 Effect of Q on $f'(\eta)$.

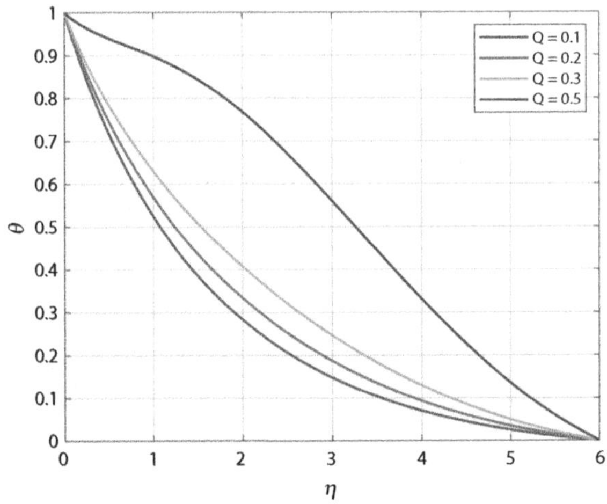

Figure 14.10 Effect of Q on $\theta(\eta)$.

14.5 Conclusion

This work investigates the boundary layer transport of a non-Newtonian Casson fluid *via* exponentially extending surface, including the influences of magnetohydrodynamic, thermal sources and sinks, and thermal radiation. This study used a verified and extremely efficient variational finite-element method, a computational fluid dynamic tool, to analyze the flow geometry. Critical parameters, such as the magnetic field, radiation, Casson fluid, Prandtl number, and thermal source/sink parameters, are examined to evaluate their influence on heat transmission coefficients, velocity profiles, skin friction values, and temperature profiles. The key findings are as follows: Radiative thermal transmission raises temperatures by maximizing exchange of energy, but the magnetic field changes fluid temperatures by modifying the alignment of magnetic dipoles. Lower temperature profiles are the result of a decrease in heat transport efficiency caused by higher Prandtl numbers. Raising the Casson fluid value makes the fluid highly resistant to flow, which in turn reduces its velocity. By increasing the fluid's resistance and thermal dissipation, the Casson fluid parameter raises the temperature of the fluid. Any change in the fluid's temperature can be ascribed to an increase or decrease in the value of the heat source or sink.

The motivation behind using the finite-element method (FEM) is that it can conserve both time and financial resources in contrast to conventional experimental testing, which tends to be costly and labor-intensive. FEM offers comprehensive findings, enabling engineers to scrutinize the behavior of structures and systems with precision. The comparative analysis of results obtained through FEM and previously published benchmark studies demonstrates that FEM offers excellent accuracy and efficiency, yielding dependable solutions and important insights for practical applications.

Acknowledgement

The authors express gratitude to the Department of Mathematics, School of Advanced Sciences, Vellore Institute of Technology, Vellore, India, for facilitating the research at the FIST Lab (SR/FST/MS-II/2023/139), which is funded by the Department of Science and Technology (DST), Government of India.

AI disclosure statement: The author confirms that no artificial intelligence (AI)-generated content or AI-assisted writing tools were used in the preparation of this chapter (including summaries and the abstract). All text, analysis, and conclusions are the result of the author's original work.

References

1. Sakiadis, B.C., Boundary-layer behavior on continuous solid surfaces: I. Boundary-layer equations for two-dimensional and axisymmetric flow. *AIChE J.*, 7, 26–8, 1961.

2. Gnaneswara Reddy, M., Unsteady radiative-convective boundary-layer flow of a Casson fluid with variable thermal conductivity. *J. Eng. Phys. Thermophys.*, 88, 240–51, 2015.
3. Pramanik, S., Casson fluid flow and heat transfer past an exponentially porous stretching surface in presence of thermal radiation. *Ain Shams Eng. J.*, 5, 205–12, 2014.
4. Khalid, A., Khan, I., Shafie, S., Exact solutions for unsteady free convection flow of Casson fluid over an oscillating vertical plate with constant wall temperature. *Abstr. Appl. Anal.*, 2015, 2015.
5. Chandra, P. and Das, R., A hybrid machine learning algorithm for studying magnetized nanofluid flow containing gyrotactic microorganisms *via* a vertically inclined stretching surface. *Int. J. Numer. Method. Biomed. Eng.*, e3780, 2023.
6. Hsiao, K.L., Micropolar nanofluid flow with MHD and viscous dissipation effects towards a stretching sheet with multimedia feature. *Int. J. Heat Mass Transf.*, 112, 983–90, 2017.
7. Chandna, O.P. and Garg, M.R., The flow of a viscous MHD fluid. *Q. Appl. Math.*, 34, 287–99, 1976.
8. Immanuel, Y., Pullepu, B., Kirubhashankar, C.K., Casson flow of MHD fluid moving steadily with constant velocity. *Appl. Math. Sci.*, 9, 1503–8, 2015.
9. Jalili, P., Azar, A.A., Jalili, B., Ganji, D.D., Study of nonlinear radiative heat transfer with magnetic field for non-Newtonian Casson fluid flow in a porous medium. *Results Phys.*, 48, 106371, 2023.
10. Mukhopadhyay, S., Casson fluid flow and heat transfer over a nonlinearly stretching surface. *Chin. Phys. B*, 22, 074701, 2013.
11. Manideep, P., Raju, R.S., Rao, T.S.N., Reddy, G.J., Unsteady MHD free convection flow of Casson fluid over an inclined vertical plate embedded in a porous media. *AIP Conf. Proc.*, 1953, 2018.
12. Salah, F. and Almohammadi, A., MHD boundary layer flow due to an exponentially stretching surface through porous medium with radiation effect. *Int. J. Adv. Appl. Sci.*, 10, 165–70, 2023.
13. Usman, M., Soomro, F.A., Ul Haq, R., Wang, W., Defterli, O., Thermal and velocity slip effects on Casson nanofluid flow over an inclined permeable stretching cylinder *via* collocation method. *Int. J. Heat Mass Transf.*, 122, 1255–63, 2018.
14. Rahman, M., Ferdows, M., Shamshuddin, M.D., Koulali, A., Eid, M.R., Aiding (opponent) flow of hybrid copper–aluminum oxide nanofluid towards an exponentially extending (lessening) sheet with thermal radiation and heat source (sink) impact. *J. Pet. Sci. Eng.*, 215, 2022.
15. Chamkha, A.J., Mujtaba, M., Quadri, A., Issa, C., Thermal radiation effects on MHD forced convection flow adjacent to a non-isothermal wedge in the presence of a heat source or sink. *Heat Mass Transf. und Stoffuebertragung*, 39, 305–12, 2003.

16. Sumalatha, C. and Bandari, S., Effects of Radiations and Heat Source/Sink on a Casson Fluid Flow over Nonlinear Stretching Sheet. *World J. Mech.*, 05, 257–65, 2015.
17. Prakash, J., Durga Prasad, P., Kumar, G.V., Kiran Kumar, R V M S S, Varma, S.V.K., Heat and Mass Transfer Hydromagnetic Radiative Casson Fluid Flow over an Exponentially Stretching Sheet with Heat Source/Sink. *Int. J. Eng. Sci. Invent.*, 2016. ISSN Online 2319–6734.
18. Seth, G.S., Sharma, R., Kumbhakar, B., Tripathi, R., MHD Stagnation Point Flow over Exponentially Stretching Sheet with Exponentially Moving Free-Stream, Viscous Dissipation, Thermal Radiation and Non-Uniform Heat Source/Sink. *Diffus. Found.*, 11, 182–90, 2017.
19. Aghalya, T. and Tamizharasi, R., Effect of heat source/sink and thermal radiation on an unsteady MHD Casson flow past over an exponentially accelerated vertical plate. *J. Therm. Anal. Calorim.*, 149, 7041–57, 2024.
20. Coorevits, P., Hild, P., Lhalouani, K., Sassi, T., Mixed finite element methods for unilateral problems: convergence analysis and numerical studies. *Math. Comput.*, 71, 1–26, 2001.
21. Hayat, T., Muhammad, K., Momani, S., Melting heat and viscous dissipation in flow of hybrid nanomaterial: a numerical study *via* finite difference method. *J. Therm. Anal. Calorim.*, 147, 6393–401, 2022.
22. Chandra, P. and Das, R., Impact of gyrotactic microorganisms in nanofluid *via* porous media along an inclined stretching plate: Finite element analysis. *Numer. Heat Transf. Part B Fundam.*, 83, 367–94, 2023.
23. Batool, S., Rasool, G., Alshammari, N., Khan, I., Kaneez, H., Hamadneh, N., Numerical analysis of heat and mass transfer in micropolar nanofluids flow through lid driven cavity: Finite volume approach. *Case Stud. Therm. Eng.*, 37, 2022.
24. Swain, K. and Mishra, S., Flow and heat transfer analysis of water-based copper nanofluid over a nonlinearly stretching sheet: a numerical approach. *Int. J. Ambient Energy*, 2021.
25. Chandra, P. and Das, R., Finite-element-based machine-learning algorithm for studying gyrotactic-nanofluid flow *via* stretching surface. *Int. J. Numer. Methods Fluids*, 1–25, 2023.

15

Non-Newtonian Fluids in Environmental Engineering

Abdulhalim Musa Abubakar[1]*, Suleiman A. Wali[1], Abubakar Mohammed[1] and Vivek Kumar Pandey[2]

[1]*Department of Chemical Engineering, Modibbo Adama University (MAU), Yola, Adamawa State, Nigeria*
[2]*Centre for Research Impact & Outcome, Chitkara University Institute of Engineering and Technology, Chitkara University, Rajpura, Punjab, India*

Abstract

Non-Newtonian fluids play a supportive role in various environmental engineering applications, due to the fact that it influences processes such as wastewater treatment, sediment transport, and pollutant dispersion. Understanding the complex flow behavior of the fluids is essential for effective modeling and simulation. In this review, key non-Newtonian fluid characteristics are examined, including shear-thinning, shear-thickening, and viscoelastic behaviors. Various models, such as Bingham plastic and power-law models, are discussed for their applicability in environmental contexts. Case studies illustrate the impact of non-Newtonian behavior on sediment transport in rivers and estuaries, and emphasizes how rheological properties affect sediment deposition and erosion processes. For that reason, the study explores the role of non-Newtonian fluids in biofilm development and the implications for nutrient and pollutant removal in wastewater treatment facilities. Advanced simulation techniques, including computational fluid dynamics (CFD), are highlighted for their ability to capture the intricacies of flow in complex environmental systems. Moreover, challenges in measuring non-Newtonian properties and the influence of environmental factors are addressed, which provided insights into improving predictive models. As environmental engineering

**Corresponding author*: abdulhalim@mau.edu.ng; ORCID: https://orcid.org/0000-0002-1304-3515
Suleiman A. Wali: ORCID: https://orcid.org/0009-0005-2745-4007
Vivek Kumar Pandey: ORCID: https://orcid.org/0009-0001-5425-2137

Dhananjay Yadav, Mukesh Kumar Awasthi and Harith Mohamed Al-Azri (eds.) Non-Newtonian Fluids for Industrial Applications: Modeling and Simulations, (319–354)

increasingly seeks sustainable solutions, understanding non-Newtonian fluid dynamics becomes essential for optimizing processes and enhancing the performance of engineered systems. Chapter 15 thus serves as a comprehensive resource for researchers and practitioners aiming to leverage the unique properties of non-Newtonian fluids to address contemporary environmental challenges.

Keywords: Non-Newtonian fluids, environmental modeling, fluid dynamics, Bingham plastic, sediment transport

15.1 Introduction

In his study, Chhabra [32] introduces readers to non-Newtonian fluids, describing it as fluids whose viscosity or flow behavior changes in response to applied shear stress, strain, or external conditions, unlike Newtonian fluids with constant viscosity. Hence, non-Newtonian fluids are often better than Newtonian fluids in applications requiring variable viscosity under different shear rates, enabling enhanced performance in processes like mixing, pumping, and pollutant removal. Tauviqirrahman *et al.* [176] noticed that the Newtonian lubricant is preferable for improving tribological performance, whereas non-Newtonian fluid is recommended for lowering bearing noise. A typical example of non-Newtonian fluids is lubricants used for improving oil life, preventing corrosion, limiting friction and wear, lowering the temperature and its effects, and reducing deposits of sludge in the engines, as comprehensively studied by Jang & Khonsari [84]. In a similar study, Chhabra & Patel [33] wrote on environmental engineering application of non-Newtonian fluids, tackling the fluid behavior, rheology, flow in pipes and the heat, momentum, and mass transfer characteristics. This study sought to add to some of the identified gaps in their study. Basically, non-Newtonian fluids play a vital role in environmental engineering, particularly in processes involving sludge treatment, waste management, and pollutant remediation, to mention a few. Ren *et al.* [147] describe how functional materials could be derived from this type of fluid. Their unique rheological properties, such as shear-thinning or shear-thickening behavior, enable efficient handling, pumping, and processing of complex materials like slurries, emulsions, and biosolids [69, 129, 162, 172]. Ideally, knowing their flow characteristics is central for designing systems that optimize energy consumption and ensure effective environmental sustainability in applications that encompass wastewater treatment and soil stabilization. In this case, the gaps filled by the current study are the provision of a comprehensive evaluation of non-Newtonian fluids in environmental engineering through advanced computational fluid dynamics (CFD) modeling. The work provides detailed insights into various types of non-Newtonian fluid behaviors plus an analysis of these behaviors' applications in environmental engineering, linking some characteristic to specific real-world processes such as sludge treatment, soil remediation, and erosion control *via* a selective case study. We move forward to divulge the environmental factors affecting the fluid as well as the hitches in weighing its properties, which, in accordance with the rigorous search we conducted, is lacking in the literature.

15.2 Characteristics of Non-Newtonian Fluids

Espinoza *et al.* [55] previously characterized specific non-Newtonian fluids for environmental applications. Non-Newtonian fluids exhibit a variety of unique flow characteristics described in Table 15.1 that makes them valuable in environmental engineering applications [77]. Unlike Newtonian fluids, their viscosity is not constant and changes with shear rate, time, or stress, enabling them to adapt to diverse environmental conditions and processes [132]. Shear-thinning (pseudoplastic) fluids are widely used in wastewater treatment to enhance the flow and settling of sludge [163]. This behavior ensures efficient transport through pipelines while maintaining stability under low-stress conditions [37, 135]. Conversely, shear-thickening (dilatant) fluids find applications in erosion control by stabilizing sediments under dynamic water flows [166]. Thixotropic fluids are beneficial in soil remediation [122, 175]. They allow easy injection into contaminated zones and regain viscosity to seal and stabilize the soil, preventing pollutant migration [122, 203]. Rheopectic fluids are less common but can be engineered for specific applications, such as creating barriers to slow the spread of contaminants, as in a study by Dewar & Joyce [47].

Viscoelastic fluids are particularly advantageous in bioreactors for wastewater treatment [61]. Their ability to maintain structure while allowing fluid flow optimizes the contact between microorganisms and pollutants, improving degradation efficiency. Yield stress fluids are useful for handling sludge and other semi-solid waste, ensuring controlled movement and deposition [108, 169]. Models like Bingham plastics are essential in designing systems for transporting dense slurries or sediments [137, 208]. Similarly, Casson fluids, with applications in blood-mimicking models, can simulate the flow of pollutants with complex rheological properties [23].

Table 15.1 Properties of non-Newtonian fluids.

Characteristic	Description	Emphasis
Shear-thinning (pseudoplastic)	Viscosity decreases with increasing shear rate. Example: paint, ketchup.	[71]
Shear-thickening (dilatant)	Viscosity increases with increasing shear rate. Example: cornstarch in water.	[107]
Thixotropy	Viscosity decreases over time under constant shear and recovers when at rest.	[24, 62]
Rheopecty	Viscosity increases over time under constant shear.	[206]
Viscoelasticity	Exhibits both solid-like and fluid-like behavior under stress.	[81]
Yield stress	Requires a minimum stress to start flowing. Example: toothpaste, mud.	[200, 213]
Time-dependent behavior	Viscosity depends on the duration of applied stress (e.g., thixotropy and rheopecty).	[109, 170]
Nonlinear stress–strain relationship	The stress–strain relationship does not follow a straight line (nonlinear behavior).	[53]
Bingham plastic behavior	Acts like a solid at low stress but flows like a viscous fluid once yield stress is exceeded. Example: drilling mud.	[182, 183]
Casson fluid behavior	Viscosity is influenced by yield stress and shear rate, commonly used for blood.	[60]
Carreau model behavior	Viscosity transitions from Newtonian at low shear to non-Newtonian at high shear.	[171]
Power–law fluids	Viscosity follows a power–law relationship with shear rate, either shear-thinning or shear-thickening.	[133]

15.3 Modeling Non-Newtonian Fluids

Key models used in non-Newtonian fluid modeling include the power-law model, which characterizes shear-thinning and shear-thickening fluids by relating shear stress to shear rate through a power–law relationship [96]. The power–law model is widely applied in modeling the transport of shear-thinning fluids like bio-polymer solutions, which are used to improve soil permeability during remediation efforts [157]. For fluids exhibiting a yield stress [213], such as sludge in environmental systems studied by Liu *et al.* [108], the Bingham plastic and Herschel–Bulkley models are often employed [143] simply because they incorporate a yield stress parameter, below which the fluid behaves as a solid and above which it flows [169, 196]. For instance, in wastewater treatment, the Bingham plastic and Herschel–Bulkley models are used to simulate the flow of sludge, which behaves as a yield-stress fluid [208]. Both models allow for the design of efficient pumping systems and sedimentation tanks by predicting flow resistance and settling dynamics [178]. The Carreau and Cross models are used for fluids transitioning between Newtonian and non-Newtonian behavior, capturing the viscosity changes over a wide range of shear rates [73]. In sediment transport and coastal management, non-Newtonian models such as the Carreau and Cross models are used to simulate the deposition and resuspension of fine sediments [91], which often exhibit non-Newtonian characteristics under varying flow conditions [158]. According to literature, these simulations are critical for maintaining reservoir capacity, preventing erosion, and managing dredging operations. In time-dependent fluids explained in Ibrahim *et al.* [80] and Bašić [16], thixotropic and rheopectic behaviors are modeled using equations that account for the change in viscosity with time under constant shear. Practically, thixotropic fluids are used to create temporary seals in contaminated [36, 149], whereas rheopectic fluids can be applied to enhance the structural integrity of pollution barriers in dynamic environments [161]. Viscoelastic fluids require more sophisticated models [157], such as the Maxwell and Kelvin–Voigt models, to capture their stress–strain responses and relaxation behavior [66]. Through advanced modeling approaches, engineers can optimize system designs, improve efficiency, and address environmental challenges by accurately predicting the behavior of non-Newtonian fluids under varying conditions [86, 101].

15.4 Case Studies

15.4.1 Sediment Transport in Rivers and Estuaries

Lately, Li *et al.* [104] investigated the formation mechanisms of fluid mud in the Yangtze Estuary and found that storm, tide, and slope-triggered sediment dynamics, including stratification effects and gravity-driven flows, significantly influence the scale, structure, and accumulation of fluid mud, offering insights to enhance sediment management and dredging strategies. Wang & Andutta [186] enhanced the understanding of sediment transport dynamics in ports, estuaries, and coastal environments, finding that sediment resuspension, deposition, and transport are strongly influenced by mechanisms like surface waves, tides, winds, and density-driven currents, as demonstrated through modeling and observations in diverse high-energy systems worldwide. Castro *et al.* [27] focused on numerically modeling bedload sediment transport using Shallow Water equations, introducing and evaluating deterministic models, unifying the solid transport discharge definition, and applying high-order finite volume methods for precise discretization, validated through comparisons with analytical solutions and experimental results. Study by Shen *et al.* [164] considered the viscosity of cohesive sediment-laden flows, finding that sediment type and concentration significantly influence relative viscosity, with clays exhibiting exponential rises, and demonstrating that a modified Costa-based viscosity model effectively predicts non-Newtonian fluid behavior and enhances large-scale sediment transport and pollutant modeling accuracy.

Li *et al.* [105] found that sediment concentration alters flow velocity distribution, with low flow rates causing asymmetric velocity profiles in pressurized pipes and vertical stratification of particles, while higher flow rates reduce asymmetry and promote uniform concentration profiles. In a quest to address knowledge gaps in sediment transport and flow dynamics in steep mountain channels, Mir & Patel [113] discovered that boulder-dominated granular beds significantly influence sediment movement and hazards, and highlighting the potential of numerical, computational, and machine learning techniques to improve understanding of channel morphology and associated risks. Nevertheless, non-Newtonian fluids in sediment transport can complicate the prediction of flow behavior, as their viscosity changes with shear rate, making it difficult to model and analyze sediment movement accurately [58, 199]. This is because the variability in viscosity could lead to unpredictable sediment deposition patterns, and may result in uneven sedimentation or erosion in riverbeds and estuaries [17, 102]. Complex interactions between the fluid and sediment particles may as well result in increased energy requirements for transportation, affecting the efficiency of waterway management and sediment control [82, 128].

15.4.2 Impact of Non-Newtonian Behavior on Deposition and Erosion

Several case studies as relates to erosion and deposition, impacted by non-Newtonian fluid, are there in the literature. Poloski *et al.* [136] found that the Waste Treatment and Immobilization Plant (WTP) design guide's correlations for slurry transport are conservative for Hanford slurry particles, are inaccurate for viscosity predictions, and require modification to address non-Newtonian systems where rheological properties significantly affect critical deposition velocities. Tong *et al.* [180] developed a semi-mechanistic model incorporating adhesion effects to predict the threshold velocity of non-Newtonian fluids required to erode solids beds, demonstrating that the model's predictions align with experimental data for water-based and oil-based muds with similar rheological properties. Adebiyi & Hu [3] developed and validated three numerical models to investigate the impacts of sediment erosion and deposition on debris flow propagation, revealing that debris flows significantly alter bed deformation due to their high shear stress and vertical deposits, while turbidity currents exhibit thinner tails, less shear stress, and horizontal deposits, with sediment size, bed slope, concentration, and porosity showing positive correlations with bed deformation. De Blasio [42] examined the rheological behavior of non-Newtonian fluids in mudflows and debris flows, highlighting their composition, potential for rapid and destructive movement, and the risks posed by volcanic lahars, particularly in regions like Mount Rainier, where ancient and current hazards coexist. Hosseini *et al.* [76] utilized the $\mu(I)$ viscoplastic model in a smoothed particle hydrodynamics framework to study the interaction of Newtonian and non-Newtonian fluids in phenomena like dam breaks and landslides, demonstrating that the model accurately captures fluid–sediment dynamics by comparing computational results with experimental data. Moreso, Dechelette *et al.* [44] developed a computational model to describe the dynamics of non-Newtonian power–law drops spreading on a flat surface, finding that variations in rheological parameters such as the non-Newtonian Reynolds and Weber numbers significantly influence film spreading and recession, with model predictions closely aligning with experimental data within a 21% discrepancy. The aforementioned studies have shown that non-Newtonian behavior significantly impacts deposition and erosion processes by altering flow dynamics and material interactions. Fluids with shear-thinning or viscoplastic properties exhibit varying resistance to deformation, influencing sediment transport, bed stability, and erosion thresholds [117]. Such behaviors can lead to non-linear sediment deposition patterns [195] and require advanced models to predict and manage environmental or industrial processes effectively [51].

15.4.3 Biofilm Development in Wastewater Treatment

In wastewater treatment, biofilm development facilitates the degradation of pollutants through microbial activity, by providing a stable and efficient process for biological treatment [12, 120]. In the context of non-Newtonian fluids, such as sludge with high solid concentrations, the rheological properties affect the mass transfer, nutrient delivery, and biofilm structure, which directly influence treatment efficiency [67]. Understanding and optimizing biofilm formation in non-Newtonian systems can enhance pollutant removal and operational stability in challenging wastewater conditions. Notably, Al-Shamary [7] investigated the hydrodynamics of membrane bioreactors (MBRs) using Newtonian and non-Newtonian fluids, optimizing tank and module geometries, aeration intensity, and bubble distribution to reduce membrane fouling, improve filtration performance, and minimize energy consumption, with experimental validation and predictive modeling showing good agreement. Prades Martell [139] advanced fixed-bed biofilm system modeling by developing and validating Eulerian multiphase computational fluid dynamics (CFD) models that integrated biofilm dynamics, fluid flow, and biochemical interactions, accurately reproducing biofilm deformation, detachment, and fluidization under laminar and turbulent conditions while introducing innovative experimental tools for interface characterization and model verification. On the other hand, Bentzen *et al.* [18] developed and validated a CFD model for rotational cross-flow MBRs, incorporating both Newtonian and non-Newtonian fluid dynamics to accurately predict shear stress on membrane surfaces, and derived empirical relationships linking shear stress, angular velocity, and total suspended solids concentration to optimize energy consumption and control fouling.

Alternatively, Edden [52] utilized nuclear magnetic resonance (NMR) to investigate biofilm growth in porous media and the flow behavior of non-Newtonian fluids, demonstrating that biofouling alters fluid flow dynamics by accelerating bulk flow through channels, while polyacrylamide exhibits shear-thinning behavior under varying shear rates, providing insights for applications in environmental and industrial processes. Karimi *et al.* [90] reviewed the interplay between physical mechanisms and biofilm processes, highlighting the role of hydrodynamics, quorum sensing, and propulsion in biofilm formation, and emphasized the potential of microfluidic techniques to unravel these mechanisms and advance biofilm research in health, environmental, and industrial contexts. Also, Swain *et al.* [173] reviewed the operational principles and performance of Aerobic Inverse Fluidized Bed Biofilm Reactors (AIFBBRs) for industrial wastewater treatment, emphasizing their high organic matter removal efficiency, operational stability, and the critical parameters influencing their effectiveness. Nevertheless, biofilm development in wastewater treatment poses challenges related to its non-Newtonian fluid behavior, such as increased flow resistance and uneven shear stress distribution, which can reduce the efficiency of bioreactors [98]. The biofilm's viscoelastic properties can lead to clogging and pressure drops in treatment systems, making maintenance more difficult and costly [174]. Heterogeneity of biofilms also complicates the modeling and prediction of their hydrodynamic interactions, limiting the optimization of treatment processes [156]. Typically, Adeboye *et al.* [4] highlighted some challenges that include the complex influence of fluid flow, shear stress, and substrate properties on biofilm initiation, structure, and stability, particularly in non-Newtonian fluids, along with difficulties in analyzing these interactions using current rheological and microbiological techniques. Accurately simulating the complex growth dynamics, detachment, reattachment, and deformation of biofilms in granular porous media, as well as the effects of shear stress, nutrient distribution, and mechanical stresses on biofilm behavior and system clogging, are additional issues pinpointed by Kapellos *et al.* [89].

15.4.4 Implications for Nutrient and Pollutant Removal

For example, heavy metals such as lead, cadmium, and mercury often exist in wastewater and require precise rheological control to enhance their adsorption onto biosolids or synthetic polymers for removal [130]. Nutrients (e.g., nitrogen and phosphorus), which contribute to eutrophication in water bodies, are effectively targeted using sludge with shear-thinning properties that enhance mixing and nutrient sequestration. Oil and grease, common in industrial effluents, can be addressed using non-Newtonian emulsifiers that stabilize and separate these pollutants for recovery or disposal [64, 183]. Suspended solids, including organic and inorganic particulates, benefit from non-Newtonian behaviors in sedimentation and flocculation processes that optimize particle aggregation and settling [68]. Pathogens, such as bacteria and viruses, can also be immobilized or inactivated using viscoelastic gels or sludge matrices designed to enhance retention times and exposure to disinfectants. Also among the list is phenolic compounds and azo dyes from industrial effluents [131], which can be adsorbed or degraded using non-Newtonian systems (viz., biosorbents and biopolymers) [116].

15.5 CFD Simulation Techniques

CFD enables the numerical simulation of these complex rheological properties by incorporating advanced mathematical models [8, 152, 202], such as the power–law [100], Herschel–Bulkley [187], and Carreau models [146], into the Navier–Stokes equations [201]. It then facilitates the accurate predictions of flow, mixing, and transport phenomena in both static and dynamic environments. In environmental engineering, CFD is widely applied to optimize processes like wastewater treatment [78], pollutant dispersion [35, 153], and sediment transport [25]. CFD simulations are used to model sludge flow in treatment plants [205], where yield stress and shear-thinning properties significantly affect pipeline transport and settling behavior in clarifiers [13, 145]. Similarly, CFD is employed to study the dispersion of pollutants in water bodies [115] or the atmosphere [126], accounting for the unique viscoelastic or time-dependent characteristics of contaminants [34]. It also plays a critical role in designing and optimizing equipment, such as wet scrubbers [40] and bioreactors [28, 41, 191–193], where non-Newtonian fluids like polymer slurries [184] or microbial suspensions are used [59]. By visualizing flow patterns, shear stress distributions, and mixing efficiency, CFD helps engineers make informed decisions to enhance performance and energy efficiency. Furthermore, CFD supports research in soil remediation, where simulations of non-Newtonian fluid injection into porous media help predict contaminant removal and fluid spread [26, 30, 39]. Table 15.2 provides detailed insights into the application of CFD in modeling non-Newtonian fluids across various environmental engineering contexts.

Other applications include the optimization of wastewater filtration processes by predicting stress–strain behaviors critical for membrane fouling prevention [185]. CFD additionally aids in modeling thixotropic fluids used in mine tailings management, ensuring the stability and controlled flow of materials during storage and reclamation [114, 159].

Table 15.2 Applications of CFD in environmental systems with non-Newtonian fluids.

Application area	Non-Newtonian fluid Type	Description	Work by
Anaerobic digester mixing	Pseudoplastic sludge	CFD simulations optimize mixing efficiency in anaerobic digesters handling pseudoplastic sludge, enhancing biogas production and waste breakdown.	[194]
Drilling fluid transport	Shear-thinning drilling mud	Modeling the flow behavior of shear-thinning drilling fluids in eccentric annuli to improve cuttings transport efficiency.	[6]
Pipeline slurry flow	Non-Newtonian slurry	Simulating three-phase non-Newtonian slurry flows in industrial pipelines to optimize transport processes.	[155]
Porous media flow	Power–law fluids	Investigating the flow of power–law non-Newtonian fluids through three-dimensional disordered porous media.	[118, 165]
Wet scrubber design	Herschel–Bulkley fluids	Applying CFD to simulate the flow of Herschel–Bulkley fluids in wet scrubbers for air pollution control.	[110]
Indoor air quality modeling	Thixotropic fluids	CFD simulations assess the distribution of airborne pollutants in indoor environments, considering the thixotropic nature of certain aerosol gels, to improve ventilation design and occupant health.	[29]
Spray atomization processes	Viscoelastic fluids	Modeling the atomization of viscoelastic fluids in spray systems to optimize droplet size distribution and enhance pollutant dispersion control.	[48]
Industrial mixing operations	Shear-thinning fluids	CFD analysis of shear-thinning non-Newtonian fluids in industrial mixers to enhance mixing efficiency and ensure uniform pollutant treatment.	[127]

15.6 Challenges in Measurement and Modeling

15.6.1 Difficulties in Assessing Non-Newtonian Properties

Assessing the non-Newtonian properties of fluids presents several challenges due to the complex and variable behavior these materials exhibit under different conditions [148, 154]. Variability in non-Newtonian fluids' viscosity makes it difficult to establish a single, definitive measurement of their properties, depending on factors such as shear rate, time, temperature, and pressure [142, 151]. Study using blood by Hong *et al.* [74] shows that the velocity profile of the non-Newtonian fluid was generally blunter than that of the Newtonian fluid. Dynamic nature of non-Newtonian behavior—ranging from shear-thinning and shear-thickening to viscoelasticity—requires advanced rheological instruments capable of simulating a wide range of operational conditions [75, 124, 167]. Interpreting the data often involves complex mathematical models that may not always accurately predict behavior across all shear rates or under transient conditions [16]. The sensitivity of non-Newtonian fluids to environmental factors such as temperature further complicates their assessment, as small variations can lead to significant changes in their rheological properties [10, 63]. Inhomogeneous or multiphase systems, such as suspensions, emulsions, or gels exemplified by Pal [123], introduce variability in measurement, as the distribution of particles or droplets can alter the flow characteristics [179, 207]. Most of these difficulties underscore the need for meticulous experimental design, sophisticated equipment, and robust modeling techniques to reliably assess the non-Newtonian properties of fluids [75, 148].

15.6.2 Environmental Factors Affecting Fluid Behavior

The behavior of non-Newtonian fluids is significantly influenced by various environmental and external factors, with some having a more severe impact than others. Among the most critical factors are temperature, particle size and distribution, shear rate, and chemical composition, shown in Table 15.3.

On the other hand, factors like humidity and pH tend to have a lesser impact, except in specific systems where the fluid is highly sensitive to these conditions. For example, gels or hydrogels may be affected by humidity [15, 125], but many other non-Newtonian fluids exhibit relatively stable behavior under varying humidity levels [1, 204]. Pressure, though it can influence certain fluids like polymers or highly compressible materials, generally has a minor effect on most non-Newtonian systems under

Table 15.3 Environmental factors influencing non-Newtonian fluid behavior.

Factor	Effect on non-Newtonian behavior	References
Temperature	Alters viscosity and can impact shear-thinning or thickening.	[9, 168]
Pressure	Affects fluid structure and flow properties in certain systems.	[88, 106, 121]
pH	Influences the stability and rheology of emulsions or suspensions.	[103]
Humidity	Can affect water-sensitive non-Newtonian fluids like gels.	[20]
Contaminants	May disrupt particle interactions, altering flow characteristics.	[177]
Shear rate	Determines whether the fluid exhibits shear-thinning, shear-thickening, or viscoelasticity.	[11]
Mechanical stress	Long-term or repetitive stresses can modify the fluid's microstructure and behavior.	[45]
Time (thixotropy/ rheopexy)	Viscosity may change over time under constant shear due to structural adjustments.	[2]
Chemical composition	Additives, solvents, or dissolved salts modify intermolecular forces and fluid behavior.	[94, 112]
Chemical reaction	Govern by reaction order index for homogeneous or heterogeneous reaction	[83, 93, 138]
Particle size and distribution	Influences flow and deformation properties in suspensions and dispersions.	[95]
Flow history	Previous shear conditions affect current rheological properties, especially in memory-like fluids.	[198]
Electromagnetic fields	External fields can drastically change properties in magnetorheological or electrorheological fluids.	[5]

typical conditions. Flow history and mechanical stress, while relevant, are secondary in impact compared to factors like temperature or shear rate, unless the fluid is memory-dependent [54, 56, 57] or subjected to long-term deformation [65, 72, 97].

15.7 Applications in Environmental Engineering

Non-Newtonian fluids play a significant role in various aspects of environmental engineering due to their unique flow behaviors, such as shear-thinning [160], shear-thickening [43, 189], or viscoelasticity [46]. In industrial wastewater treatment, non-Newtonian fluids such as xanthan gum or guar gum solutions exemplified by Zhao *et al.* [209] and Espinoza *et al.* [55] are utilized in settling tanks to promote the flocculation and sedimentation of heavy particles. In wastewater treatment processes, as contained in Table 15.4, the fluids can adapt their viscosity under varying flow conditions, aiding in the efficient removal of suspended solids [134]. Additionally, non-Newtonian fluids are used in bioreactors for the treatment of organic pollutants [78]. For example, viscoelastic microbial suspensions enhance mixing efficiency and improve contact between microorganisms and pollutants, leading to faster degradation rates [19]. Tables 15.5 and 15.6 present other environmental application of non-Newtonian fluids.

In air pollution control systems, non-Newtonian fluids find application in wet scrubbers [210]. Shear-thickening slurries, often containing fine particulates like calcium carbonate [21, 99], improve the capture efficiency of acidic gases such as sulfur dioxide [79]. Interestingly, this approach is particularly effective in industrial emission control systems, as contained in McManus [111]. Finally, in coastal protection and marine pollution mitigation, non-Newtonian fluids are applied to create barriers against oil spills [64, 211]. Gel-like, viscoelastic polymers can be deployed to contain or divert the spread of oil slicks, minimizing ecological damage and facilitating cleanup operations [38, 211]. Table 15.6 exemplifies the use of the fluids to stabilize marine sediments in areas prone to erosion or contamination. According to Richards *et al.* [150], non-Newtonian fluids can be used for the protection of flexible laminates.

15.8 Conclusion

The current comprehensive review examined the unique characteristics and applications of non-Newtonian fluids in environmental engineering, in order to address critical gaps in their modeling, behavior, and practical utility. Some key rheological properties such as shear-thinning, shear-thickening, thixotropy, and viscoelasticity were analyzed, so as to highlight their significant contributions to processes that include sludge treatment, sediment transport, and pollutant remediation. Advanced mathematical

Table 15.4 Non-Newtonian fluid in wastewater treatment processes.

Application	Non-Newtonian fluid role	Examples	Reference
Sludge dewatering	Improves solid–liquid separation efficiency.	Sludge in wastewater treatment.	[190]
Flocculation and coagulation	Enhances particle aggregation for sedimentation or filtration.	Polymer-based flocculants.	[212]
Aeration processes	Optimizes oxygen transfer rates in non-Newtonian media.	Activated sludge systems	[49]
Hydraulic transport	Facilitates the transport of sludge and viscous effluents.	Sludge pipelines.	[87]
Heavy metal removal	Enhances adsorption using viscoelastic or shear-thinning fluids	Biopolymer-based adsorbents.	[181]
Membrane filtration	Reduces fouling by controlling flow behavior near the membrane.	Shear-thinning feed solutions.	[144]

Table 15.5 Pollution dispersion modeling application of non-Newtonian fluid.

Application	Non-Newtonian fluid role	Examples	Reference
Oil spill modeling	Simulates the spread and behavior of viscous oil pollutants.	Crude oil spills in marine systems.	[38, 119]
Sediment transport	Models the movement of suspended particles in non-Newtonian flow.	Mudflows in polluted rivers.	[197]
Groundwater contamination	Predicts the transport of viscous contaminants underground.	Leachates from industrial waste.	[50]
Air pollution dispersion	Models the spread of aerosol-based non-Newtonian pollutants.	Industrial particulate emissions.	[14]
Chemical spill dispersion	Simulates the flow and spread of viscous chemical leaks.	Polymer and slurry spills.	[22]
Urban runoff analysis	Analyzes non-Newtonian stormwater carrying pollutants.	Runoff from urbanized areas.	[85]

Table 15.6 Applications in sediment management strategies.

Application	Non-Newtonian fluid role	Examples	Reference
Sediment transport modeling	Simulates the flow of fine-grained, cohesive sediments.	Riverbed sediment movement.	[13, 140]
Dredging operations	Enhances efficiency in the handling and pumping of sediments.	Sludge and sediment dredging.	[141]
Erosion control	Predicts sediment stability and erosion in non-Newtonian media.	Coastal and riverbank protection.	[188]
Reservoir sedimentation management	Models the deposition and resuspension of sediments.	Dam and reservoir maintenance.	[92]
Wetland restoration	Aids in sediment redistribution for habitat recovery.	Restoring degraded wetlands.	[31]
Slurry/sludge transport	Facilitates the movement of sediment-laden fluids.	Slurry pipelines for mining or dredging.	[70, 85]

models (to wit, Bingham plastic and power–law equations) are emphasized for their role in accurately predicting non-Newtonian fluid behavior under varying environmental conditions. Moreover, the study stresses the potential of rarely explored fluid behaviors, such as rheopexy, in innovative applications like pollution barriers and soil stabilization. Principally, the present research not only advanced the theoretical understanding of non-Newtonian fluid dynamics but also offers practical insights into optimizing energy use and improving the efficiency of engineered systems. In order to bridge the knowledge gaps existing in this area, this work contributes to the development of sustainable, high-performance environmental engineering solutions, to pave way to studies and applications that leverage the versatility of non-Newtonian fluids. A study specifically discussing 'nano-rheology and its environmental applications' is absent in the literature (based on Google Scholar search before 2025).

AI disclosure statement: The author acknowledges the use of AI-based tools for minor language editing or grammatical correction. However, all intellectual contributions, analyses, and conclusions presented in this chapter are entirely the author's own.

References

1. Abusharha, A., Pearce, E.I., Afsar, T., Razak, S., Protecting tear-film stability under adverse environmental conditions using a mucomimetic with a Non-Newtonian viscosity agent. *Medicina*, *59*, 1862, 2023. https://doi.org/10.3390/medicina59101862.
2. Adamu, G.T., Kwami, A.M., Abdulhameed, M., Yakubu, G., Effects of retardation time on non-Newtonian electro-osmotic flow in a micro-channel. *Diffus. Found.*, *26*, 39–52, 2020. https://doi.org/10.4028/www.scientific.net/DF.26.39.
3. Adebiyi, A.A. and Hu, P., A numerical study on impacts of sediment erosion/deposition on debris flow propagation. *Water*, *13*, 1698, 2021. https://doi.org/10.3390/w13121698.
4. Adeboye, A., Onyeaka, H., Al-Sharify, Z., Nnaji, N., Understanding the influence of rheology on biofilm adhesion and implication for food safety. *Int. J. Food Sci.*, 2024. https://doi.org/10.1155/ijfo/2208472.
5. Adıgüzel, O.B. and Atalık, K., Magnetic field effects on Newtonian and non-Newtonian ferrofluid flow past a circular cylinder. *Appl. Math. Modell.*, *42*, 161–174, 2017. https://doi.org/10.1016/j.apm.2016.10.014.
6. Ahsan, M., Fahad, S., Butt, M.S., Computational Fluid Dynamics simulation and analysis of non-Newtonian drilling fluid flow and cuttings transport in an eccentric annulus. *Mathematics*, *13*, 101, 2024. https://doi.org/10.3390/math13010101.
7. Al-Shamary, L., *Hydrodynamics of membrane bioreactors with Newtonian and nonNewtonian fluids*, S. Enders, M. Kraume, F. Ziegler (Eds.), Doktorin der Ingenieurwissenschaften, genehmigte Dissertation, von der Fakultät III - Prozesswissenschaften, der Technischen Universität Berlin, 2013, https://api-depositonce.tu-berlin.de/server/api/core/bitstreams.
8. Alade, O., Mahmoud, M., Al-Nakhli, A., Rheological studies and numerical investigation of barite sag potential of drilling fluids with thermochemical fluid additive using computational fluid dynamics (CFD). *J. Pet. Sci. Eng.*, *220*, Part A, 2023. https://doi.org/10.1016/j.petrol.2022.111179.

9. Ali, D.H. and Mohammed, M.A.R., Studying the rheological properties of non-Newtonian fluids under the effect of temperature using different chemical additives. *Iraqi J. Chem. Pet. Eng.*, *21*, 2, 47–56, 2020. https://doi.org/10.31699/IJCPE.2020.2.6.
10. Ali, F., Souayeh, B., Hou, Y., Usman, M., Bhattacharyya, S., Sarfraz, M., Theoretical investigation and sensitivity analysis of non-Newtonian fluid during roll coating process by response surface methodology. *Open Phys.*, *22*, 1, 1–22, 2024. https://doi.org/10.1515/phys-2024-0105.
11. Amaratunga, M., Rabenjafimanantsoa, H.A., Time, R.W., Estimation of shear rate change in vertically oscillating non-Newtonian fluids: Predictions on particle settling. *J. Non-Newtonian Fluid Mech.*, *277*, 104236, 1–13, 2020. https://doi.org/10.1016/j.jnnfm.2020.104236.
12. Andersson, S., Rajarao, G.K., Land, C.J., Dalhammar, G., Biofilm formation and interactions of bacterial strains found in wastewater treatment systems. *FEMS Microbiol. Lett.*, *283*, 1, 83–90, 2008. https://doi.org/10.1111/j.1574-6968.2008.01149.x.
13. Balhoff, M.T., Lake, L.W., Bommer, P.M., Lewis, R.E., Weber, M.J., Calderin, J.M., Rheological and yield stress measurements of non-Newtonian fluids using a Marsh Funnel. *J. Pet. Sci. Eng.*, *77*, 3–4, 393–402, 2011. https://doi.org/10.1016/j.petrol.2011.04.008.
14. Barros, P.L., Ein-Mozaffari, F., Lohi, A., Gas dispersion in non-Newtonian fluids with mechanically agitated systems: A review. *Processes*, *10*, 275, 2022. https://doi.org/10.3390/pr10020275.
15. Barry, R.A. and Wiltzius, P., Humidity-sensing inverse opal hydrogels. *Langmuir*, *22*, 3, 1369–1374, 2006. https://doi.org/10.1021/la0519094.
16. Bašić, M., *An overview of numerical methods for non-Newtonian flows*, Qualification Exam, Postgraduate Study of Mechanical Engineering, Faculty of Electrical Engineering, Mechanical Engineering and Naval Architecture, University of Split, 2021, https://data.fesb.hr/documents/public/news/Martina.
17. Beck, J.S., *Sediment transport dynamics in South African estuaries*, University of Stellenbosch, 2005, http://hdl.handle.net/10019.1/1465.
18. Bentzen, T.R., Ratkovich, N., Madsen, S., Jensen, J.C., Bak, S.N., Rasmussen, M.R., Analytical and numerical modelling of Newtonian and non-Newtonian liquid in a rotational cross-flow MBR. *Water Sci. Technol.*, *66*, 11, 2318–2327, 2012. https://doi.org/10.2166/wst.2012.443.
19. Bhat, S., Jun, D., Paul, B.C., Dahms, T.E.S., Viscoelasticity in biological systems: A special focus on microbes, in: *Viscoelasticity - From Theory to Biological Applications*, J. De Vicente (Ed.), InTech Open, 2012, https://doi.org/10.5772/49980.
20. Binesh, A.R., Mousavi, S.M., Kamali, R., Effect of temperature-dependency of Newtonian and non-Newtonian fluid properties on the dynamics of droplet impinging on hot surfaces. *Int. J. Mod. Phys. C*, *26*, 9, 2015. https://doi.org/10.1142/S0129183115501065.

21. Bosco, A., Paiva, F., Odenbach, S., Calado, V., Influence of particle shape and sample preparation on shear thickening behavior of precipitated calcium carbonate suspensions. *Adv. Powder Technol.*, *32*, 3, 802–809, 2021. https://doi.org/10.1016/j.apt.2021.01.023.
22. Bouzid, I., Maire, J., Laurent, F., Broquaire, M., Fatin-Rouge, N., Controlled treatment of a high velocity anisotropic aquifer model contaminated by hexachlorocyclohexanes. *Environ. Pollut.*, *268*, Part A, 2021. https://doi.org/10.1016/j.envpol.2020.115678.
23. Boyd, J., Buick, J.M., Green, S., Analysis of the Casson and Carreau-Yasuda non-Newtonian blood models in steady and oscillatory flows using the lattice Boltzmann method. *Phys. Fluids*, *19*, 9, 2007. https://doi.org/10.1063/1.2772250.
24. Bui, C. and Ho, T., Influence of thixotropic parameters on a non-Newtonian fluid flow. *AIP Adv.*, *10*, 035303, 2020. https://doi.org/10.1063/1.5143620.
25. Busch, A., *On particle transport and turbulence in wellbore flows of non-Newtonian fluids - Findings from a cuttings transport process analysis by means of computational fluid dynamics, rheometry, and dimensional analysis*, NTNU, 2020, https://hdl.handle.net/11250/2649308.
26. Callahan, T.P., *Non-Newtonian fluid injection into granular media*, H. Huang, C. Santamarina, L. Germanovich (Eds.), Degree of Master of Science in Civil Engineering, Georgia Institute of Technology, 2011, https://d1wqtxts1xzle7.cloudfront.net/33322285.
27. Castro, M.J., Fernández-Nieto, E.D., Ferreiro, A.F., Sediment transport models in Shallow Water equations and numerical approach by high order finite volume methods. *Comput. Fluids*, *37*, 3, 299–316, 2008. https://doi.org/10.1016/j.compfluid.2007.07.017.
28. Chen, J. and Xiao, W., Solids suspension study in a side-entering stirred tank through CFD modeling. *Int. J. Chem. Reactor Eng.*, 2013. https://doi.org/10.1515/ijcre-2012-0062.
29. Chen, Q. and Srebric, J., The use of CFD for indoor environmental design, in: *The Use of CFD Tools for Indoor Environmental Design-Advanced Building Simulation*, A. Malkawi and G. Augenbroe (Eds.), pp. 119–140, Spon Press, New York, 2004, https://engineering.purdue.edu/~yanchen/paper/2004-7.pdf.
30. Cheng, Z., Lien, F.-S., Zhang, J.H., Gu, G.X., Blind zones in radiating dispersion at high Péclet number driven by non-Newtonian fluids in porous media. *J. Fluid Mech.*, *986*, 2024. https://doi.org/10.1017/jfm.2024.306.
31. Chhabra, R.P., Hydrodynamics of nonspherical particles in non-Newtonian fluids, in: *Handbook of Applied Polymer Processing Technology*, 1st, pp. 1–45, CRC Press, 1996, https://doi.org/10.1201/9781003067528-1.
32. Chhabra, R.P., Non-Newtonian fluids: An introduction, in: *Rheology of Complex Fluids*, J.M. Krishnan, A.P. Deshpande, P.B.S. Kumar (Eds.), pp. 3–34, Springer, New York, NY, 2010, https://doi.org/10.1007/978-1-4419-6494-6_1.

33. Chhabra, R.P. and Patel, S.A., *Non-Newtonian flow and applied rheology: Engineering applications*, 3rd, Elsevier Ltd, 2025, https://shop.elsevier.com/books/.
34. Chinyoka, T. and Makinde, O.D., Viscoelastic modeling of the diffusion of polymeric pollutants injected into a pipe flow. *Acta Mech. Sin.*, *29*, 166–178, 2013. https://doi.org/10.1007/s10409-013-0016-3.
35. Chu, A.K.M., Kwok, R.C.W., Yu, K.N., Study of pollution dispersion in urban areas using Computational Fluid Dynamics (CFD) and Geographic Information System (GIS). *Environ. Modell. Software*, *20*, 3, 273–277, 2005. https://doi.org/10.1016/j.envsoft.2004.05.007.
36. Ciriello, V. and Federico, V. Di, Similarity solutions for flow of non-Newtonian fluids in porous media revisited under parameter uncertainty. *Adv. Water Resour.*, *43*, 38–51, 2012. https://doi.org/10.1016/j.advwatres.2012.03.028.
37. Collivignarelli, M.C., Miino, M.C., Bellazzi, S., Caccamo, F.M., Durante, A., Abbà, A., Review of rheological behaviour of sewage sludge and its importance in the management of wastewater treatment plants. *Water Pract. Technol.*, *17*, 1, 483–491, 2022. https://doi.org/10.2166/wpt.2021.098.
38. Cunningham, R.J., *Gel stability in waves: breakage, behaviour, and implications for oil spill remediation*, G.A. Lawrence and M. Isaacson (Eds.), Thesis/Dissertation, University of British Columbia, 2020, https://doi.org/10.14288/1.0392486.
39. Dallagi, H., *Numerical and experimental investigations of the rheological behavior of foam flow: Application to the cleaning of surfaces contaminated by microorganisms in the food industries*, M.J. Legrand, M.K. Whitehead, M.A. Saint-Jalmes, M.I. Sampers, M.H. Timmerman, M.C. Faille, M.F. Aloui, M.T. Bénézech (Eds.), Doctorate Thesis in Chemical Engineering, Université de Lille, 2023, https://theses.hal.science/tel-03921788v1.
40. Danzomo, B.A., Salami, M.-J.E., Khan, R.M., Nor, M.I.B.M., CFD based parametric analysis of gas flow in a counter-flow wet scrubber system. *Int. J. Environ. Prot. Policy*, *1*, 2, 16–23, 2013. https://doi.org/10.11648/j.ijepp.20130102.11.
41. Darbandi, T., Risberg, M., Westerlund, L., CFD modeling of the forces in the wet scrubber acting on particulate matter released from biomass combustion. *Therm. Sci. Eng. Prog.*, *25*, 100997, 2021. https://doi.org/10.1016/j.tsep.2021.100997.
42. De Blasio, F.V., Non-Newtonian fluids, mudflows, and debris flows: A rheological approach, in: *Introduction to the Physics of Landslides*, 1st, pp. 89–130, Springer, Dordrecht, 2011, https://doi.org/10.1007/978-94-007-1122-8_4.
43. De Goede, T.C., De Bruin, K.G., Bonn, D., High-velocity impact of solid objects on non-Newtonian fluids. *Sci. Rep.*, *9*, 1250, 2019. https://doi.org/10.1038/s41598-018-37543-1.
44. Dechelette, A., Sojka, P.E., Wassgren, C.R., Non-Newtonian drops spreading on a flat surface. *J. Fluids Eng.*, *132*, 10, 2010. https://doi.org/10.1115/1.4002281.

45. Deshpande, N.S. and Barigou, M., Vibrational flow of non-Newtonian fluids. *Chem. Eng. Sci.*, *56*, 12, 3845–3853, 2001. https://doi.org/10.1016/S0009-2509(01)00059-8.
46. Deville, M.O. and Gatski, T.B., Non-Newtonian and viscoelastic fluids, in: *Mathematical Modeling for Complex Fluids and Flows*, 1st, pp. 95–147, Springer, Berlin, Heidelberg, 2012, https://doi.org/10.1007/978-3-642-25295-2_5.
47. Dewar, R.J. and Joyce, M.J., The thixotropic and rheopectic behaviour of maize starch and maltodextrin thickeners used in dysphagia therapy. *Carbohydr. Polym.*, *65*, 3, 296–305, 2006. https://doi.org/10.1016/j.carbpol.2006.01.018.
48. Di Martino, M., Ahirwal, D., Maffettone, P.L., Computational fluid dynamics characterization of the hollow-cone atomization: Newtonian and non-Newtonian spray comparison. *Phys. Fluids*, *34*, 9, 2022. https://doi.org/10.1063/5.0104658.
49. Durán, C., Fayolle, Y., Pechaud, Y., Cockx, A., Gillot, S., Impact of suspended solids on the activated sludge non-Newtonian behaviour and on oxygen transfer in a bubble column. *Chem. Eng. Sci.*, *141*, 154–165, 2016. https://doi.org/10.1016/j.ces.2015.10.016.
50. Eberhard, U., Seybold, H.J., Secchi, E., Jiménez-Martínez, J., Rühs, P.A., Ofner, A., Andrade, J.S., Holzner, M., Mapping the local viscosity of non-Newtonian fluids flowing through disordered porous structures. *Sci. Rep.*, *10*, 11733, 2010. https://doi.org/10.1038/s41598-020-68545-7.
51. Ebtehaj, I. and Bonakdari, H., Assessment of evolutionary algorithms in predicting non-deposition sediment transport. *Urban Water J.*, *13*, 5, 499–510, 2016. https://doi.org/10.1080/1573062X.2014.994003.
52. Edden, A.S., *Characterization of non-Newtonian fluids and fluid flow through biofilms in porous media using nuclear magnetic resonance*, Montana State University - Bozeman, College of Engineering, 2013, https://scholarworks.montana.edu/handle/1/2895.
53. Ellmenreich, J., *A mass conserving mixed stress-strain rate finite element method for non-Newtonian fluid simulations*, J. Schöberl and P.L. Lederer (Eds.), Diploma Thesis, Technische Universität Wien, 2021, https://doi.org/10.34726/hss.2021.95386.
54. Eringen, A.C., Memory-dependent orientable nonlocal micropolar fluids. *Int. J. Eng. Sci.*, *29*, 12, 1515–1529, 1991. https://doi.org/10.1016/0020-7225(91)90123-K.
55. Espinoza, I., Hauswirth, S., Cerda, C.C., Sadeghi, S., *Characterization of non-Newtonian fluids for environmental applications*, American Geophysical Union, 2017, https://ui.adsabs.harvard.edu/abs/2017AGUFM.H43I1767E/abstract.
56. Ezzat, M.A., El-Karamany, A.S., El-Bary, A.A., Generalized thermo-viscoelasticity with memory-dependent derivatives. *Int. J. Mech. Sci.*, *89*, 470–475, 2014. https://doi.org/10.1016/j.ijmecsci.2014.10.006.

57. Ezzat, M.A., El Karamany, A.S., El-Bary, A.A., Thermoelectric viscoelastic materials with memory-dependent derivative. *Smart Struct. Syst.*, *19*, 5, 1–13, 2017. https://doi.org/10.12989/sss.2017.19.5.000.
58. Fiorot, G.H., *On unsteady open-channel flows: a contribution to non-stationary sediment transport in runoff flows and to unstable non-Newtonian mudflow studies*, G. de F. Maciel and P. Dupont (Eds.), Universidade Estadual Paulista (Unesp), 2016, http://hdl.handle.net/11449/143758.
59. Gachelin, J., Miño, G., Berthet, H., Lindner, A., Rousselet, A., Clément, E., Non-Newtonian viscosity of Escherichia coli suspensions. *Phys. Rev. Lett.*, *110*, 268103, 2013. https://doi.org/10.1103/PhysRevLett.110.268103.
60. Gandhi, R. and Sharma, B.K., Modelling pulsatile blood flow using Casson fluid model through an overlapping stenotic artery with Au-Cu hybrid nanoparticles: Varying viscosity approach, in: *Advances in Mathematical Modelling, Applied Analysis and Computation*, vol. 666, J. Singh, G.A. Anastassiou, D. Baleanu, D. Kumar (Eds.), pp. 155–176, Springer, Cham, 2023, https://doi.org/10.1007/978-3-031-29959-9_10.
61. Gao, C., Habibi, M., Hendrickx, T.L.G., Rijnaarts, H.H.M., Temmink, H., Sudmalis, D., Variation of viscoelastic properties of extracellular polymeric substances and their relation to anaerobic granule's mechanical strength in full-scale treatment plants. *Bioresour. Technol.*, *411*, 131233, 2024. https://doi.org/10.1016/j.biortech.2024.131233.
62. Gholami, A., Mansoori, Z., Avval, M.S., Ahmadi, G., Study of thixotropic behavior of non-Newtonian fluids in tool-joints of oil wells. *ASME 2022 Fluids Engineering Division Summer Meeting*, pp. 1–7, 2022, https://doi.org/10.1115/FEDSM2022-87150.
63. Giannetti, F., Luchini, P., Marino, L., Stability and sensitivity analysis of non-Newtonian flow through an axisymmetric expansion. *J. Phys. Conf. Ser.*, *318*, 032015, 2011. https://doi.org/10.1088/1742-6596/318/3/032015.
64. Gogarty, W.B. and Poettmann, F.H., The use of non-Newtonian fluids in oil recovery, in: *Paper Presented at the 8th World Petroleum Congress*, Moscow, USSR, June 1971, pp. 287–297, 1971, https://onepetro.org/WPCONGRESS/proceedings-abstract/WPC08.
65. Greco, F., Drop deformation for non-Newtonian fluids in slow flows. *J. Non-Newtonian Fluid Mech.*, *107*, 1–3, 111–131, 2002. https://doi.org/10.1016/S0377-0257(02)00144-1.
66. Greco, R. and Marano, G.C., Identification of parameters of Maxwell and Kelvin–Voigt generalized models for fluid viscous dampers. *J. Vib. Control*, *21*, 2, 2013. https://doi.org/10.1177/1077546313487937.
67. Greener, J., Gashti, M.P., Eslami, A., Zarabadi, M.P., Taghavi, S.M., A microfluidic method and custom model for continuous, non-intrusive biofilm viscosity measurements under different nutrient conditions. *Biomicrofluidics*, *10*, 6, 2016. https://doi.org/10.1063/1.4968522.

68. Grigoryeva, A. and Abiev, R.S., Experimental and theoretical study of the flocculants mixing in water. *E3S Web Conferences*, *266*, 266, 1–8, 2021. https://doi.org/10.1051/e3sconf/202126602012.
69. Grosso, M., Mei, R., Baratti, R., Tronci, S., On-line control of the rheological properties for a continuous production of a non-Newtonian fluid, in: *EuroPACT 2017 - 4th European Conference on Process Analytics and Control Technology at Potsdam*, Germany, 2017, https://www.researchgate.net/publication/320417130.
70. Gruszczyński, M.F., Kostecki, S., Zieliński, S., Skrzypczak, Z., Stefanek, P., Czaban, S., Popczyk, M., A simple and effective method for measuring the density of non-Newtonian thickened tailings slurry during hydraulic transport. *Sensors*, *22*, 7857, 2022. https://doi.org/10.3390/s22207857.
71. Guedda, M. and Kersner, R., Non-Newtonian pseudoplastic fluids: Analytical results and exact solutions. *Int. J. Non-Linear Mech.*, *46*, 7, 949–957, 2011. https://doi.org/10.1016/j.ijnonlinmec.2011.04.009.
72. Ha, J.-W. and Yang, S.-M., Deformation and breakup of Newtonian and non-Newtonian conducting drops in an electric field. *J. Fluid Mech.*, *405*, 131–156, 2000. https://doi.org/10.1017/S0022112099007223.
73. Hassan, M., Issakhov, A., Khan, S.U.-D., Assad, M.E.H., Hani, E.H.B., Rahimi-Gorji, M., Nadeem, S., Khan, S.U.-D., The effects of zero and high shear rates viscosities on the transportation of heat and mass in boundary layer regions: A non-Newtonian fluid with Carreau model. *J. Mol. Liq.*, *317*, 113991, 2020. https://doi.org/10.1016/j.molliq.2020.113991.
74. Hong, H., Song, J.M., Yeom, E., Variations in pulsatile flow around stenosed microchannel depending on viscosity. *PLoS One*, *14*, e0210993, 2019. https://doi.org/10.1371/journal.pone.0210993.
75. Hosen, H.F., Shahmardi, A., Brandt, L., Solsvik, J., Dynamics of a single bubble in Newtonian and non-Newtonian fluids: Experimental and simulation approaches. *Int. J. Multiphase Flow*, *174*, 104789, 2024. https://doi.org/10.1016/j.ijmultiphaseflow.2024.104789.
76. Hosseini, K., Omidvar, P., Kheirkhahan, M., Farzin, S., Smoothed particle hydrodynamics for the interaction of Newtonian and non-Newtonian fluids using the μ(I) model. *Powder Technol.*, *351*, 325–337, 2019. https://doi.org/10.1016/j.powtec.2019.02.045.
77. Hoyt, J.W., Some applications of non-Newtonian fluid flow. *Rheol. Ser.*, *8*, 797–826, 1999. https://doi.org/10.1016/S0169-3107(99)80008-2.
78. Huang, F., Kuang, S., Ismail, N.I., Zou, R., Yu, A., Modeling and analysis of particle behavior in fluidized bed bioreactors during non-Newtonian sewage treatment. *J. Water Process Eng.*, *60*, 105203, 2024. https://doi.org/10.1016/j.jwpe.2024.105203.
79. Huglen Bratten, S., *An experimental study on the rise and absorption of CO2 bubbles in Newtonian and non-Newtonian fluids*, R.W. Time and A.H. Rabenjafimanantsoa (Eds.), Department of Energy and Petroleum

Engineering (TN-IEP), Faculty of Science and Technology, Universitetet i Stavanger, 2023, https://hdl.handle.net/11250/3089890.
80. Ibrahim, M., Saeed, T., Zeb, S., Numerical simulation of time-dependent two-dimensional viscous fluid flow with thermal radiation. *Eur. Phys. J. Plus, 137*, 609, 2022. https://doi.org/10.1140/epjp/s13360-022-02813-5.
81. Ionescu, C.M., Birs, I.R., Copot, D., Muresan, C.I., Caponetto, R., Mathematical modelling with experimental validation of viscoelastic properties in non-Newtonian fluids. *Philos. Trans. R. Soc. A, 378*, 2172, 2020. https://doi.org/10.1098/rsta.2019.0284.
82. Iwuoha, P.O., Adiela, P.U., Nwannah, C.C., Okeke, O.C., Sediment source and transport in river channels: Implications for river structures. *Int. J. Of Eng. And Sci. (IJES), 5*, 5, 19–26, 2016. https://www.theijes.com/papers/v5-i5/D0505019026.pdf.
83. Jaber, K.K. and Al-Athari, F.M., Non Newtonian fluid flow under the effect of chemical reaction and Ion and Hall currents over A moving cylinder. *J. Purity Util. React. Environ., 2*, 1, 1–13, 2013. https://www.researchgate.net/publication/282709814.
84. Jang, J.Y. and Khonsari, M.M., Lubrication with a non-Newtonian fluid, in: *Encyclopedia of Tribology*, Q.J. Wang and Y.-W. Chung (Eds.), pp. 2146–2151, Springer, Boston, MA, 2013, https://doi.org/10.1007/978-0-387-92897-5_151.
85. Jankowska, H., Dzido, A., Krawczyk, P., Determination of rheological parameters of non-Newtonian fluids on an example of biogas plant substrates. *Energies, 16*, 1128, 2023. https://doi.org/10.3390/en16031128.
86. Jharwal, A.K., Challenges in classical and non-Newtonian fluid mechanics-A mathematical perspective. *Afr. Diaspora J. Math., 25*, 1, 100–112, 2022. https://newjournalzone.in/index.php/ijmfsmr/article/view/95.
87. Jiang, M., *Dilute species transport in non-Newtonian, single-fluid, porous medium systems*, C.T. Miller (Ed.), Master of Science in Environmental Engineering Thesis, Department of Environmental Sciences and Engineering, Gillings School of Global Public Health, 2019, https://doi.org/10.17615/y4nv-7h26.
88. Kaczmarczyk, K., Kruk, J., Ptaszek, P., Ptaszek, A., Pressure drop method as a useful tool for detecting rheological properties of non-Newtonian fluids during flow. *Appl. Sci., 11*, 14, 1–11, 2021. https://doi.org/10.3390/app11146583.
89. Kapellos, G.E., Alexiou, T.S., Payatakes, A.C., Hierarchical simulator of biofilm growth and dynamics in granular porous materials. *Adv. Water Resour., 30*, 6–7, 1648–1667, 2007. https://doi.org/10.1016/j.advwatres.2006.05.030.
90. Karimi, A., Karig, D., Kumar, A., Ardekani, A.M., Interplay of physical mechanisms and biofilm processes: review of microfluidic methods. *Lab Chip, 15*, 23–42, 2015. https://doi.org/10.1039/C4LC01095G.
91. Kaveh, K. and Malcherek, A., Enhancing non-Newtonian fluid modeling: A novel extension of the cross flow curve model. *J. Hydro-environ. Res., 56*, 17–27, 2024. https://doi.org/10.1016/j.jher.2024.08.001.

92. Khahledi, M., Haldenwang, R., Chhabra, R., Fester, V., Non-Newtonian fluid flow from bottom of tank using orifices of different shapes. *Chem. Eng. Res. Des.*, *157*, 34–45, 2020. https://doi.org/10.1016/j.cherd.2020.02.015.
93. Khan, Z., Khan, I., Ullah, M., Tlili, I., Effect of thermal radiation and chemical reaction on non-Newtonian fluid through a vertically stretching porous plate with uniform suction. *Results Phys.*, *9*, 1086–1095, 2018. https://doi.org/10.1016/j.rinp.2018.03.041.
94. Khatibi, M., Potokin, N., Time, R.W., Experimental investigation of effect of salts on rheological properties of non-Newtonian fluids. *Annu. Trans. Nord. Rheol. Soc.*, *24*, 117–126, 2016. https://nrs.blob.core.windows.net/pdfs/nrspdf.
95. Khushbu, S. and Sunil, C.K., Effect of particle size, concentration, temperature and pH on rheological properties of shallots flour. *J. Food Sci. Technol.*, *57*, 10, 3601–3610, 2020. https://doi.org/10.1007/s13197-020-04392-8.
96. Kim, G., Hyun, J.M., Kwak, H.S., Transient buoyant convection of a Power-law non-Newtonian fluid in an enclosure. *Int. J. Heat Mass Transfer*, *46*, 19, 3605–3617, 2003. https://doi.org/10.1016/S0017-9310(03)00149-2.
97. Kiran, R.M., Chakraborty, J., DasGupta, S., Chakraborty, S., Flow-induced deformation in a microchannel with a non-Newtonian fluid. *Biomicrofluidics*, *12*, 3, 2018. https://doi.org/10.1063/1.5036632.
98. Krsmanovic, M., Biswas, D., Ali, H., Kumar, A., Ghosh, R., Dickerson, A.K., Hydrodynamics and surface properties influence biofilm proliferation. *Adv. Colloid Interface Sci.*, *288*, 102336, 2021. https://doi.org/10.1016/j.cis.2020.102336.
99. Kugge, C. and Daicic, J., Shear response of concentrated calcium carbonate suspensions. *J. Colloid Interface Sci.*, *271*, 1, 241–248, 2004. https://doi.org/10.1016/j.jcis.2003.09.048.
100. Lavrov, A., Numerical modeling of steady-state flow of a non-Newtonian power-law fluid in a rough-walled fracture. *Comput. Geotech.*, *50*, 101–109, 2013. https://doi.org/10.1016/j.compgeo.2013.01.004.
101. Lavrov, A., Flow of non-Newtonian fluids in single fractures and fracture networks: Current status, challenges, and knowledge gaps. *Eng. Geol.*, *321*, 107166, 2023. https://doi.org/10.1016/j.enggeo.2023.107166.
102. Leupi, C., *Numerical modeling of cohesive sediment transport and bed morphology in estuaries*, M. Deville, M. Altinakar, J. Bühler, E. Miglio, A. Quarteroni (Eds.), Laboratoire d'Ing´enierie Num´erique-LIN, Facult´e des Sciences et Technique de l'Ing´enie, Institut des sciences de l'´energie, Ecole Polytechnique F´ed´erale de Lausanne, 2005, https://www.theijes.com/papers/v5-i5/D0505019026.pdf.
103. Li, S., Fan, J., Xu, S., Li, R., Luan, J., The influence of pH on gas-liquid mass transfer in non-Newtonian fluids. *Chem. Ind. Chem. Eng. Q.*, *23*, 3, 321–327, 2017. https://doi.org/10.2298/CICEQ160705046L.
104. Li, W., Li, M., Zhang, X., Li, J., Characteristics of fluid mud in the Yangtze Estuary: Storm, tide, and slope-triggered sediment dynamics and effects.

Estuarine Coastal Shelf Sci., *281*, 108194, 2023. https://doi.org/10.1016/j.ecss.2022.108194.

105. Li, Y., Zhang, J., Xu, H., Bai, Y., Experimental study on the characteristics of sediment transport and sorting in pressurized pipes. *Water*, *13*, 2782, 2021. https://doi.org/10.3390/w13192782.
106. Lin, J., Chu, L., Liang, L., Effects of viscosity pressure dependency on the non-Newtonian squeeze film of parallel circular plate. *Lubr. Sci.*, *25*, 1, 2013. https://doi.org/10.1002/ls.1188.
107. Liu, H., Fu, K., Cui, X., Zhu, H., Yang, B., Shear thickening fluid and its application in impact protection: A review. *Polymers*, *15*, 2238, 2023. https://doi.org/10.3390/polym15102238.
108. Liu, W.-H., Zhang, H., Sun, P., Zeng, Y.-P., Gao, Y.-Y., Wang, H.-F., Zeng, R.J., Yield stress Measurement of municipal sludge: A comprehensive evaluation of testing methods and concentration effects using a rotational rheometer. *Environ. Res.*, *250*, 118554, 2024. https://doi.org/10.1016/j.envres.2024.118554.
109. Loenko, D.S., Shenoy, A., Sheremet, M.A., Effect of time-dependent wall temperature on natural convection of a non-Newtonian fluid in an enclosure. *Int. J. Therm. Sci.*, *166*, 106973, 2021. https://doi.org/10.1016/j.ijthermalsci.2021.106973.
110. Lovato, S., Toxopeus, S., Settels, J., Keetels, G., Application of a maritime CFD code to a benchmark problem for non-Newtonian fluids: the flow around a sphere, in: *International Shipbuilding Progress*, 2022, https://doi.org/10.3233/ISP-220002.
111. McManus, T.N., Non-Newtonian fluids and uncontrolled emission of toxic gases: A major threat to worker safety. *Open J. Saf. Sci. Technol.*, *13*, 2, 2023. https://doi.org/10.4236/ojsst.2023.132004.
112. Melton, L.L. and Malone, W.T., Fluid mechanics research and engineering application in non-Newtonian fluid systems. *SPE J.*, *4*, 1, 56–66, 1964. https://doi.org/10.2118/739-PA.
113. Mir, A.A. and Patel, M., A comprehensive review on sediment transport, flow dynamics, and hazards in steep channels. *J. Water Manage. Model. (JWMM)*, 2024. https://doi.org/10.14796/JWMM.C517.
114. Mizani, S., *Experimental study and surface deposition modelling of amended oil sands tailings products*, P. Simms (Ed.), Doctor of Philosophy (Ph.D.) Thesis in Environmental Engineering, Carleton University, 2017, https://doi.org/10.22215/etd/2017-11931.
115. Modenesi, K., Furlan, L.T., Tomaz, E., Guirardello, R., Núnez, J.R., A CFD model for pollutant dispersion in rivers. *Braz. J. Chem. Eng.*, *21*, 4, 2004. https://doi.org/10.1590/S0104-66322004000400005.
116. Moghimipour, E., Kouchak, M., Salimi, A., Bahrampour, S., Handali, S., The effect of polymer content on the non-Newtonian behavior of acetaminophen suspension. *J. Drug Delivery*, *2013*, 907471, 1–5, 2013. https://doi.org/10.1155/2013/907471.

117. Moon, J.H., Lee, J.B., Lee, S.H., Dynamic behavior of non-Newtonian droplets impinging on solid surfaces. *Mater. Trans.*, *54*, 2, 2013. https://doi.org/10.2320/matertrans.M2012215.
118. Morais, A., Seybold, H., Herrmann, H.J., Andrade, J.J.S., Non-Newtonian fluid flow through three-dimensional disordered porous media. *Phys. Rev. Lett.*, *103*, 194502, 2009. https://doi.org/10.1103/PhysRevLett.103.194502.
119. Moreno, E., Dialami, N., Cervera, M., Modeling of spillage and debris floods as Newtonian and viscoplastic Bingham flows with free surface with mixed stabilized finite elements. *J. Non-Newtonian Fluid Mech.*, *290*, 104512, 2021. https://doi.org/10.1016/j.jnnfm.2021.104512.
120. Morgan-Sagastume, F., Biofilm development, activity and the modification of carrier material surface properties in moving-bed biofilm reactors (MBBRs) for wastewater treatment. *Crit. Rev. Environ. Sci. Technol.*, *48*, 5, 439–470, 2018. https://doi.org/10.1080/10643389.2018.1465759.
121. Mori, N., Eguchi, T., Nakamura, K., Horikawa, A., Pressure flow of non-Newtonian fluids between eccentric cylindes with the inner cylinder rotating. *J. Text. Mach. Soc. Japan Trarisactioris*, *38*, 2, T37–T45, 1987. https://www.jstage.jst.go.jp/article/jte1955/33/2/33_2_46/_pdf/-char/ja.
122. Nalezny, C.L. and Li, M.C., Effect of soil structure and thixotropic hardening on the swelling behavior of compacted clay soils, in: *Paper Sponsored by Committee on Physico-Chemical Phenomena in Soils and Presented at the 46th Annual Meeting*, pp. 1–22, 1967, https://onlinepubs.trb.org/Onlinepubs/hrr/1967/209/209-001.pdf.
123. Pal, R., Non-Newtonian behaviour of suspensions and emulsions: Review of different mechanisms. *Adv. Colloid Interface Sci.*, *333*, 103299, 1–37, 2024. https://doi.org/10.1016/j.cis.2024.103299.
124. Palaksha, C.P., Shivaprakash, S., Jagadish, H.P., Dynamic characteristics of Non Newtonian fluid Squeeze film damper, in: *International Conference on Advances in Materials and Manufacturing Applications (IConAMMA-2016)*, Bangalore, India, 14–16 July 2016, vol. *149*, 2016, https://doi.org/10.1088/1757-899X/149/1/012199.
125. Pandiyarajan, C.K., Prucker, O., Rühe, J., Humidity driven swelling of the surface-attached poly(N-alkylacrylamide) hydrogels. *Macromolecules*, *49*, 21, 8254–8264, 2016. https://doi.org/10.1021/acs.macromol.6b01379.
126. Pantusheva, M., Mitkov, R., Hristov, P.O., Petrova-Antonova, D., Air pollution dispersion modelling in urban environment using CFD: A systematic review. *Atmosphere*, *13*, 1640, 2022. https://doi.org/10.3390/atmos13101640.
127. Panunzi, A., Moroni, M., Mazzelli, A., Bravi, M., Industrial case-study-based Computational Fluid Dynamic (CFD) modeling of stirred and aerated bioreactors. *ACS Omega*, *7*, 29, 25152–25163, 2022. https://doi.org/10.1021/acsomega.2c01886.
128. Papanicolaou, A. (Thanos), N., Elhakeem, M., Krallis, G., Prakash, S., Edinger, J., Sediment transport modeling review—Current and future

developments. *J. Hydraul. Eng.*, *134*, 1, 2008. https://doi.org/10.1061/(ASCE)0733-9429(2008)134:1(1).

129. Park, N.A. and Park, H.L., Rheological property measurements of Newtonian and non-Newtonian fluids with the falling needle viscometer, in: *Symposia Paper*, pp. 1–13, 2013, https://doi.org/10.1520/STP156420120068.
130. Park, Y.-G. and Kim, D., Heavy metal removal by non-Newtonian emulsion liquid membrane dispersed in Taylor-Couette flow: experiments and modeling. *Desalin. Water Treat.*, *61*, 58–67, 2017. https://doi.org/10.5004/dwt.2016.1710.
131. Park, Y., Skelland, A.H.P., Forney, L.J., Kim, J.-H., Removal of phenol and substituted phenols by newly developed emulsion liquid membrane process. *Water Res.*, *40*, 9, 1763–1772, 2006. https://doi.org/10.1016/j.watres.2006.03.005.
132. Partal, P. and Franco, J.M., Non-Newtonian fluids, in: *Rheology*, vol. 1, C. Gallegos (Ed.), pp. 96–119, Encyclopedia of Life Support Systems (EOLSS, 2010, https://books.google.com.
133. Patnana, V.K., Bharti, R.P., Chhabra, R.P., Two-dimensional unsteady flow of power-law fluids over a cylinder. *Chem. Eng. Sci.*, *66*, 12, 2978–2999, 2009. https://doi.org/10.1016/j.ces.2009.03.029.
134. Peker, S.M., Helvacı, S.S., Yener, H.B., Alparslan, A., İkizler, B., Non-Newtonian behavior of solid–liquid suspensions, in: *Solid-Liquid Two Phase Flow*, S.M. Peker, S.S. Helvacı, H.B. Yener, A. Alparslan, B. İkizler (Eds.), pp. 71–166, Elsevier Science, 2008, https://doi.org/10.1016/B978-044452237-5.50004-7.
135. Pinho, F.T. and Whitelaw, J.H., Flow of non-newtonian fluids in a pipe. *J. Non-Newtonian Fluid Mech.*, *34*, 2, 129–144, 1990. https://doi.org/10.1016/0377-0257(90)80015-R.
136. Poloski, A.P., Adkins, H.E., Abrefah, J., Casella, A.M., Hohimer, R.E., Nigl, F., Minette, M.J., Toth, J.J., Tingey, J.M., Yokuda, S.T., *Deposition velocities of Newtonian and non-Newtonian slurries in pipelines*, G.H. Beeman (Ed.), 2009, https://doi.org/10.2172/963206.
137. Poloski, A.P., Nigl, F., Adkins, H.E., Minette, M.J., Abrefah, J., Toth, J.J., Casella, A.M., Tingey, J.M., Hohimer, R.E., Yokuda, S.T., *Deposition velocities of Newtonian and non-Newtonian slurries in pipelines*, 2009. https://www.pnnl.gov/rpp-wtp/documents/WTP-RPT-175.pdf.
138. Ponalagusamy, R., Murugan, D., Priyadharshini, S., Effects of rheology of non-Newtonian fluid and chemical reaction on a dispersion of a solute and implications to blood flow. *Int. J. Appl. Comput. Math.*, *8*, 109, 2022. https://doi.org/10.1007/s40819-022-01312-6.
139. Prades Martell, L., *Computational fluid dynamics techniques for fixed-bed biofilm systems modeling: Numerical simulations and experimental characterization*, J. Gamisans Noguera, D. Castaño, A. David (Eds.), Doctoral Thesis, Universitat Politècnica de Catalunya, 2018, https://doi.org/10.5821/dissertation-2117-125845.

140. Pullum, L., Chryss, A., Graham, L., Matoušek, V., Pěník, V., Modelling turbulent transport of solids in non-Newtonian carrier fluids applicable to tailings disposal, in: *12th International Conference on Transport and Sedimentation of Solid*, pp. 229–240, 2015, http://http//ts.upwr.edu.pl/ts17.
141. Randall, R.E., Fluid dynamics of dredging. *Phys. Fluids*, *34*, 1, 2022. https://doi.org/10.1063/5.0076904.
142. Rao, P.S. and Rahul, A.K., Combined effect of viscosity variation and non-Newtonian Rabinowitsch fluid in wide parallel rectangular-porous plate with squeeze-film characteristics. *Meccanica*, *54*, 2399–2409, 2019. https://doi.org/10.1007/s11012-019-01092-2.
143. Rashaida, A.A., *Flow of a non-Newtonian Bingham plastic fluid over a rotating disc*, Degree of Doctor of Philosophy in Department of Mechanical Engineering, College of Graduate Studies and Research, University of Saskatchewan, 2005, https://www.collectionscanada.gc.ca/obj/s4/f2/dsk3/SSU/TC-SSU-08172005120844.pdf.
144. Raspati, G.S. and Leiknes, T., Applicability of blocking laws in non-Newtonian fluid membrane filtration. *Desalin. Water Treat.*, *54*, 7, 1771–1781, 2015. https://doi.org/10.1080/19443994.2014.891465.
145. Reddy, N.V.K., Pothal, J.K., Barik, R., Senapati, P.K., Pipeline slurry transportation system: An overview. *J. Pipeline Syst. Eng. Pract.*, *14*, 3, 2023. https://doi.org/10.1061/JPSEA2.PSENG-139.
146. Rehman, S., Hashim, Shah, S.I., Numerical simulation for heat and mass transport of non-Newtonian Carreau rheological nanofluids through convergent/divergent channels. *Proc. Inst. Mech. Eng., Part C: J. Mech. Eng. Sci.*, *236*, 11, 2022. https://doi.org/10.1177/09544062211065688.
147. Ren, Y., Koh, K.S., Zhang, Y., Synthesis of functional materials by non-Newtonian microfluidic multiphase system, in: *Advances in Microfluidics - New Applications in Biology, Energy, and Materials Sciences*, X.-Y. Yu (Ed.), InTech Open, 2016, https://doi.org/10.5772/64521.
148. Renardy, M., Current issues in non-Newtonian flows: a mathematical perspective. *J. Non-Newtonian Fluid Mech.*, *90*, 2–3, 243–259, 2000. https://doi.org/10.1016/S0377-0257(99)00081-6.
149. Rezaee, D., Numerical study on miscible viscous fingering in thixotropic fluids. *Nonlinear Sci.*, *2*, 100005, 2025. https://doi.org/10.1016/j.nls.2024.100005.
150. Richards, J.A., Hodgson, D.J.M., O'Neill, R.E., Poon, W.C.K., Optimizing non-Newtonian fluids for impact protection of laminates. *PNAS*, *121*, 10, 2024. https://doi.org/10.1073/pnas.2317832121.
151. Rojas, R.O., Quijano, J.C., Ruiz, C.P.T., López, A.F.E., Development and experimental validation of a viscosity meter for Newtonian and non-Newtonian fluids, in: *Preprints.Org*, pp. 1–9, 2020, https://doi.org/10.48550/arXiv.2008.05287.
152. Rooki, R., Ardejani, F.D., Moradzadeh, A., Norouzi, M., CFD simulation of rheological model effect on cuttings transport. *J. Dispersion Sci. Technol.*, *36*, 3, 402–410, 2015. https://doi.org/10.1080/01932691.2014.896219.

153. Sabatino, S., Buccolieri, R., Pulvirenti, B., Britter, R., Simulations of pollutant dispersion within idealised urban-type geometries with CFD and integral models. *Atmos. Environ.*, *41*, 37, 8316–8329, 2007. https://doi.org/10.1016/j.atmosenv.2007.06.052.
154. Saccomandi, G. and Vergori, L., Old problems revisited from new perspectives in implicit theories of fluids, in: *Non-Newtonian Fluid Mechanics and Complex Flows*, 1st, A. Farina, L. Fusi, A. Mikelić, G. Saccomandi, A. Sequeira, E.F. Toro (Eds.), pp. 45–90, Springer Cham, 2018, https://doi.org/10.1007/978-3-319-74796-5.
155. Sadeghi, M., Li, S., Zheng, E., Sontti, S.G., Esmaeili, P., Zhang, X., CFD simulation of turbulent non-Newtonian slurry flows in horizontal pipelines. *Ind. Eng. Chem. Res.*, *61*, 15, 5324–5339, 2022. https://doi.org/10.1021/acs.iecr.1c04471.
156. Sadino-Riquelme, M.C., Donoso-Bravo, A., Zorrilla, F., Valdebenito-Rolack, E., Gómez, D., Hansen, F., Computational fluid dynamics (CFD) modeling applied to biological wastewater treatment systems: An overview of strategies for the kinetics integration. *Chem. Eng. J.*, *466*, 143180, 2023. https://doi.org/10.1016/j.cej.2023.143180.
157. Safari, A., Emami, N., Cervantes, M.J., Viscoelasticity and shear-thinning effects on bio-polymer solution and suspended particle behaviours under oscillatory curve Couette flow conditions. *Biosurf. Biotribol.*, *4*, 1, 1–17, 2018. https://doi.org/10.1049/bsbt.2018.0002.
158. Salahuddin, T. and Awais, M., A comparative study of Cross and Carreau fluid models having variable fluid characteristics. *Int. Commun. Heat Mass Transfer*, *139*, 106431, 2022. https://doi.org//10.1016/j.icheatmasstransfer.2022.106431.
159. Sánchez-Peralta, J.A., Beltrán-Rodríguez, L.N., Trujillo-Vela, M.G., Larrahondo, J.M., Flows of liquefied filtered tailings: Laboratory-scale physical and numerical modeling. *Int. J. Civ. Eng.*, *18*, 393–404, 2020. https://doi.org/10.1007/s40999-019-00482-7.
160. Sanchez, E. and Bhatta, D., Flow patterns for Newtonian and non-Newtonian fluids in a cylindrical pipe. *Appl. Appl. Mathematics An Int. J. (AAM)*, *18*, 1, 1–11, 2023. https://digitalcommons.pvamu.edu/aam/vol18/iss1/4.
161. Schröder, A., *Extensional viscosity of curtain coating colors evaluated as Euler number*, M. Toivakka (Ed.), Master´s thesis in Chemical and Process Engineering, Laboratory of Natural Materials Technology, Faculty of Science and Engineering, Åbo Akademi University, 2023, https://www.doria.fi/bitstream/handle/10024/188143/schroder_andreas.pdf?sequence=3.
162. Sedaghat, A., A novel and robust model for determining rheological properties of Newtonian and non-Newtonian fluids in a marsh funnel. *J. Pet. Sci. Eng.*, *156*, 3, 2017. https://doi.org/10.1016/j.petrol.2017.06.057.
163. Seyssiecq, I., Ferrasse, J.-H., Roche, N., State-of-the-art: Rheological characterisation of wastewater treatment sludge. *Biochem. Eng. J.*, *16*, 1, 41–56, 2003. https://doi.org/10.1016/S1369-703X(03)00021-4.

164. Shen, X., Lin, M., Pan, F., Maa, J.P.-Y., Ha, H.K., Bi, Q., Shao, Y., Zhang, J., Wu, Z., Viscosity of cohesive sediment-laden flows: Experimental and empirical methods. *J. Geophys. Res. (JGR) Oceans*, *129*, 2, 2024. https://doi.org/10.1029/2023JC020043.
165. Shenoy, A.V., Non-Newtonian fluid heat transfer in porous media. *Adv. Heat Transfer*, *24*, 101–190, 1994. https://doi.org/10.1016/S0065-2717(08)70233-8.
166. Shetty, A., *Understanding the behavior of shear-thickening dilatant non-Newtonian fluids in a microgravity (μg) environment*, 2020. https://www.researchgate.net/publication/363885687.
167. Shin, D.H., Moon, J.H., Choi, C.K., Lee, S.H., Dynamics of a non-Newtonian droplet impact on a micro-textured surface. *J. Heat Transfer*, *134*, 080905–1, 1–1, 2012. https://doi.org/10.1115/1.4006719.
168. Soto-Castruita, E., Ramírez-González, P.V., Martínez-Cortés, U., Quiñones-Cisneros, S.E., Effect of the temperature on the non-Newtonian behavior of heavy oils. *Energy Fuels*, *29*, 5, 2883–2889, 2015. https://doi.org/10.1021/ef502171d.
169. Spinosa, L. and Lotito, V., A simple method for evaluating sludge yield stress. *Adv. Environ. Res.*, *7*, 3, 655–659, 2003. https://doi.org/10.1016/S1093-0191(02)00041-2.
170. Sun, H., Jiang, Y., Zhang, Y., Jiang, L., A review of constitutive models for non-Newtonian fluids. *Fract. Calc. Appl. Anal.*, *27*, 1483–1526, 2024. https://doi.org/10.1007/s13540-024-00294-0.
171. Sun, J., Guo, L., Jing, J., Tang, C., Lu, Y., F, J., Ullmann, A., Brauner, N., Investigation on laminar pipe flow of a non-Newtonian Carreau-Extended fluid. *J. Pet. Sci. Eng.*, *205*, 108915, 2021. https://doi.org/10.1016/j.petrol.2021.108915.
172. Suzuki, K. and Hagura, Y., Experimental comparison of static rheological properties of non-Newtonian food fluids with dynamic viscoelasticity. *Nihon Reoroji Gakkaishi*, *46*, 1, 1–12, 2018. https://doi.org/10.1678/rheology.46.1.
173. Swain, A.K., Sahoo, A., Jena, H.M., Patra, H., Industrial wastewater treatment by Aerobic Inverse Fluidized Bed Biofilm Reactors (AIFBBRs): A review. *J. Water Process Eng.*, *23*, 61–74, 2018. https://doi.org/10.1016/j.jwpe.2018.02.017.
174. Tang, B. and Zhang, Z., Essence of disposing the excess sludge and optimizing the operation of wastewater treatment: Rheological behavior and microbial ecosystem. *Chemosphere*, *105*, 1–13, 2014. https://doi.org/10.1016/j.chemosphere.2013.12.067.
175. Tang, B., Zhou, B., Xie, L., Yin, J., Evaluation method for thixotropy of clay subjected to unconfined compressive test. *Front. Earth Sci.*, *9*, 2021. https://doi.org/10.3389/feart.2021.683454.
176. Tauviqirrahman, M., Jamari, J., Susilowati, S., Pujiastuti, C., Setiyana, B., Pasaribu, A.H., Ammarullah, M.I., Performance comparison of Newtonian and non-Newtonian fluid on a heterogeneous slip/no-slip journal

bearing system based on CFD-FSI method. *Fluids*, *7*, 225, 2022. https://doi.org/10.3390/fluids7070225.
177. Theodoropoulou, M., Karoutsos, V., Tsakiroglou, C., Investigation of the contamination of fractured formations by non-Newtonian oil pollutants. *Environ. Forensics*, *2*, 4, 321–334, 2001. https://doi.org/10.1006/enfo.2001.0070.
178. Thomas, A.D. and Wilson, K.C., Rough-wall and turbulent transition analyses for Bingham plastics. *J. South. Afr. Inst. Min. Metall.*, *107*, 6, 359–364, 2006. https://www.saimm.co.za/Journal/v107n06p359.pdf.
179. Thorvaldsen, G.S., *The effect of the particle size distribution on non-Newtonian turbulent slurry flow in pipes*, M Tech Degree Thesis in Chemical Engineering, School of Mechanical and Process Engineering, Cape Technikon, 1996, https://core.ac.uk/download/pdf/148364823.pdf.
180. Tong, T.A., Liu, Y., Ozbayoglu, E., Yu, M., Ettehadi, R., May, R., Threshold velocity of non-Newtonian fluids to initiate solids bed erosion in horizontal conduits. *J. Pet. Sci. Eng.*, *199*, 108256, 2021. https://doi.org/10.1016/j.petrol.2020.108256.
181. Tosco, T. and Sethi, R., Transport of non-Newtonian suspensions of highly concentrated micro- and nanoscale iron particles in porous media: A modeling approach. *Environ. Sci. Technol.*, *44*, 23, 9062–9068, 2010. https://doi.org/10.1021/es100868n.
182. Trivedi, M., Nirmalkar, N., Gupta, A.K., Chhabra, R.P., Effect of non-Newtonian fluid behavior on forced convection from a cluster of four circular cylinders in a duct, Part II: Bingham plastic fluids. *Heat Transfer Eng.*, *43*, 1, 2022. https://doi.org/10.1080/01457632.2020.1844432.
183. Uwaezuoke, N., Perspective Chapter: Rheological considerations for drilling and enhanced oil recovery fluids, in: *Advances in Rheology of Materials*, A. Dutta and H.M. Ali (Eds.), InTech Open, 2022, https://doi.org/10.5772/intechopen.106897.
184. Visintainer, R., Matoušek, V., Pullum, L., Sellgren, A., Non-Newtonian slurries and suspensions, in: *Slurry Transport Using Centrifugal Pumps*, pp. 159–207, Springer, Cham, 2023, https://doi.org/10.1007/978-3-031-25440-6_6.
185. Vroman, T., Mechanisms of hollow-fiber membrane fouling removal used in water treatment: Critical backwash fluxes and membrane deformation for an enhanced backwash efficiency [Doctorate Thesis, Université Paul Sabatier - Toulouse III], in: *Polymers*, Rabiller-Baudry, M., Moulin, M.P., Heran, M.M., Bacchin, M.P., Guigui, M.C., Remigy, M.J.-C., Beaume, M.F. (Eds.), 2020, https://theses.hal.science/tel-02978055/.
186. Wang, X.H. and Andutta, F.P., Sediment transport dynamics in ports, estuaries and other coastal environments, in: *Sediment Transport Processes and Their Modelling Applications*, A.J. Manning (Ed.), InTech Open, 2013, https://doi.org/10.5772/51022.
187. Wang, Z., Guo, W., Ding, W., Liu, K., Qin, W., Wang, C., Wang, Z., Numerical study on the hydrodynamic properties of bentonite slurries with

Herschel-Bulkley-Papanastasiou rheology model. *Powder Technol., 419,* 118375, 2023. https://doi.org/10.1016/j.powtec.2023.118375.

188. Wang, Z., Zhang, J., Shirazi, S.A., Dou, Y., Experimental and numerical study of erosion in a non-Newtonian hydraulic fracturing fluid. *Wear, 422–423,* 1–8, 2019. https://doi.org/10.1016/j.wear.2018.12.080.
189. Wei, M., Lin, K., Sun, L., Shear thickening fluids and their applications. *Mater. Des., 216,* 110570, 2022. https://doi.org/10.1016/j.matdes.2022.110570.
190. Weiss, M., Plósz, B.G., Essemiani, K., Meinhold, J., Suction-lift sludge removal and non-Newtonian flow behaviour in circular secondary clarifiers: Numerical modelling and measurements. *Chem. Eng. J., 132,* 1–3, 241–255, 2007. https://doi.org/10.1016/j.cej.2007.01.004.
191. Wu, B., CFD simulation of gas and non-Newtonian fluid two-phase flow in anaerobic digesters. *Water Res., 44,* 13, 3861–3874, 2010a. https://doi.org/10.1016/j.watres.2010.04.043.
192. Wu, B., Computational Fluid Dynamics investigation of turbulence models for Non-Newtonian fluid flow in anaerobic digesters. *Environ. Sci. Technol., 44,* 23, 8989–8995, 2010b. https://doi.org/10.1021/es1010016.
193. Wu, B. and Chen, S., CFD simulation of non-Newtonian fluid flow in anaerobic digesters. *Biotechnol. Bioeng., 99,* 3, 700–711, 2008. https://doi.org/10.1002/bit.21613.
194. Wu, B. and Shulin, C., CFD simulation of non-Newtonian fluid flow in anaerobic digesters. *Biotechnol. Bioeng., 99,* 3, 700–711, 2008. https://doi.org/10.1002/bit.21613.
195. Wu, X., Cai, C., Li, D., Zhou, J., Zhang, W., Non-linear response of sediment size characteristics and associated transport patterns to soil structural stability in sheet erosion under field rainfall simulation. *CATENA, 228,* 107120, 2023. https://doi.org/10.1016/j.catena.2023.107120.
196. Wu, Y.-S., Pruess, K., Witherspoon, P.A., *Flow and displacement of Bingham non-Newtonian fluids in porous media,* Director, Office of Energy Research, Office of Basic Energy Sciences, Engineering and Geosciences Division, of the U.S. Department of Energy under Contract No. DE-AC03- 76SF00098, 1990, https://www.osti.gov/servlets/purl/7081417.
197. Xia, B. and Krueger, P.S., Rheology of particulate suspensions with non-Newtonian fluids in capillaries. *Proc. R. Soc. A, 478,* 20210615, 2022. https://doi.org/10.1098/rspa.2021.0615.
198. Yamaguchi, H., Non-Newtonian fluid and flow, in: *Engineering Fluid Mechanics,* 1st, vol. 85, pp. 399–496, Springer, Dordrecht, 2008, https://doi.org/10.1007/978-1-4020-6742-6_7.
199. Yan, Z., Li, Z., Cheng, S., Wang, X., Zhang, L., Zheng, L., Zhang, J., From Newtonian to non-Newtonian fluid: Insight into the impact of rheological characteristics on mineral deposition in urine collection and transportation. *Sci. Total Environ., 823,* 153532, 2022. https://doi.org/10.1016/j.scitotenv.2022.153532.

200. Yang, F., Bick, A., Shandalov, S., Brenner, A., Oron, G., Yield stress and rheological characteristics of activated sludge in an airlift membrane bioreactor. *J. Membr. Sci.*, *334*, 1–2, 83–90, 2009. https://doi.org/10.1016/j.memsci.2009.02.022.
201. Yang, H., Moresi, L., Weng, H., Giordani, J., Numerical modeling of earthquake cycles based on Navier-Stokes equations with viscoelastic-plasticity rheology. *Geochemistry Geophysics Geosytems*, *24*, 9, 2023. https://doi.org/10.1029/2023GC010872.
202. Yang, L., Tian, J.-L., Yang, Z., Li, Y., Fu, C.-H., Zhu, Y.-H., Pang, X.-L., Numerical analysis of non-Newtonian rheology effect on hydrocyclone flow field. *Pet. Chem.*, *1*, 1, 68–74, 2015. https://doi.org/10.1016/j.petlm.2015.05.001.
203. Yang, S., Ren, Y., Andersen, K.H., Effects of thixotropy and reconsolidation on the undrained shear characteristics of remoulded marine clays. *Ocean Eng.*, *239*, 109888, 2021. https://doi.org/10.1016/j.oceaneng.2021.109888.
204. Ye, J., Gao, L., Shao, E., Kovács, B., Li, J., Gu, Y., The effect of non-Newtonian fluid midsole footwear on lower limb biomechanics after 5 km of running in high temperature. *Appl. Sci.*, *13*, 8024, 2023. https://doi.org/10.3390/app13148024.
205. Yousuf, N., Kurukulasuriya, N., Chryss, A., Rudman, M., Rees, C., Usher, S., Farno, E., Lester, D., Eshtiaghi, N., An accurate and robust method for intensification of wastewater sludge pipe flow. *Sci. Total Environ.*, *949*, 175143, 2024. https://doi.org/10.1016/j.scitotenv.2024.175143.
206. Yusof, F.A.A., Yamaki, M., Kawai, M., Okajima, M.K., Kaneko, T., Mitsumata, T., Rheopectic behavior for aqueous solutions of megamolecular polysaccharide sacran. *Biomolecules*, *10*, 155, 1–11, 2020. https://doi.org/10.3390/biom10010155.
207. Zade, S., *Experimental studies of large particles in Newtonian and non-Newtonian fluids*, 2019. https://www.diva-portal.org/smash/get/diva2:1347711/FULLTEXT01.pdf.
208. Zengeni, B.T., *Bingham yield stress and Bingham plastic viscosity of homogeneous non-Newtonian slurries*, G.J. Oliver (Ed.), Master of Technology Mechanical Engineering Dissertation in the Faculty of Engineering at the Cape Peninsula University of Technology, 2016, https://core.ac.uk/download/pdf/148366243.pdf.
209. Zhao, C., Zhang, X., Si, W., Luan, D., Chen, S., Cavern development characteristics of pseudoplastic fluid in a stirred tank with dual-impellers. *J. Taiwan Inst. Chem. Eng.*, *150*, 105057, 2023. https://doi.org/10.1016/j.jtice.2023.105057.
210. Zhao, L., You, R., Liu, J., Chen, Q., A real-time measurement and analysis method for gas holdup in a wet scrubber with the use of image information entropy. *Sep. Purif. Technol.*, *345*, 127255, 2024. https://doi.org/10.1016/j.seppur.2024.127255.

211. Zhao, M., Cao, M., Hu, Y., Yang, Q., Wu, Y., Chu, Z., Hu, X., Dai, C., Investigating breakup behaviors of the non-Newtonian fluid: A case study of oil droplet using 3-D pore throat structured microchannels. *Colloids Surf., A*, *587*, 124330, 2020. https://doi.org/10.1016/j.colsurfa.2019.124330.
212. Zhao, S.H. and Luo, T.N., Study on the model of magnetic flocculation of magneto-rheological fluid. *Adv. Mater. Research*, 335–336, 994–999, 2011. https://doi.org/10.4028/www.scientific.net/AMR.335-336.994.
213. Zhu, H., Kim, Y.D., Kee, D.D., Non-Newtonian fluids with a yield stress. *J. Non-Newtonian Fluid Mech.*, *129*, 3, 177–181, 2005. https://doi.org/10.1016/j.jnnfm.2005.06.001.

16

Non-Newtonian Fluid Dynamics in Additive Manufacturing and 3D Printing

Gandhimathi G.[1], Chellaswamy C.[2*], Geetha T. S.[3] and Awad M. M.[4]

[1]Department of Electronics and Communication Engineering, SRM TRP Engineering College, Tiruchirappalli, India
[2]Electronics and Communication Engineering, SRM TRP Engineering College, Tiruchirappalli, India
[3]Department of Electronics and Communication Engineering, SA Engineering College, Chennai, India
[4]Department of Mechanical Engineering, Mansoura University, Mansoura, Egypt

Abstract

Additive manufacturing, also known as 3D printing, is a rapidly growing technology used in industries like healthcare, aerospace, construction, and electronics. Many of the materials used in 3D printing, such as polymer melts, hydrogels, and cement pastes, do not behave like normal liquids. These materials are called non-Newtonian fluids because their flow properties change under different conditions. This chapter explores how non-Newtonian fluids behave during the 3D printing process. It covers key properties such as shear-thinning, viscoelasticity, and thixotropy, which affect how materials flow through nozzles, form layers, and solidify. Different printing methods, including extrusion-based printing, inkjet printing, and photopolymerization, are discussed, along with how fluid behavior impacts their performance. The chapter also introduces computer simulations used to model fluid movement and improve print quality. Process optimization techniques, namely adjusting printing speed, nozzle pressure, and material composition, are highlighted to ensure better accuracy and reliability in printed

**Corresponding author*: drchellaswamy@gmail.com; ORCID: 0000-0002-2473-6042
Gandhimathi G.: ORCID: 0000-0003-4576-3338
Geetha T. S.: ORCID: 0000-0003-0034-3225
Awad M. M.: ORCID: 0000-0001-6238-5872

Dhananjay Yadav, Mukesh Kumar Awasthi and Harith Mohamed Al-Azri (eds.) Non-Newtonian Fluids for Industrial Applications: Modeling and Simulations, (355–382) © 2026 Scrivener Publishing LLC

objects. Finally, challenges and future trends, including multimaterial printing, bio-based materials, and AI-driven optimizations, are discussed. By understanding non-Newtonian fluid dynamics in 3D printing, researchers and engineers can develop better materials and processes, leading to stronger, more precise, and more sustainable printed products.

***Keywords*:** Non-Newtonian fluids, additive manufacturing, 3D printing, shear-thinning, viscoelasticity, extrusion-based printing, process optimization, AI-driven printing

16.1 Introduction to Non-Newtonian Fluids in Additive Manufacturing

16.1.1 Overview of Additive Manufacturing and 3D Printing Technologies

Additive manufacturing (AM), commonly known as 3D printing, is a modern production technique where objects are built layer by layer from digital designs. Unlike traditional manufacturing methods that often involve cutting away material from a solid block (subtractive manufacturing), additive processes add material only where needed, making them more efficient and often less wasteful. This technology starts with a digital 3D model created using computer-aided design (CAD) software. The model is then sliced into thin horizontal layers, and the 3D printer follows this blueprint to deposit or solidify material—such as plastic, resin, metal, or even biomaterials—layer by layer until the final object is formed [1].

The type of 3D printing technology used depends on what kind of material is being printed, how precise the object needs to be, and how much it will cost. As the technology improves, we are now able to print with more types of materials, including those that are soft, flexible, or biodegradable [2]. Many of these materials do not flow like regular liquids—they behave in more complex ways, which is why understanding how they move and change during printing is becoming more important. This is especially true for materials like hydrogels, polymer melts, and cement pastes that are commonly used in modern 3D printing.

Additive manufacturing is widely used across industries—from prototyping and custom tooling to producing end-use parts in aerospace, healthcare, automotive, and consumer goods. Its key benefits include design flexibility, rapid prototyping, reduced material waste, and the ability to create complex geometries that are difficult or impossible to produce with conventional methods.

16.1.2 Importance of Non-Newtonian Fluid Behavior in 3D Printing

Non-Newtonian fluid (NNF) behavior plays a vital role in the success of many 3D printing processes, especially those that involve complex or soft materials such as polymer melts, hydrogels, pastes, slurries, and cement-based mixtures. Unlike Newtonian fluids, whose viscosity remains constant regardless of the applied force or movement, NNFs change their flow behavior depending on the stress or shear rate they experience. This property is extremely useful in 3D printing [3]. For instance, many 3D printing materials are shear-thinning, meaning their viscosity decreases as the material is forced through the nozzle, making them easier to print. Once the shear force is removed—after the material is deposited—the viscosity increases again, helping the material hold its shape and form stable layers. This behavior is especially important for printing detailed structures, tall or overhanging parts, and soft or delicate materials.

Another key property found in some NNFs is viscoelasticity, where the material behaves like both a liquid and a solid. This dual behavior is helpful because it allows the material to stretch or flow during extrusion, but then recover and maintain structural integrity after deposition. Similarly, some printing materials exhibit thixotropy, a time-dependent property where the material becomes more fluid when shaken or moved but slowly thickens again when left at rest. This is important in preventing printed layers from spreading out too much or collapsing under their own weight. Without these non-Newtonian behaviors, many advanced 3D printing applications—like bioprinting tissues, printing concrete structures, or creating functional soft electronics—would not be possible or would lack precision and quality. As 3D printing technology continues to grow, the importance of mastering NNF behavior becomes even more critical for innovation and industrial success.

16.1.3 Comparison of Newtonian vs. NNF in Printing Applications

The success of the 3D printing process is heavily influenced by the kind of material or fluid utilized. Both Newtonian and NNFs are used in different types of additive manufacturing, but they behave in very different ways under pressure or during movement [4]. Newtonian fluids, like water or simple oils, have a constant viscosity. This means their thickness (or resistance to flow) does not change when force or shear is applied. These fluids are predictable and easier to model mathematically, making them suitable for printing processes like inkjet printing, where consistent droplet formation is crucial. However, because Newtonian fluids do not thicken after deposition, they may spread out or lose shape easily after being printed. This limits their use in applications where shape stability, layer stacking, or structure support is important.

NNFs, on the other hand, change their viscosity depending on how much force is applied or how long they are moved. This behavior is highly beneficial in 3D printing. For example, many 3D printing materials like hydrogels, polymer melts, cement pastes, or biological inks are shear-thinning, meaning they flow easily through a nozzle under pressure but quickly thicken or solidify when the force stops. This allows for better control of the material, more stable structures, and higher accuracy in the final print. NNFs are essential for extrusion-based printing, bioprinting,

Table 16.1 Comparison between Newtonian vs. NNF.

Aspect	Newtonian fluids	NNFs
Viscosity	Constant (independent of shear rate)	Variable (changes with shear rate or stress)
Examples	Water, air, alcohol	Polymer melts, hydrogels, cement paste, toothpaste
Printing suitability	Inkjet printing, spray-based applications	Extrusion printing, bioprinting, direct ink writing
Flow behavior	Predictable, steady flow	Complex, adjustable flow depending on conditions
Shape retention after printing	Poor (may spread or drip)	Good (material thickens or solidifies post-deposition)
Main advantage	Easy to control droplet formation	Enables precise structures and stability
Main limitation	Lacks structural stability after deposition	Requires deeper understanding of fluid dynamics

and construction-scale printing, where precision and shape retention are critical [5]. NNFs offer more advanced control, making them ideal for complex 3D printing applications that require precise geometry, layer stability, and the use of smart or functional materials. Comparison between Newtonian and NNF is listed in Table 16.1.

16.2 Rheology and Material Behavior in 3D Printing

16.2.1 Shear-Thinning and Shear-Thickening Effects in Printing Fluids

In 3D printing, it is important to understand how materials flow when we apply pressure or force to them. This is where a science called rheology helps us. Rheology looks at how liquids and soft materials behave when they are pushed, pulled, or moved. Two common flow behaviors in 3D printing materials are called shear-thinning and shear-thickening. Shear-thinning means that a material becomes thinner and flows more easily when we apply force. This is very useful in 3D printing. For example, when the printer pushes a material through a nozzle, the force makes it thinner so it can come out smoothly. But once it is printed and the force stops, it becomes thick again and stays in place. This helps the material hold its shape and form strong, stable layers. Many 3D printing materials, like gels, pastes, polymer melts, and bio-inks, show this behavior [6]. Shear-thinning is especially helpful in bioprinting, where soft materials with living cells need to flow gently through the nozzle but still keep their shape after printing.

Shear-thickening is the opposite. These materials become thicker and harder to move when force is applied. This can be a problem in 3D printing because the material might clog the nozzle or be difficult to control. However, in some special cases, shear-thickening is useful—like in materials that need to become stronger or stiffer when under pressure. But in most 3D printing applications, shear-thinning is the better and more commonly used behavior [7]. The shear-thinning materials are very helpful in 3D printing because they flow easily when needed and become firm after printing. This makes printing more accurate and reliable. Understanding how these materials behave helps engineers choose the right materials and improve the printing process.

16.2.2 Viscoelasticity and Its Impact on Printability

Viscoelasticity means that a material can act like both a liquid and a solid. This is important in 3D printing because the material needs to do two things: it must flow like a liquid when it is pushed through the nozzle, and it must stay in shape like a solid once it is printed. Many materials used in 3D printing, like gels, pastes, and bio-inks, show viscoelastic behavior [8].

When printing, if the material is too much like a liquid, it may drip, spread out, or lose its shape after being printed. If it is too much like a solid, it may not flow properly through the nozzle and could clog or break. Viscoelastic materials are just right because they flow well when force is applied but quickly become firm when the force stops. This helps the printed layers stay in place and hold their shape. Viscoelasticity is especially useful in areas like bioprinting, where soft materials with living cells need to flow gently and form stable structures. It also helps when printing flexible or soft objects that need to be both strong and easy to shape. The viscoelasticity helps materials print smoothly and stay in shape, which improves the quality and success of 3D printing.

16.2.3 Yield Stress Behavior in Paste-Like Printing Materials

Yield stress is a property that tells us how much force is needed to make a thick material start to flow. In simple words, a material with yield stress will not move until you push it hard enough. This is very important in 3D printing when using paste-like materials, such as cement paste, clay, food-based inks, or thick gels. These materials do not flow like water. Instead, they stay in place until enough pressure is applied [9]. In 3D printing, this behavior is useful because we want the material to stay still after it comes out of the nozzle. If the material starts to flow too easily (like a liquid), it might spread, drip, or lose shape. But with the right yield stress, the paste will only flow when pushed through the nozzle, and then stop flowing and hold its shape when printing is done. This helps build strong and stable layers, especially in large or complex prints.

If the yield stress is too low, the material might slump or collapse after printing. If it is too high, it can be hard to push the material through the nozzle, which may cause clogs or printing errors. So, it is important to find a good balance—enough yield stress to hold the shape, but not so much that it becomes too hard to print. The yield stress helps control when and how paste-like materials flow during 3D printing. By understanding and adjusting this property, we can print thicker, more stable materials for use in construction, food printing, or soft robotics.

16.2.4 Thixotropy and Structural Recovery During Deposition

Thixotropy is a special property of some materials where they become thinner and flow more easily when stirred or moved, but then slowly become thick again when left still. This behavior is very helpful in 3D printing, especially when using materials like gels, pastes, and certain clays or bio-inks. When a thixotropic material is pushed through a nozzle, the movement makes it soft and easy to flow [10]. This helps the material come out smoothly and form the printed shape. But once the material is placed on the printing surface, it stops moving, and its structure starts to recover—it becomes thick and firm again. This process is called structural recovery, and it helps the material hold its shape after printing, without dripping or flattening.

Thixotropy is very useful for building strong, tall, or detailed structures because it allows the material to flow when needed, but then solidify enough to support itself right after printing. This is important in applications like concrete printing, food printing, and bioprinting, where materials must stay in place after each layer is deposited. If the material does not recover quickly enough, the printed layers may collapse or lose shape. If it recovers too fast, it might block the nozzle or not spread properly [11]. So, finding the right thixotropic behavior is important for smooth printing and strong, accurate structures. The thixotropy helps printing materials behave like a liquid when moving, and like a solid when still, making 3D printing more reliable and precise.

16.2.5 Types of Non-Newtonian Materials in Additive Manufacturing

In 3D printing, many materials are Non-Newtonian, meaning their flow changes when pressure or force is applied. These materials are helpful because they can flow easily when needed and then quickly become stable after printing [12, 13]. Below are the main types of non-Newtonian materials used in additive manufacturing:

Polymer-based materials: These include polymer melts and thermoplastic pastes. They are often shear-thinning, so they flow easily under pressure but become firm when the pressure stops. They are widely used in extrusion-based printing like FDM.

Hydrogels: Soft, water-filled materials commonly used in bioprinting. Hydrogels are viscoelastic and shear-thinning, which means they are gentle

when printing cells but can still hold their shape. Ideal for printing tissues and soft structures.
Cementitious materials and ceramic pastes: These are used in construction and structural printing. They have yield stress and thixotropic behavior. They do not flow until pushed but firm up quickly after printing—great for printing walls, bricks, or large parts.
Food-based inks: Materials like chocolate, dough, and vegetable pastes used in food 3D printing. They often show shear-thinning and thixotropy, which helps them be shaped easily and hold that shape afterward.
Silicone and elastomers: These are flexible, rubber-like materials used for soft robotics, medical devices, or wearable parts. They usually have visco-elastic behavior, allowing them to stretch and return to shape after printing.

Each of these materials is chosen for its special flow behavior and how well it can form strong, accurate, and stable printed parts. Understanding these types helps in selecting the right material for different 3D printing applications.

16.3 Deposition Techniques for Non-Newtonian Fluids

16.3.1 Flow Behavior and Nozzle Design Considerations

In 3D printing, the way a material flows through the nozzle is very important. This is known as flow behavior, and it depends on the type of material being used—especially if it is an NNF. These materials, like gels, pastes, or polymer melts, change their thickness when pressure is applied [13]. Some become thinner (shear-thinning), some thicker (shear-thickening), and others may have a yield stress, meaning they only start to flow after a certain amount of force is applied. To get smooth and accurate printing, the nozzle must be designed to match the flow behavior of the material.

If the nozzle is too small, thick materials might not flow well and could clog the nozzle. If the nozzle is too wide, the printed lines may be rough or lose detail. So, it is important to choose the right nozzle size and shape based on how easily the material flows. For shear-thinning materials, a narrow nozzle can often be used because the material becomes thin when pressure is applied and flows easily [14]. But for thicker pastes or materials with yield stress, a wider nozzle may be needed to allow smooth extrusion.

The shape of the nozzle tip also matters. A sharp, conical tip helps focus the flow and improves accuracy. A round tip may be better for materials that need gentle flow. The length of the nozzle affects the pressure needed to push material through—it must be long enough to guide the material smoothly, but not so long that it increases resistance. Temperature is another factor. Some materials need to be heated to flow properly, so heated nozzles are used in processes like FDM. For materials that are sensitive to heat, the nozzle design should keep the temperature low and steady. A good nozzle design works hand-in-hand with the material's flow behavior. Matching the right nozzle size, shape, and temperature control to the material ensures better flow, reduces clogs, and helps create clean, precise printed layers.

16.3.2 Resin Viscosity and Curing Dynamics

In some types of 3D printing—like stereolithography and digital light processing—liquid resins are used instead of solid or paste-like materials. These resins are special liquids that harden (or "cure") when exposed to light, usually UV light. Two important things to understand about these materials are resin viscosity and curing dynamics, because they affect how well the printing process works. Viscosity is how thick or thin a liquid is. A resin with low viscosity is very thin and flows easily, like water. A resin with high viscosity is thicker and flows more slowly. The viscosity of a resin affects how it spreads across the printing area and how quickly new layers can be formed. If the resin is too thick, it might not flow smoothly or refill the printing area fast enough, which can lead to poor layer bonding or defects in the printed object. If it is too thin, it may spread too much or not stay where it is needed [15].

Curing dynamics refer to how fast and how completely the resin hardens when exposed to light. For accurate printing, the resin must cure just the right amount—not too fast and not too slow. If the resin cures too quickly, it may over-harden or create rough surfaces. If it cures too slowly, the printed layers might stay soft, causing poor detail or structural problems. The speed and depth of curing depend on several things, such as the resin formula, light intensity, and exposure time.

Balancing viscosity and curing dynamics is key to getting good results. Engineers and researchers often adjust the chemical makeup of the resin to control how it flows and how fast it cures. They also adjust the printer settings, like light power and exposure time, to match the resin's behavior. The resin viscosity and curing dynamics work together to control how smoothly the resin flows and how well it hardens into solid layers. Understanding these properties helps ensure strong, accurate, and high-quality prints in resin-based 3D printing methods.

16.3.3 Droplet Formation and Spreading for High-Precision Deposition

In some 3D printing methods—like inkjet printing or material jetting—the material is printed as tiny droplets instead of continuous lines. These droplets are very small, and they are released one by one from a nozzle onto the printing surface. For high-precision printing, like in electronics, fine structures, or tiny medical parts, it is very important to control how each droplet forms, how it moves, and how it spreads after landing [16]. First, the droplet formation must be smooth and consistent. If a droplet is too big, it can cause the printed details to look blurry or thick. If it is too small or uneven, parts of the design might not print at all. The shape, size, and speed of each droplet depend on the material's viscosity, surface tension, and the pressure or heat used to push it out of the nozzle.

After the droplet lands on the surface, it starts to spread out. This spreading must be carefully controlled. If it spreads too much, it can mix with nearby droplets and ruin the sharpness of the design. If it spreads too little, the layers might not bond properly, or gaps may form between droplets. The surface temperature, material properties, and drying speed all affect how much spreading occurs. The goal is to create droplets that land in the right spot, stay the right size, and stick properly without moving too much. This helps build sharp, clean, and detailed printed parts. In high-precision applications like micro-patterns, sensors, or printed circuits, controlling droplet behavior is essential for getting perfect results. The droplet formation and spreading are key to making accurate, tiny structures in 3D printing. By understanding how droplets behave, we can print very fine details with high quality and precision.

16.3.4 Interaction of Binders and Powder Flowability

In some types of 3D printing, like binder jetting, the process uses powder materials (such as metal, ceramic, or sand powders) and a liquid binder. The binder is sprayed onto thin layers of powder to glue the particles together and form solid shapes, layer by layer. For this method to work well, both the binder and the powder must interact properly. This interaction affects how smoothly the powder spreads and how strong the printed part becomes. First, the flowability of the powder is very important. Good flowability means the powder can spread easily and evenly to form flat, uniform layers. If the powder is too sticky or clumpy, it will not spread well, and the layers will be uneven. This can lead to defects in the final part.

The size, shape, and dryness of the powder particles all affect flowability—smaller and rounder particles usually flow better.

Next, the binder must be able to wet the powder particles and hold them together without making the powder too wet or sticky. If too much binder is used, the powder might clump or shift, ruining the layer. If too little binder is used, the particles will not stick properly, and the printed object will be weak or fall apart. The binder also needs to dry or cure quickly so that each layer becomes stable before the next one is added. The right balance between powder flowability and binder behavior is the key to printing parts that are strong, smooth, and accurate. Scientists and engineers often adjust the binder formula or add flow-enhancing agents to the powder to improve this balance. The interaction between binder and powder plays a big role in how well the 3D printer builds solid parts from loose powder. Good flow and proper binding lead to better quality prints with fewer errors.

16.4 Computational Modeling and Simulation

16.4.1 Governing Equations for Non-Newtonian Fluid Flow in 3D Printing

The continuity equation is one of the most important rules in fluid mechanics. It is based on the idea of mass conservation, which means that mass cannot be created or destroyed. In 3D printing, this rule helps us understand how fluid (such as a Non-Newtonian material) flows through the nozzle and forms layers without any loss or gain in material.

The continuity equation makes sure that whatever material enters a space must either stay there or exit—nothing disappears. It keeps track of how much material is flowing and ensures that the flow is balanced. For fluids, this equation is written as

$$\nabla.\mathrm{u} = 0(\text{for incompressible fluids}) \tag{16.1}$$

Here, $\nabla \cdot \mathrm{u}$ means the divergence of velocity, which shows how the fluid is moving in or out of a small region. If the result is zero, it means the fluid is incompressible, and the amount flowing in is equal to the amount flowing out.

For incompressible NNFs, which are very common in 3D printing (like pastes, gels, or molten polymers), this simplified form is usually enough.

It tells us that the fluid is not shrinking or expanding, and the volume remains constant as it moves. In some special cases, like when using compressible materials (where the density changes), the more general form of the continuity equation is used:

$$\frac{\partial \rho}{\partial t} + \nabla.(\rho V) = 0 \tag{16.2}$$

This version includes ρ, the fluid density, and accounts for how both the speed and density of the fluid can change over time. The continuity equation helps ensure that the 3D printer deposits the correct amount of material, without gaps or extra buildup.

In many 3D printing situations, especially when analyzing a flat layer of material or looking at flow in a thin region, we can simplify the problem to two dimensions (2D). This means we only consider how the fluid moves in the x-direction (left to right) and y-direction (front to back) and ignore the vertical (z) direction for that case. For incompressible flow (where the fluid's density stays the same), the continuity equation in 2D is

$$\frac{\partial u}{\partial x} + \frac{\partial v}{\partial y} = 0 \tag{16.3}$$

where u and v are the velocity of the fluid in the x-direction and y-direction, respectively. $\frac{\partial u}{\partial x}$ and $\frac{\partial v}{\partial y}$ are the rate of change of velocity in the x-direction and y-direction, respectively. Figure 16.1 shows the visualization of velocity field and its divergence using continuity equation. Assume a square domain of length L=1.0 discretized into a grid of N=10 points and N=100 points resulting in a grid spacing dx=L/N. The velocity field can be depicted using arrows (quiver plot), where each arrow indicates the direction and magnitude of the velocity vector at a grid point. This visualization provides insight into the flow patterns, such as identifying regions of uniform flow, vortices, or recirculation zones. For the fine grid (100×100) in Figure 16.1(a), the velocity field is highly detailed, accurately capturing flow structures, boundary layers, and vortices. The smooth velocity contours ensure that velocity gradients are well resolved, leading to a more realistic flow representation. In contrast, the coarse grid (10×10) in Figure 16.1(b) results in a much less detailed velocity field, with significant numerical diffusion and poor resolution of velocity variations.

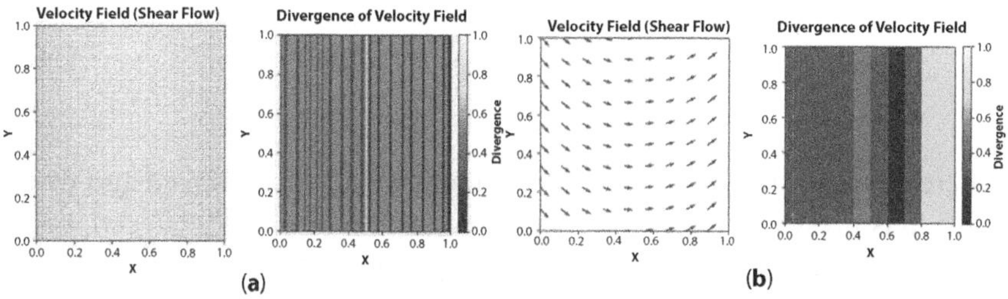

Figure 16.1 Velocity visualization and continuity equation verification. (a) Nano fluid grid size 100×100. (b) Nano fluid grid size 10×10.

The coarser mesh averages out small-scale structures, reducing the accuracy of the flow representation.

In Figure 16.1, heat map can represent the divergence values across the domain. In incompressible flow, according to the continuity equation, this divergence should ideally be zero everywhere. Areas where the divergence deviates from zero may indicate numerical errors or regions where the incompressibility assumption is violated. In computational fluid dynamics (CFD), these visualizations are essential for verifying that numerical simulations adhere to physical laws like mass conservation. The fine grid (100×100) ensures a better approximation of the continuity equation, as smaller control volumes reduce discretization errors. The mass conservation is well maintained, leading to minimal errors in divergence calculations. On the other hand, the coarse grid (10×10) introduces larger discretization errors, making it more challenging to maintain mass conservation accurately. The larger cell sizes may lead to artificial sources or sinks in the flow field, reducing the precision of the numerical solution.

16.4.2 Momentum Equation (Navier–Stokes for NNF)

The momentum equation is one of the most important tools used to describe how fluids move. It is founded on Newton's second law of motion, which states that force is the product of mass and acceleration. In fluid mechanics, this becomes the Navier–Stokes equation. This equation shows how fluid velocity changes due to forces like pressure, viscosity, and gravity.

For Newtonian fluids (like water or air), the viscosity is constant. But for NNFs, which are often used in 3D printing (like gels, pastes, and polymer melts), the viscosity changes with the flow—for example, it may become thinner or thicker when stirred. So, the standard Navier–Stokes equation must be modified to include variable viscosity or other stress behaviors. The general form of the Navier–Stokes equation for a fluid is

$$\rho\left(\frac{\partial u}{\partial t}+u.\nabla u\right)=\nabla.\tau-\nabla p+F \tag{16.4}$$

where ρ is the fluid density, u is the velocity vector (fluid speed and direction), p is pressure, τ is the stress tensor (includes viscosity effects), and F are body forces (like gravity or electromagnetic forces).

Figure 16.2 shows the fluid flow using a basic numerical solution to the Navier–Stokes equations. The grid spacing 50×50 (Figure 16.2(a)) and 10×10 (Figure 16.2(b)) are determined by dividing the total length by the number of grid points, ensuring uniform spacing. The time step is small to maintain numerical stability when solving the equations. This represents the kinematic viscosity of the fluid (nu = 0.1), which affects the diffusion of momentum. The velocity fields (U, V) and pressure field (P) are initialized, and a loop iterates through time steps (3000) to update the velocity fields based on pressure gradients and viscosity effects. A Laplacian function is used to compute diffusion effects, and boundary conditions enforce no-slip conditions at the domain edges.

The analysis of nano-fluid flow at different grid sizes (50×50 and 10×10) reveals significant differences in velocity visualization, pressure distribution, and the verification of the Navier–Stokes equations. For the fine grid (50×50) in Figure 16.2(a), the velocity field is smooth and accurately captures small flow structures, including vortices. The pressure distribution is well resolved, with minimal numerical diffusion, ensuring that pressure gradients are correctly represented. Additionally, the Navier–Stokes equations are better verified, as the discretization errors are significantly reduced. However, this finer resolution comes at the cost of increased computational time and memory usage. In contrast, the coarse grid (10×10) in Figure 16.2(b) results in a rough and less detailed velocity field, where small-scale flow structures are not well resolved. The pressure distribution is less

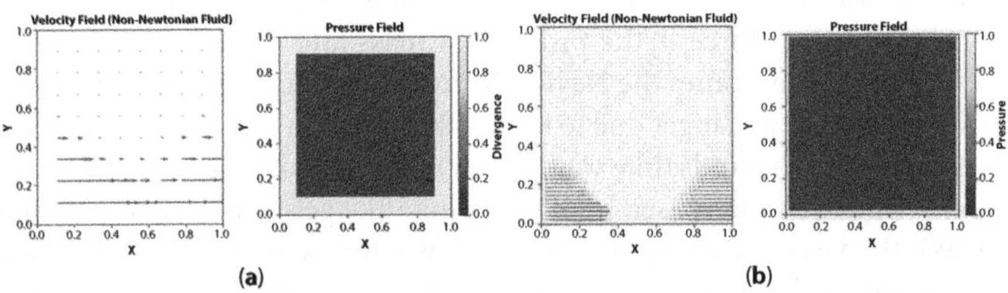

Figure 16.2 Velocity visualization and Navier–Stokes equations verification. (a) Nano fluid grid size 50×50. (b) Nano fluid grid size 10×10.

accurate due to larger discretization errors, leading to potential numerical diffusion. The verification of the Navier–Stokes equations is less precise, as the approximation errors in convective and diffusive terms become more pronounced. However, the computational cost is significantly lower, making it a faster but less accurate approach. Inclusive, while the fine grid provides better accuracy in flow visualization and equation verification, the coarse grid offers a computationally efficient alternative at the expense of precision. For a balance between accuracy and efficiency, an adaptive meshing approach or a moderate grid resolution may be preferable.

Figure 16.3 illustrates the effect on the velocity field and its divergence at different time steps (1000, 1500, and 2000) in the simulation using the

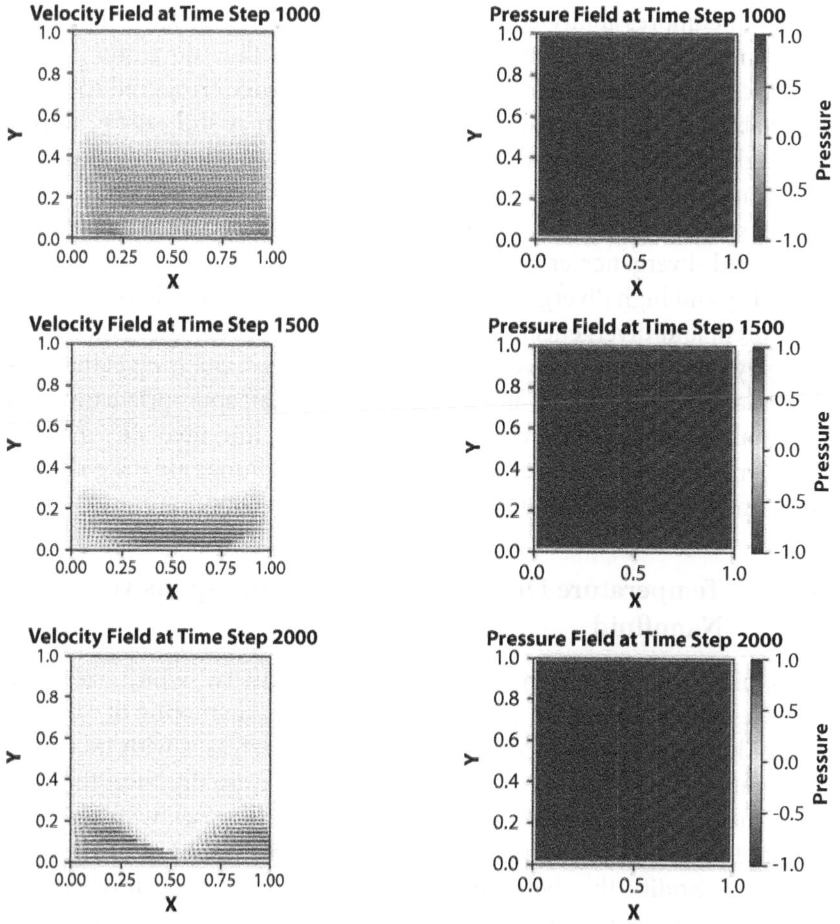

Figure 16.3 Evolution of velocity field and its divergence at different time stamps (1000, 1500, 2000) using Navier–Stokes equations.

Navier–Stokes equations. As time progresses, the velocity field evolves due to the combined effects of pressure gradients and viscosity. At time step 1000, the flow structure begins to develop, showing initial variations in velocity. By time step 1500, the velocity field exhibits more pronounced changes, with stronger directional flow and visible divergence patterns indicating the redistribution of momentum. At time step 2000, the flow becomes more refined, demonstrating the effects of viscosity in smoothing out sharp gradients while maintaining the underlying fluid motion. The divergence of the velocity field highlights regions of expansion and contraction, revealing the balance between convection and diffusion in the evolving fluid dynamics.

The divergence of the velocity field indicates how well the flow satisfies the incompressibility condition ($\nabla \cdot u = 0$). At time step 1000, higher divergence errors are expected due to transient flow development and numerical fluctuations. At time step 1500, divergence errors reduce as the flow stabilizes, but minor numerical inaccuracies may still exist in regions with high velocity gradients. At time step 2000, divergence errors are minimized, and the flow reaches a near-steady or fully developed state, improving mass conservation. As time progresses, the velocity field becomes more stable, and divergence errors reduce. At $t = 1000$, transient effects cause instability and high divergence errors. By $t = 1500$, the flow becomes more structured, and divergence errors decrease. At $t = 2000$, the velocity field is mostly stabilized, and divergence approaches zero, ensuring better mass conservation. For practical applications, selecting an appropriate time step is crucial. If real-time flow dynamics are required, intermediate time steps (1500) may be useful. However, if steady-state conditions are the goal, running simulations until $t = 2000$ ensures better accuracy.

16.4.3 Temperature Distribution in Thermoresponsive Nanofluid

Nanofluids exhibit enhanced thermal conductivity, which generally increases with temperature due to improved Brownian motion of nanoparticles. The specific heat of nanofluids typically decreases with increasing temperature, as the heat capacity of nanoparticles is usually lower than that of the base fluid. Viscosity, on the other hand, decreases with temperature due to reduced intermolecular forces, but the presence of nanoparticles can modify this behavior depending on their concentration and size. Heat transfer in nanofluids occurs through conduction, convection, and radiation. Conduction is the primary mechanism at the microscopic level, where enhanced thermal conductivity of nanoparticles improves heat transfer. Convection plays a crucial role in macroscopic transport, with nanoparticles modifying the boundary layer characteristics and fluid motion. Radiation becomes significant at high temperatures, as energy is transferred through electromagnetic waves, and nanoparticles can enhance radiative heat transfer due to their optical properties. The dynamic response of nanofluid properties to temperature changes is crucial in applications like cooling systems and heat exchangers. As temperature varies, the interaction between nanoparticles and the base fluid changes dynamically, affecting viscosity, density, thermal conductivity, and specific heat in real time. This dynamic behavior influences the stability and performance of nanofluids in practical applications.

The basic nanofluid structure is designed to simulate heat transfer in a nanofluid using a one-dimensional finite difference method while incorporating a temperature-dependent thermal conductivity model. The simulation domain is set to a length of one meter, divided into discrete spatial intervals of 0.01 meter. Initially, the temperature of the entire domain is set to 25°C, except for a heat source placed at the center, which is maintained at 100°C as shown in Figure 16.4(a). The base fluid and nanoparticles have predefined thermal conductivities of 0.6 W/mK and 1.0 W/mK, respectively. Thermal diffusivity values for the base fluid and nanoparticles are also specified to govern the rate of heat propagation. The key aspect of the model is the effective thermal conductivity function, which dynamically adjusts based on the local temperature. This function assumes a linear increase in thermal conductivity from the base fluid value to the nanoparticle value as the temperature rises from 25°C to 100°C. This variation is clamped within the specified limits to ensure stability and to prevent extreme values. During the simulation, a time-stepping loop iterates 1000 times to update the temperature profile across the domain. At each step, the finite difference method is used to solve the heat conduction equation, where the effective thermal conductivity at each spatial point is considered. The temperature update formula accounts for the influence of neighboring points, allowing heat to propagate throughout the domain. Additionally, a simple convection model is incorporated, which redistributes heat by introducing a correction term based on the difference between the boundary and interior temperatures. A radiation term is also included, simulating energy exchange with the surroundings, ensuring that heat dissipates realistically over time. To prevent numerical instabilities, temperature values are clamped between the initial and source temperature limits. This ensures that the temperature distribution remains physically realistic throughout the simulation as shown in Figure 16.4(b). The combined effects of conduction, convection, and radiation in the model provide an

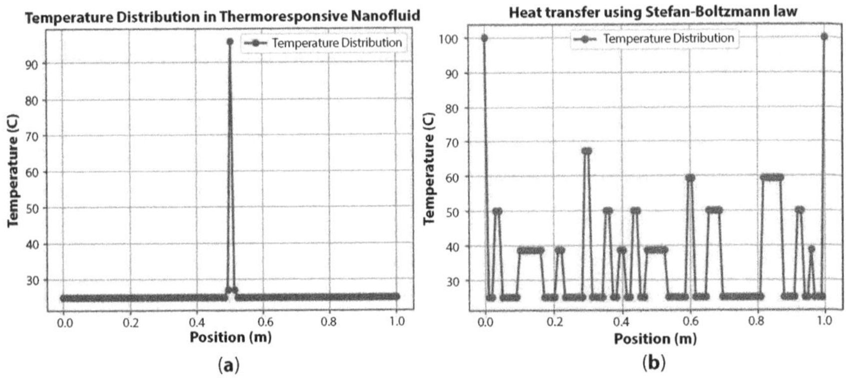

Figure 16.4 Thermal conductivity in nanofluid. (a) Temperature distribution. (b) Heat transfer.

accurate representation of nanofluid heat transfer, where the presence of nanoparticles enhances thermal conductivity and improves heat dissipation efficiency. This approach makes the simulation applicable to various practical scenarios, such as thermal management in electronic systems and heat exchangers.

This model treats the effective thermal conductivity as a linear function of temperature. The values are constrained within the range of the base fluid and nanoparticle conductivities to prevent numerical instabilities. This dynamic adjustment allows the simulation to represent the enhanced heat transfer properties of nanofluids more accurately. The simulation iterates through 1000 time steps, updating the temperature profile in each iteration using the finite difference method. The heat conduction equation is solved using an explicit scheme, where the new temperature at each spatial point is computed based on the temperatures of neighboring points and the effective thermal conductivity at that location, then incorporates convection effects using Newton's law of cooling. The convective heat transfer coefficient is set to 10 W/m^2K, and heat transfer due to convection is modeled as proportional to the difference between the heat source temperature and the local temperature. This ensures that heat is dissipated efficiently from the heated region to the surrounding fluid.

Stefan–Boltzmann principle is applied by computing the radiation heat loss for each temperature point in the domain. The Stefan–Boltzmann constant is set to $5.67 \times 10{-}85.67$ W/m^2K^4, and the radiation heat flux is calculated based on the difference in the fourth power of the absolute temperatures of the source and the local temperature. This addition enhances the realism of the simulation, especially at higher temperatures where radiation becomes a dominant heat transfer mechanism. The refined model now provides a more accurate representation of heat transfer in nanofluids, considering the combined effects of conduction, convection, and radiation. This approach is particularly useful for analyzing heat dissipation in cooling systems, heat exchangers, and other thermal management applications where nanofluids are employed to enhance performance.

The finite volume method (FVM) is an advanced numerical approach used to solve the governing heat transfer equations in nanofluids. In FVM, the computational domain is discretized into small control volumes, and the conservation equations for mass, momentum, and energy are applied to each volume. This method ensures accurate treatment of fluxes at control volume boundaries and is widely used for simulating complex heat transfer phenomena in nanofluids. Figure 16.5 shows variations in flow velocity, temperature distribution, or heat transfer rates under different magnetic field strengths. A strong magnetic field can influence nanoparticle alignment, altering viscosity and convective heat transfer. The Lorentz force generated by the field can either enhance or restrict movement, modifying the overall heat transport efficiency. If the graph incorporates convection modelling, it may compare forced and natural convection scenarios. Forced convection, driven by an external force (e.g., a pump or fan), tends to create more uniform heat transfer, while natural convection relies on buoyancy forces and temperature gradients. Figure 16.6 enhancing convection models ensures more accurate simulations of real-world heat transfer applications by incorporating dynamic parameters as forced or natural convection

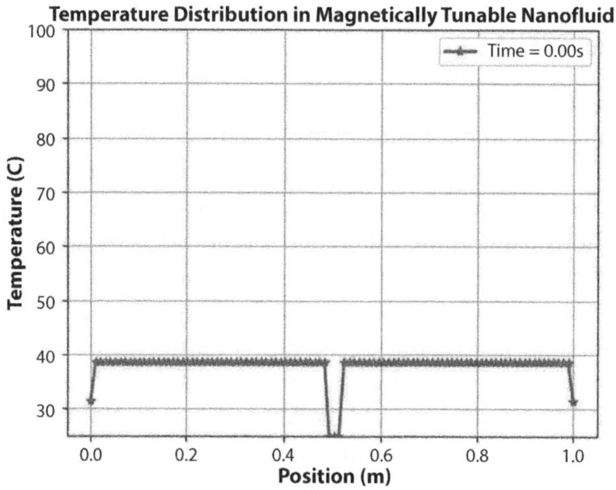

Figure 16.5 Temperature distribution in different magnetic field strengths.

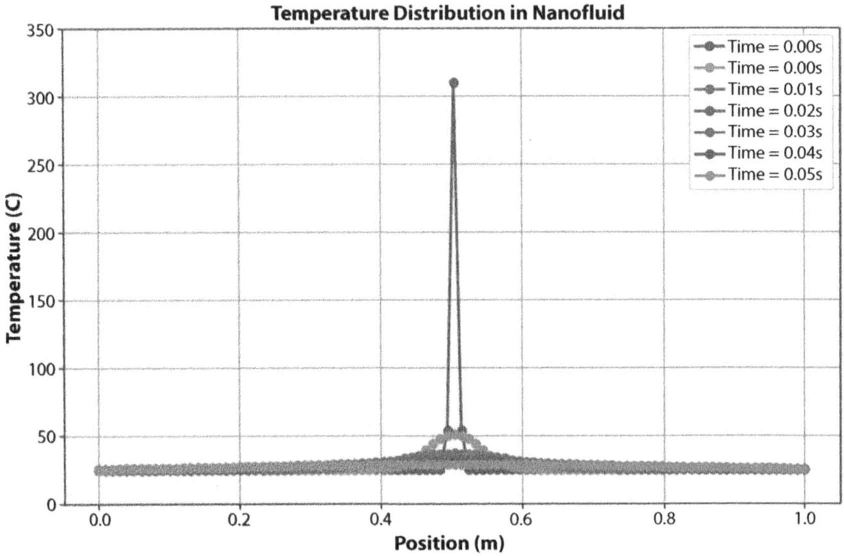

Figure 16.6 Temperature distribution under dynamic parameters in nanofluid.

and more complex boundary conditions, such as insulated boundaries. The graph may highlight differences in heat transfer efficiency under varying boundary conditions, such as insulated boundaries, constant heat flux, or spatially varying heat sources. These enhancements improve the accuracy of predictions, optimize heat transfer performance, and enable better control over fluid dynamics in advanced thermal applications.

16.4.4 Case Study 1

Implementing the Governing Equations for NNF Flow in 3D Using the FDM
CFD is essential for understanding material flow, deposition patterns, and defects in 3D printing. Figures 16.8 and 16.10 represent the evolution of a three-dimensional velocity field for an NNF governed by the Navier–Stokes equations using finite difference method (FDM) and finite volume method (FVM). The simulation incorporates the effects of viscosity, shear rate, and pressure corrections to analyze fluid motion over time. The computational domain is structured as a 50×50×50 grid and 20×20×20 with uniform spacing, and the velocity components (u,v,w) are initialized to zero. A power-law model is used to define the non-Newtonian viscosity, which varies dynamically with shear rate. The time-stepping loop iterates over 100 steps, updating velocity fields based on the continuity equation and the momentum equations. Throughout the simulation, shear rates are

computed using velocity gradients, and viscosity is adjusted accordingly. The continuity equation ensures mass conservation, while the momentum equations incorporate convection and viscosity effects to determine the updated velocity values. Pressure is also corrected iteratively to maintain a physically consistent flow.

Figure 16.7 illustrates the individual velocity components u, v, and w (corresponding to the x, y, and z directions, respectively) of the flow field, computed on a uniform 20×20×20 grid. Each component is represented spatially across the 3D domain, revealing the distribution and directional tendencies of the fluid motion. The visualization highlights localized variations in flow magnitude and direction, essential for understanding the behavior of NNFs where velocity gradients significantly affect viscosity.

Figure 16.8(a) depicted here is the three-dimensional velocity field of an NNF, obtained by numerically solving the Navier-Stokes equations using the FDM over a 20×20×20 discretized grid. The non-Newtonian behavior is captured through appropriate constitutive models integrated into the Navier-Stokes formulation. The resulting velocity vectors represent the complex interactions within the flow, including shear-thinning or shear-thickening effects, and the impact of non-linear viscosity on momentum transport. Despite the relatively coarse grid, the major flow features such as vortices, recirculation zones, and directional asymmetries are clearly discernible.

Figure 16.8(b) extends the analysis to a refined grid of 50×50×50, offering higher spatial resolution and improved accuracy. The increased number

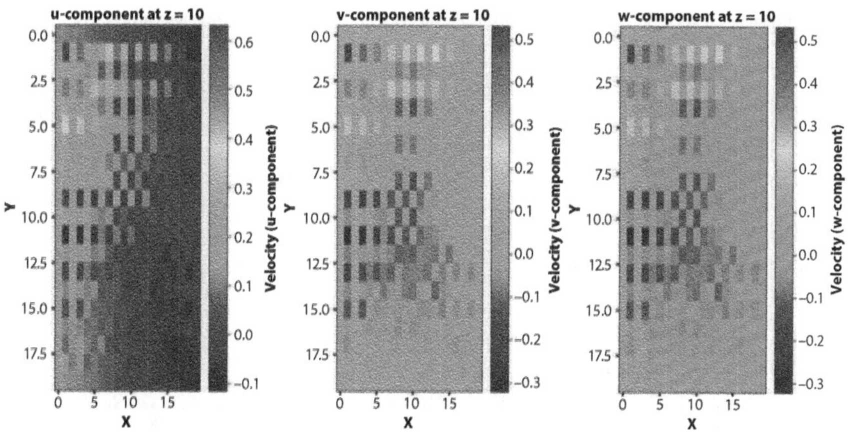

Figure 16.7 u, v, and w components view for grid size 20×20×20 governed by the Navier-Stokes equations using a finite difference method.

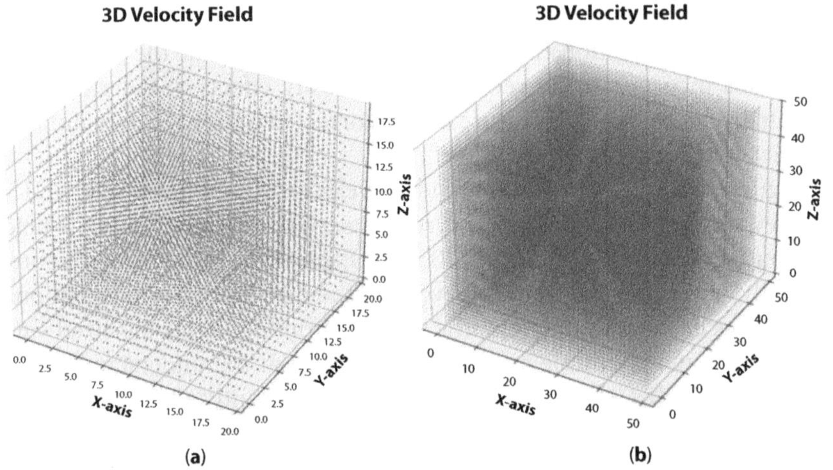

Figure 16.8 3D velocity field for a non-Newtonian fluid governed by the Navier-Stokes equations using a finite difference method (a) grid size 20×20×20 (b) grid size 50×50×50.

of grid points enhances the ability to resolve finer-scale flow structures and more accurately capture steep velocity gradients, especially near boundaries or regions of high shear. Compared to the 20×20×20 grid, the results demonstrate smoother vector transitions, better-defined flow paths, and improved depiction of local instabilities or turbulence-like behavior. The use of a finer grid is particularly important in non-Newtonian flow simulations, where small-scale variations in shear rate can significantly alter local viscosity and thereby affect the overall flow dynamics.

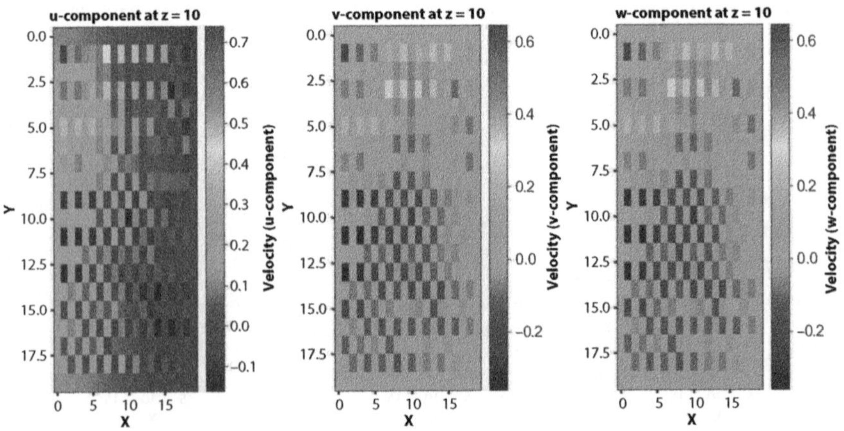

Figure 16.9 u, v, and w components view for grid size 20×20×20 governed by the Navier-Stokes equations using a finite volume method.

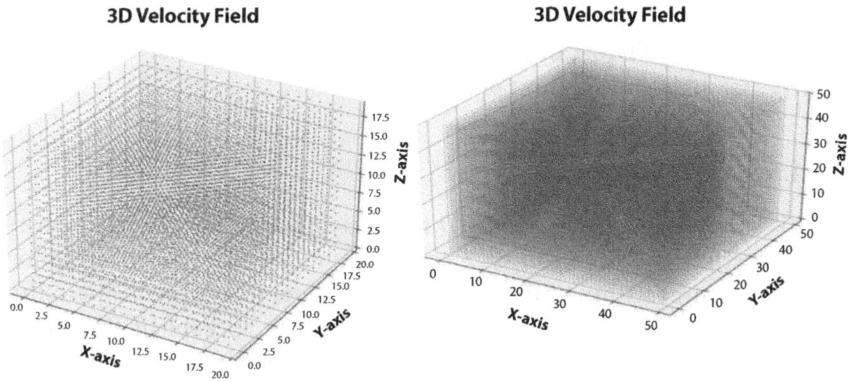

Figure 16.10 3D velocity field for a non-Newtonian fluid governed by the Navier–Stokes equations using a FVM - grid size 20×20×20 (b) grid size 50×50×50.

Implementing the Governing Equations for NNF Flow in 3D Using FVM
Similarly Figure 16.9 presents the individual velocity components u, v, and w of the flow field, computed on a uniform 20×20×20 grid under FVM. The FVM framework ensures conservation of mass and momentum at the control volume level, making it suitable for simulating complex fluid behaviors. The velocity field highlights key flow features such as recirculation zones, boundary-layer development, and non-linear effects introduced by the fluid's rheology.

Figures 16.10(a) and (b) depicts the same simulation conducted on a 20×20×20 and refined 50×50×50 grid, enabling a higher spatial resolution. FVM shown in Figures 16.10 (a)-(b) is inherently conservative, ensuring local conservation of mass, momentum, and energy within each control volume. It is well suited for structured grids and particularly popular in CFD applications due to its intuitive physical interpretation. FVM's strength lies in its ability to accurately capture fluxes across control volume faces, which is beneficial in resolving sharp gradients and non-linearities typical of NNF behavior. Using FVM, the results demonstrate strong conservation properties and are computationally efficient for structured grids, although less flexible for unstructured or highly distorted domains.

The NNF flow governed by the Navier–Stokes equations, FDM and FVM offer robust numerical approaches but differ significantly in formulation, implementation, and outcomes. FDM provides higher flexibility in handling complex geometries and irregular meshes, making it ideal for problems involving curved domains or intricate boundary conditions. FDM discretizes the domain using elements and approximates the solution using basis functions. In the case of non-Newtonian flows, FDM offers smoother and more continuous interpolation of velocity and pressure fields. However, FDM can be computationally more intensive due to its matrix assembly and solution procedures.

16.4.5 Case Study: 2

In this simulation, an enhanced 3D finite-element model (FEM) is developed to study the diffusion of a scalar quantity (such as concentration or temperature) within a 3D domain. The model incorporates the diffusion equation, which is numerically solved using an explicit time stepping method. To ensure a realistic and mathematically accurate solution, boundary conditions are implemented. Two types of boundary conditions are commonly used in FEM for diffusion problems. Dirichlet boundary condition specifies the value of the function (e.g., temperature or concentration) on the boundary. It is commonly used to simulate fixed conditions like a constant temperature. Neumann boundary condition specifies the derivative (flux) across the boundary and is used to model heat flow or diffusion through surfaces. In this particular implementation, Dirichlet boundary conditions are applied, where the concentration is fixed to zero at all boundaries of the domain. This means that the scalar quantity is assumed to be zero at the edges of the 3D grid, mimicking a scenario where the domain is surrounded by a medium that absorbs or does not allow the diffusion of the scalar. The function apply_ boundary _condition (c) explicitly sets the values at the six faces of the 3D domain left, right, front, back, bottom, and top to zero. This simulates a boundary where the value of the diffusing property is held constant (zero), which is characteristic of Dirichlet conditions. Following this, the diffusion process is computed using the function diffuse (C, D, dt). This function calculates the updated concentration values at each internal point of the domain using a FEM approximation of the Laplacian operator in 3D. The new value at each point depends on the concentration at that point and its six neighboring points in the x, y, and z directions. The parameters D, dt, dx, dy, and dz represent the diffusion coefficient, time step, and spatial resolution in the three respective directions.

By iteratively applying these updates over time while maintaining the boundary conditions, the model simulates how the scalar quantity diffuses through the domain. This enhanced 3D FEM provides a foundational numerical approach for solving diffusion-dominated problems in physics, engineering, and environmental science. The finer grid in Figure 16.11(a) captures the diffusion process more accurately, especially near boundary layers, but requires more memory and time. The coarser grid (10×10×10)

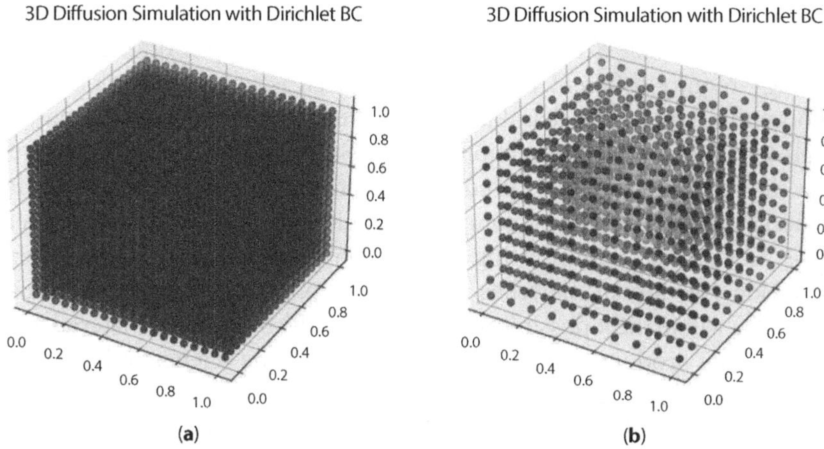

Figure 16.11 Enhanced 3D FEM with diffusion: Application of Dirichlet and Neumann boundary conditions. (a) Grid size 20×20×20. (b) Grid size 10×10×10.

in Figure 16.11(b) is faster but less detailed. The impact of Dirichlet conditions is more sharply defined in (a), while in (b), the transition appears more diffuse. The 20×20×20 case can serve as a benchmark to validate results from coarser meshes or different numerical schemes. The figure demonstrates that mesh refinement significantly affects the fidelity of FEM simulations. While both grid sizes implement the same boundary conditions, only the finer mesh (20×20×20) can resolve the complex interactions between diffusion and mixed boundary effects with high precision. Hence, mesh resolution should be chosen based on the required level of detail, available computational resources, and the physical phenomena being modelled.

16.5 Conclusion and Future Scope

In this chapter, we learned how important NNFs are in 3D printing and additive manufacturing. Unlike regular fluids (called Newtonian fluids), non-Newtonian materials behave in special ways—they can become thinner or thicker when they are moved or pressed. Many of the materials used in 3D printing, like polymer melts, hydrogels, pastes, and bio-inks, are non-Newtonian. Understanding how these materials flow and behave helps improve the print quality, accuracy, and strength of printed objects. We also looked at key properties like shear-thinning, viscoelasticity, yield stress, and thixotropy, which all affect how materials are pushed through nozzles, how they spread, and how they hold their shape. Different printing methods—like extrusion-based printing, inkjet printing, and resin curing—require different fluid behaviors. That is why it is important to choose the right material and understand its flow behavior. Mathematical tools like the continuity equation, Navier–Stokes equations, and constitutive models help engineers predict and control the flow of these special materials. With the help of computer simulations and better process settings, we can make 3D printing more reliable and efficient. In the future, new Non-Newtonian materials and smarter printing systems (like those using AI) will make it possible to create more advanced, multimaterial, and eco-friendly products. By learning and applying the science of NNFs, we can push 3D printing to new levels of innovation.

The study of NNFs in 3D printing has a lot of potential for growth in the coming years. As technology and materials continue to improve, new opportunities are opening up for more advanced and specialized applications. One major area of future development is the use of bio-based and sustainable Non-Newtonian materials. These include plant-based gels, biodegradable polymers, and other eco-friendly substances that reduce waste and environmental impact. These materials are especially important in industries like biomedical printing and tissue engineering, where safe and natural ingredients are preferred. Another exciting direction is multimaterial 3D printing, where different types of materials are printed together in one object. This often involves combining fluids with different flow behaviors, so understanding how NNFs interact becomes even more important.

AI disclosure statement: The author acknowledges the use of AI-based tools for minor language editing or grammatical correction. However, all intellectual contributions, analyses, and conclusions presented in this chapter are entirely the author's own.

References

1. Ali, B., Zhou, Y.-T., Jubair, S., Tariq, M.H., Kumar, A., Siddiqui, M.I.H., Entropy optimization and Prandtl-Eyring non-Newtonian fluid flow with second-order slip conditions past a curved Riga sheet; numerical simulation. *Int. J. Therm. Sci.*, 214, 2025. Art. no. 109916. doi: 10.1016/j.ijthermalsci.2025.109916.
2. Gupta, A., Rana, M., Mondal, N., Determining the optimal design parameters for gyroids using computational fluid dynamics analysis under a non-Newtonian perfusion system. *Biomed. Eng. Adv.*, 9, 2025. Art. no. 100153. doi: 10.1016/j.bea.2025.100153.

3. Zhang, K., Cai, C., Sun, H., Jia, N., Liu, J., Xie, Z., Flexible concentration control of Newtonian and non-Newtonian fluids in a microfluidic device *via* AC electrothermal flow. *Sens. Actuators A Phys.*, 380, 2024. Art. no. 116043. doi: 10.1016/j.sna.2024.116043.
4. Patel, S.A. and Chhabra, R.P., Non-Newtonian Fluids, in: *Kirk-Othmer Encyclopedia of Chemical Technology*, 2020, doi: 10.1002/0471238961.koe00057.
5. Hao, X., Yang, K., Wang, H., Peng, F., Yang, H., Biocatalytic Feedback-Controlled Non-Newtonian Fluids, in: *Angew. Chem*, 2019, doi: 10.1002/ange.201914398.
6. Bijender, B., Soni, A., Boora, A., Verma, D., Investigation of Tensile Strength Performance of 3D Printed PLA, ABS and PET-G Materials for Various Industrial Applications, in: *Proc. 2024 Int. Conf. Intell. Innov. Pract. Eng. Manag. (IIPEM)*, Singapore, pp. 1–6, 2024, doi: 10.1109/IIPEM62726.2024.10925687.
7. Kopřiva, J., Trnka, P., Michal, O., Hornak, J., The Evaluation of 3D Printed Layered Insulation Consisting of PET-G and TPE Materials, in: *Proc. 2024 Int. Conf. Diagnostics Elect. Eng. (Diagnostika)*, Pilsen, Czech Republic, pp. 1–4, 2024, doi: 10.1109/Diagnostika61830.2024.10693879.
8. Marconi, S., Alaimo, G., Mauri, V., Torre, M., Auricchio, F., Impact of graphene reinforcement on mechanical properties of PLA 3D printed materials, in: *2017 IEEE MTT-S Int. Microw. Workshop Series Adv. Mater. Process. RF THz Appl. (IMWS-AMP)*, Pavia, Italy, pp. 1–3, 2017, doi: 10.1109/IMWS-AMP.2017.8247414.
9. Bugday, Z.Y., Venkatachalam, A., Anderson, P.D., van der Sman, R.G.M., Rheology of paste-like food inks for 3D printing: Effects of nutrient and water content. *Curr. Res. Food Sci.*, 9, 2024. Art. no. 100847. doi: 10.1016/j.crfs.2024.100847.
10. Chen, D., Zhao, L., Diao, H., *et al.*, Rheological properties and related screen-printing performance of low-temperature silver pastes for a-Si:H/c-Si heterojunction solar cells. *J. Mater. Sci. - Mater. Electron.*, 25, 5322–5330, 2014. doi: 10.1007/s10854-014-2308-1.
11. Li, S., Wei, M., Hao, J., *et al.*, Rheological characterization and jet printing performance of Sn–58Bi solder pastes. *J. Mater. Sci. - Mater. Electron.*, 29, 4575–4582, 2018. doi: 10.1007/s10854-017-8408-y.
12. Liravi, F., Darleux, R., Toyserkani, E., Additive manufacturing of 3D structures with non-Newtonian highly viscous fluids: Finite element modeling and experimental validation. *Addit. Manuf.*, 13, 113–123, 2017. doi: 10.1016/j.addma.2016.10.008.
13. Pricci, A. and Percoco, G., Towards a digital twin of extrusion-based additive manufacturing: an experimentally validated numerical model of ironing process, in: *Prog. Addit. Manuf*, 2025, doi: 10.1007/s40964-025-00988-8.

14. Chatterjee, A., Heat transfer enhancement in laminar impinging flows with a non-Newtonian inelastic fluid. *J. Non-Newtonian Fluid Mech.*, 211, 50–61, 2014. doi: 10.1016/j.jnnfm.2014.06.009.
15. Wang, D., Wang, R., Chen, S., *et al.*, Low viscosity and highly flexible stereolithographic 3D printing resins for flexible sensors. *Mater. Des.*, 243, 2024. Art. no. 113052. doi: 10.1016/j.matdes.2024.113052.
16. Karakurt, I. and Lin, L., 3D printing technologies: techniques, materials, and post-processing. *Curr. Opin. Chem. Eng.*, 28, 134–143, 2020. doi: 10.1016/j.coche.2020.04.001.

About the Editors

Dr. Dhananjay Yadav received his Ph.D. degree from the Department of Mathematics, Indian Institute of Technology (IIT) Roorkee, India in 2013 and post-graduation (M.Sc.) in Mathematics from DDU University Gorakhpur, India in 2007. Currently, he is working as an Associate Professor with the Department of Mathematics at the University of Nizwa, Oman. Prior to his appointment to the University of Nizwa, Oman, he had worked as a Principal Research Scientist at Athabasca University, Canada; Yonsei University, South Korea; and Jeju National University, South Korea. He is a leading expert in CO_2 capture, storage and oil recovery, computational sustainability and environmental analytics, fluid mechanics, numerical analysis, hydrodynamic and hydromagnetic stability, nanofluids, and fluid flow in porous media. He has published more than 100 research articles (high impact factor) in various reputed international journals. He is also listed in the top 2% influential researchers in the world prepared by Stanford University based on Scopus data.

Dr. Mukesh Kumar Awasthi has done his Ph.D. on the topic "Viscous Correction for the Potential Flow Analysis of Capillary and Kelvin–Helmholtz Instability". He is working as an Assistant Professor with the Department of Mathematics at Babasaheb Bhimrao Ambedkar University, Lucknow. Dr. Awasthi is specialized in the mathematical modeling of flow problems. He has taught courses of fluid mechanics, discrete mathematics, partial differential equations, abstract algebra, mathematical methods, and measure theory to postgraduate students. He has acquired excellent knowledge in the mathematical modeling of flow problems, and he can solve these problems analytically as well as numerically. He has a good grasp of the subjects like viscous potential flow, electro-hydrodynamics, magneto-hydrodynamics, heat, and mass transfer. He has excellent communication skills and leadership qualities. He is self-motivated and responds to suggestions in a more convincing manner. Dr. Awasthi has qualified National Eligibility Test (NET) conducted on all India level in the year 2008 by the Council of Scientific and Industrial Research (CSIR) and got Junior

Research Fellowship (JRF) and Senior Research Fellowship (SRF) for doing research. He has published **135 plus** research publications (journal articles/books/book chapters/conference articles) in Elsevier, Taylor & Francis, Springer, Emerald, World Scientific, and many other national and international journals and conferences. Also, he has published **16** books. He is also a series editor of ***Artificial Intelligence and Machine Learning for Intelligent Engineering Systems*** published by CRC Press (Taylor & Francis USA). He has attended many symposia, workshops, and conferences in mathematics as well as fluid mechanics. He has received the "Research Awards" consecutively four times from 2013 to 2016 by the University of Petroleum and Energy Studies, Dehradun, India. He has also received the start-up research fund for his project "Nonlinear study of the interface in multilayer fluid system" from UGC, New Delhi. He is listed among the **top 2% of influential researchers** in the world, as prepared by Stanford University based on Scopus data for the years 2022 and 2023. His Orcid is ***0000-0002-6706-5226***, his Google Scholar web link is ***https://scholar.google.co.in/citations?user=Dj3ktGAAAAAJ***, and his research gate web link is ***https://www.researchgate.net/profile/Mukesh-Awasthi-2***.

Dr. Harith Mohamed Al-Azri has done his M.Sc. in radiation physics at the University of Liverpool, UK in 2005 and obtained his Ph.D. in nuclear physics at the University of York, UK in 2012. Between 2001 and 2021, he has worked on the colleges of applied science that belongs to the Ministry of Higher Education in the Sultanate of Oman. He has participated in publishing papers in high impact factor journals like *Nature* and *Physical Review C*. Currently, he is the head of the Department of Mathematical and Physical Science at the University of Nizwa. His main interest is nuclear and radiation physics. He is a member of the IAEA to establish an international nuclear science academy in Asia.

Index

Also of Interest

Check out these other related titles from Scrivener Publishing

Nanofluid Heat Transfer, Edited by Mukesh Kumar Awasthi and Reshu Gupta, ISBN: 9781394336371. Future-proof your thermal system designs with this essential guide, providing a comprehensive, cutting-edge exploration of nanofluids to drive innovation and efficiency in heat transfer.

Mechanics of Fluid Flow, By Kaplan S. Basniev, Nikolay M. Dmitriev, and George V. Chilingar, ISBN: 9781118385067. The mechanics of fluid flow is one of the most important fundamental engineering disciplines explaining both natural phenomena and human-induced processes. A group of some of the best-known petroleum engineers in the world give a thorough understanding of this important discipline, central to the operations of the oil and gas industry.

CHEMICAL PROCESS ENGINEERING VOLUME 1: Design, Analysis, Simulation, Integration, and Problem Solving with Microsoft Excel-UniSim Software for Chemical Engineers Computation, Physical Property, Fluid Flow, Equipment & Instrument Sizing, Pumps & Compressors, Mass Transfer, By A. Kayode Coker and Rahmat Sotudeh-Gharebagh, ISBN: 9781119510185. Written by one of the most prolific and respected chemical engineers in the world and his co-author, also a well-known and respected engineer, this two-volume set is the "new standard" in the industry, offering engineers and students alike the most up-do-date, comprehensive, and state-of-the-art coverage of processes and best practices in the field today.

CHEMICAL PROCESS ENGINEERING VOLUME 2: Design, Analysis, Simulation, Integration, and Problem Solving with Microsoft Excel-UniSim Software for Chemical Engineers, Heat Transfer and Integration, Process Safety, Chemical Kinetics and Reactor Design, Engineering Economics, Optimization, By A. Kayode Coker and Rahmat Sotudeh-Gharebagh, ISBN: 9781119853992. Written by one of the most prolific and respected chemical engineers in the world and his co-author, also a well-known and respected engineer, this two-volume set is the "new standard" in the industry, offering engineers and students alike the most up-do-date, comprehensive, and state-of-the-art coverage of processes and best practices in the field today.

Liquid Biofuels: Fundamentals, Characterization, and Applications, Edited by Krushna Prasad Shadangi, ISBN: 9781119791980. From the selection and pretreatment of raw materials to design of reactors, methods of conversion, selection of process parameters, optimization, and production of various types of biofuels to the industrial applications for the technology, this is the most up-to-date and comprehensive coverage of liquid biofuels for engineers and students.

PETROLEUM REFINING DESIGN AND APPLICATIONS HANDBOOK VOLUME 1, By A. Kayode Coker, ISBN: 9781118233696. The most comprehensive and up-to-date coverage of the advances of petroleum refining designs and applications, written by one of the world's most well-known process engineers, this is a must-have for any chemical, process, or petroleum engineer.

PETROLEUM REFINING DESIGN AND APPLICATIONS HANDBOOK VOLUME 2: Rules of Thumb, Process Planning, Scheduling, and Flowsheet Design, Process Piping Design, Pumps, Compressors, and Process Safety Incidents, By A. Kayode Coker, ISBN: 9781119476412. The second of a three-volume set of the most comprehensive and up-to-date coverage of the advances of petroleum refining designs and applications, written by one of the world's most well-known process engineers, this is a must-have for any chemical, process, or petroleum engineer.

PETROLEUM REFINING DESIGN AND APPLICATIONS HANDBOOK VOLUME 3: Mechanical Separation, Distillation, Packed Towers, Liquid-Liquid Extraction, Heat Transfer and Process Safety Incidents, By A. Kayode Coker, ISBN: 9781119794868. The third volume of a four-volume set of the most comprehensive and up-to-date coverage of the advances of petroleum refining designs and applications, written by one of the world's most well-known process engineers, this is a must-have for any chemical, process, or petroleum engineer.

PETROLEUM REFINING DESIGN AND APPLICATIONS HANDBOOK VOLUME 4: Heat Transfer, Pinch Analysis and Process Safety Incidents, By A. Kayode Coker, ISBN: 9781119827528. The fourth volume of a multi-volume set of the most comprehensive and up-to-date coverage of the advances of petroleum refining designs and applications, written by one of the world's most well-known process engineers, this is a must-have for any chemical, process, or petroleum engineer.

Printed and bound by CPI Group (UK) Ltd, Croydon, CR0 4YY
17/09/2026
14964266-0004